Decoding the Hand

DECODING *the* HAND

A HISTORY OF SCIENCE, MEDICINE, AND MAGIC

Alison Bashford

The University of Chicago Press ☞ *Chicago and London*

The University of Chicago Press, Chicago 60637
The University of Chicago Press, Ltd., London

Published 2025

Printed in the United States of America

34 33 32 31 30 29 28 27 26 25   1 2 3 4 5

ISBN-13: 978-0-226-83115-2 (cloth)
ISBN-13: 978-0-226-83116-9 (ebook)
DOI: https://doi.org/10.7208/chicago/9780226831169.001.0001

Library of Congress Cataloging-in-Publication Data

Names: Bashford, Alison, 1963– author.
Title: Decoding the hand : a history of science, medicine, and magic / Alison Bashford.
Description: Chicago : The University of Chicago Press, 2025. | Includes bibliographical references and index.
Identifiers: LCCN 2025013884 | ISBN 9780226831152 (cloth) | ISBN 9780226831169 (ebook)
Subjects: LCSH: Palmistry—History. | Fortune-telling—History. | Occultism—History. | Medicine—History. | Science—History.
Classification: LCC BF921 .B295 2025 | DDC 133.609—dc23/eng/20250428
LC record available at https://lccn.loc.gov/2025013884

♾ This paper meets the requirements of ANSI/NISO Z39.48-1992 (Permanence of Paper).

Authorized Representative for EU General Product Safety Regulation (GPSR) queries: **Easy Access System Europe**—Mustamäe tee 50, 10621 Tallinn, Estonia, gpsr.requests@easproject.com
Any other queries: https://press.uchicago.edu/press/contact.html

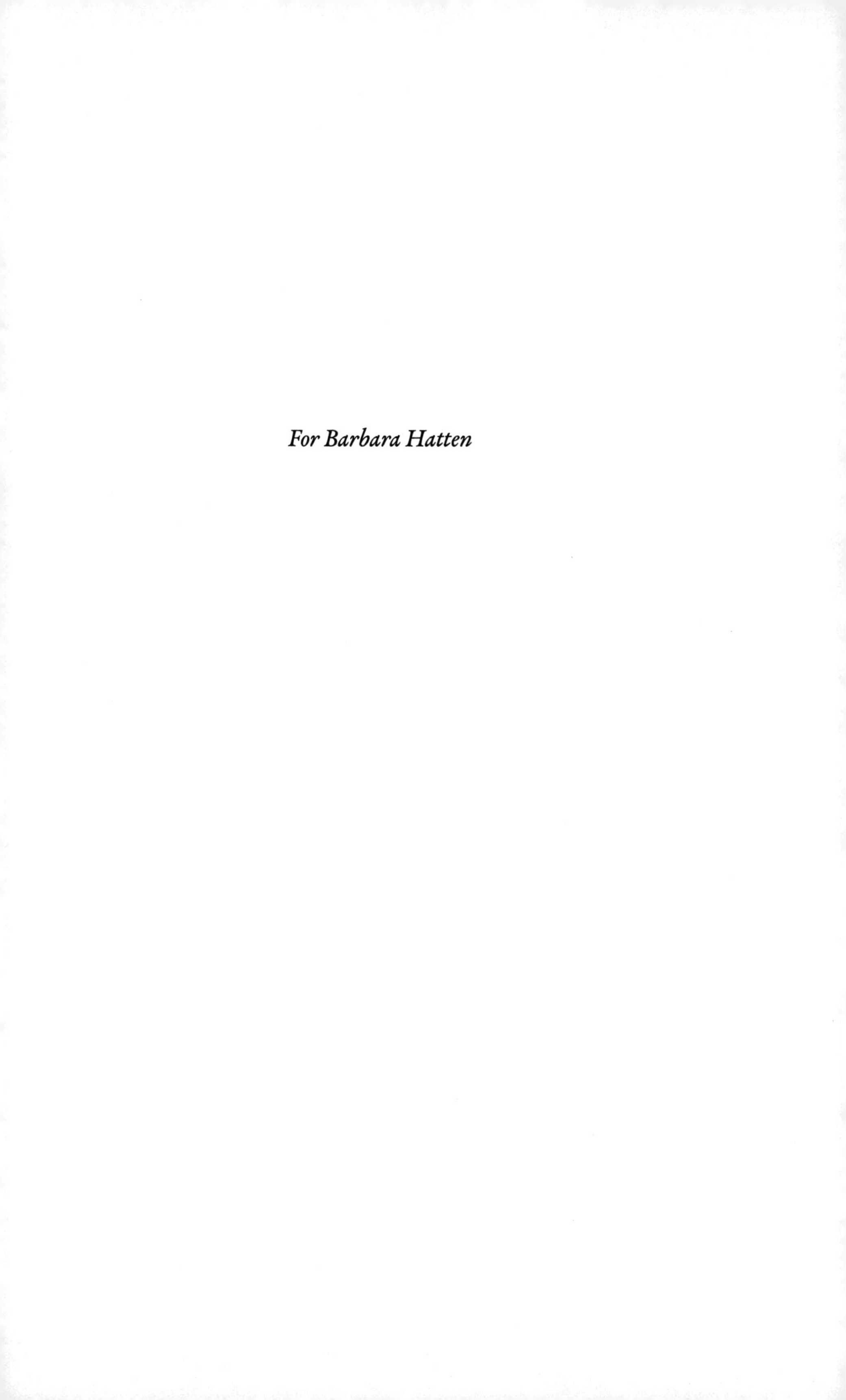

For Barbara Hatten

Nothing is so secret in a human being that
it does not have an outward sign.

PARACELSUS, C. 1530

Contents

Illustrations

Abbreviations

BL	The British Library
BMJ	*The British Medical Journal*
CWP	Charlotte Wolff Papers (the Wellcome Collection, London)
EHA	Edward Heron-Allen Collection (West Sussex Record Office)
LPP	Lionel Penrose Papers (University College London Special Collections)
TNA	The National Archives (Kew, UK)
UCL	University College London
UCLSC	University College London Special Collections
WSRO	West Sussex Record Office

Author's Note on Terminology

In a study of the past, I generally retain historical and archaic terms, including medical terms and nominations of people. On occasion I retain "Mongolism," a historical term for Down syndrome. On occasion I retain "Gipsy," deliberately an antiquated spelling, to signal a historical usage. This book tells the history of these words themselves, including the history of who they offended, and of how they changed.

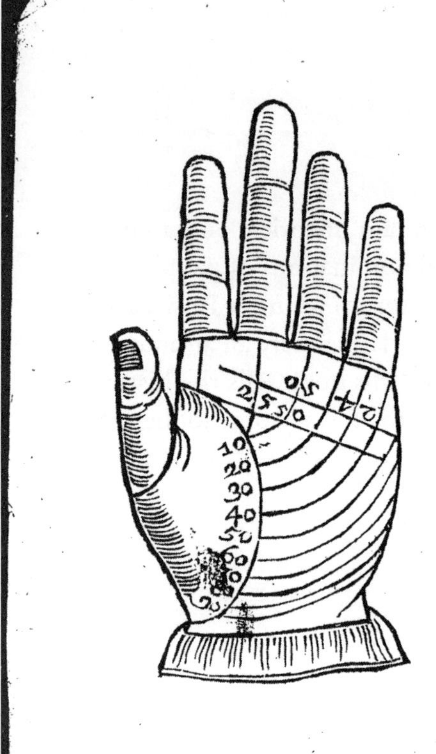

PALMISTRY,
The SECRETS thereof
DISCLOSED,
Or a *Familiar, Eaſy* and *New* Method, whereby to *judge* of the moſt *General* Accidents of mans *Life* from the *Lines* of the hand, withal its *Dimentions* and *Significations.*

As alſo that *Moſt Uſeful Piece* of Aſtrology (long ſince promiſed) concerning *ELECTIONS* for every *particular* Occaſion, now *plainly* manifeſted from *Rational Principles* of *ART*, not Publiſhed till now.

By *RICHARD SAUNDERS*, Author of the former Book of *Chyromancy* and *Phiſiognomy.*

Cuique ſua eſt Tempeſtas, et tempus cuique voluntati ſub Cælis, Eccle. Cap. 3. verſe 1. to 12.
Tempus eſt potentius Legibus.

LONDON,
Printed by *H. B.* for *G. Sawbridge*, at the Signe of the *Bible* upon *Lud-Gate Hill.* 1663.

I.1 Richard Saunders, *Palmistry, the secrets thereof disclosed* (London, 1663). Copy in author's possession.

Secrets Disclosed

TAKEN BY THE HAND

In high summer 1663, young Isaac Newton strolled over to the fair on Cambridge's Stourbridge Common and bought himself a new book, *Palmistry, the secrets thereof disclosed*. It had been written that year by the physician and astrologer Richard Saunders, who described himself as a "student in the Physical-Coelestial Sciences," not unlike the mature Newton. The future giant of astronomy and physics read about "chiromancy," from the Greek *cheír*, how health and fortunes might be told from marks on the hand. He was interested in part because those lines made shapes and patterns. A geometry of the hand was set out, one that Saunders indicated should be read strictly and precisely with a measuring compass. This chiromancy was perhaps old-fashioned for the seventeenth century, but still not implausible. Reading signs of the body—"physiognomy"—had even been described by one of Newton's contemporaries as a "seventh part of physics." No wonder he spent time on this book of the body, lines, angles, shapes, and numbers (fig. I.1).[1]

Yet in explaining and interpreting this geometry of the hand, the book that Newton bought was also occult—it disclosed secrets—because it was inspired by the sacred Jewish texts of Kabbalah, originally medieval knowledge that had been revived in the Renaissance with great scholarly impact. Newton was to become fascinated with that as well. Behind Saunders's *Palmistry* was a long and learned tradition in which signs on the hand were decoded both geometrically and astrologically, mathematically and cosmologically.

I own a precious copy of Richard Saunders's book too. But why would it be that the great physicist Newton, and an ordinary historian of the modern sciences, share an interest in the hand? A counterintuitive answer is because both Newton and I find the calculations within incomprehensible: needless to say, for entirely different reasons.

For Newton, this astrology and numerology made little sense mathematically. It didn't add up, so to say. But that didn't stop him from reading more books on chiromancy by Christian, Jewish, and Islamic scholars. For me, however, the mathematics of Saunders's *Palmistry*, and especially his more expansive prior work on *Physiognomie and Chiromancie*, doesn't add up simply because it is too unfamiliar, too complex, too difficult. I'm neither a physicist nor a mathematician, nor anything but a passing scholar of Jewish Kabbalah or the Christianized Kabbalah in which Newton and Saunders were so interested.[2]

I am, however, a historian of health, medicine, and human bodies, and now—still to my own surprise—a historian of the hand. If chiromancers discern the future, historians discern the past, including that past's own future, up to our present. In this case, I am interested in where chiromancy came from, and where it went to, over the centuries. The latter is my special focus in this book. What happened to this seventeenth-century geometry of the hand in late modernity? Did the learned chiromancy authored by Richard Saunders and read by Isaac Newton end up as curiosity, leisure and pleasure, the palmistry of piers and parlors? Yes, quite directly: already an interesting story. But that is just the beginning. It also ended up as human genetics. To jump to the last chapter of this chronologically arranged book, a strict geometry of the hand became the twentieth-century business of another

Cambridge-trained mathematician, the brilliant medical geneticist, chess theorist, and pediatrician Lionel Penrose. Between the 1930s and 1970s, Penrose correlated lines and marks on the hand to a range of chromosomal syndromes. Just before his death in 1972, he wrote his final article, "Fingerprints and Palmistry," published in the top medical journal, *The Lancet*, no less.[3]

What is the connection between these decoders of hands, the physicians Richard Saunders and Lionel Penrose, the first astrologically trained, the second genetically trained? To render the core business of this book pictorial, what is the relation between their two hand illustrations, Saunders's from 1663, Penrose's from 1963 (figs. I.2 and I.3)? If the human hand itself looks more or less the same over the centuries, might it be that the systems for reading its signs also form a more continuous history than these two representations initially suggest?

Between Richard Saunders's seventeenth century and Lionel Penrose's twentieth century, all manner of anatomists, physicians, statisticians, primatologists, and psychoanalysts were decoding hand signs. Many were fringe dwellers of their disciplines, yet not a few were distinguished enough to be Fellows of the Royal Society, the world's oldest academy of eminent scientists. Were they "palmists" or "chiromancers"? In fact, several did self-identify as just that, but as a rule, no. Were they readers of hands? Unequivocally yes, and in this regard they were of a piece with better-known hand readers; occult revivalists of hermetic magic, brazen celebrity palmists, and the people known in Isaac Newton's time as "Egyptians" or "Gipsies," women who read fortunes; *dukkerin* in the Romani tongue. This book places them all together, if not joining hands, joined by hands.

I became a historian of the hand between one instant and another in early 2020, in the time it takes to open a manila folder. Researching the files of a Berlin-trained medical doctor, Charlotte Wolff, delicate papers within revealed the ink handprints of Aldous

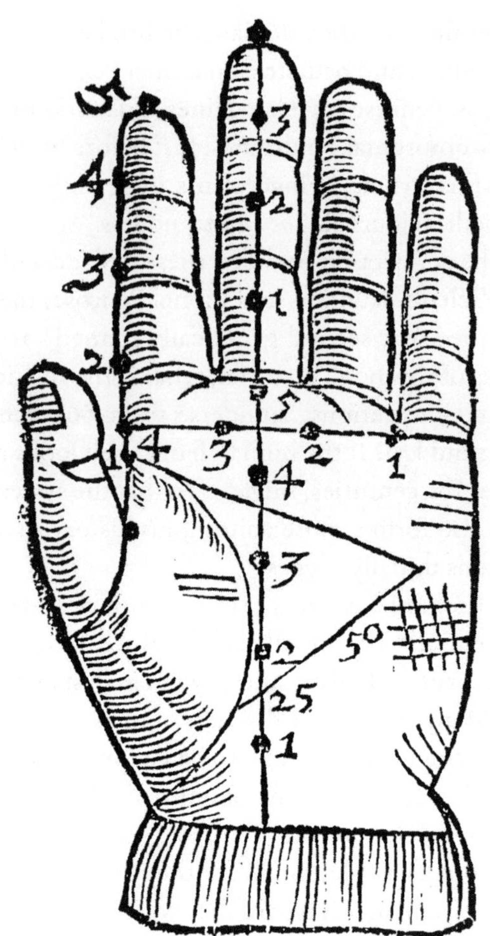

1.2 Richard Saunders, the hermetic hand, 1663. From *Palmistry, the secrets thereof disclosed* (London, 1663), 30. Copy in author's possession.

Huxley, Virginia Woolf, Vaslav Nijinsky, Maurice Ravel, André Gide, and more. It was both unnerving and magical, a too-intimate intrusion into the interiors of these interwar cultural luminaries, as if uninvited and unseen at a psychoanalytic session. Now I know the extent to which clever Charlotte Wolff's so-called "palmistry" was just that: she called them her *analysands*. If that wasn't extraordinary enough, in the next folder appeared the palm prints

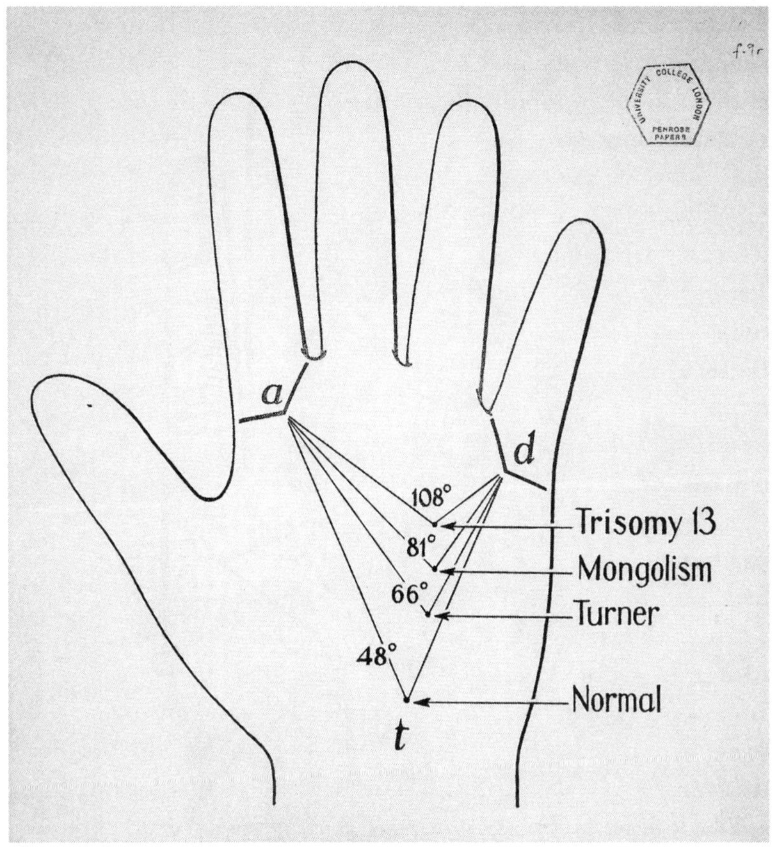

I.3 Lionel Penrose, the genetic hand, 1963. Plate prepared for L. S. Penrose, "Finger-Prints, Palms and Chromosomes," *Nature* (1963). Lionel S. Penrose Papers, University College London, Special Collections PENROSE 2/42/3/1. By kind permission, Professor Shirley Hodgson and UCL Library Services, Special Collections.

of creatures even more special: the monkeys and apes in London Zoo, 1938. Many were named—Peter, Daisy, Mok—as individual as Aldous, Virginia, and Maurice. Mok was a lowland gorilla, and Wolff had been called to the zoo's morgue to take his prints two days after death. Capturing his mark, his soul, Mok's handprint is like a strange primate death mask, the magnificent beast both dead and still living-in-sleep. We can touch him. With these fragile prints before me—then inscrutable, now comprehensible—I was entirely taken by the hand (fig. I.4).

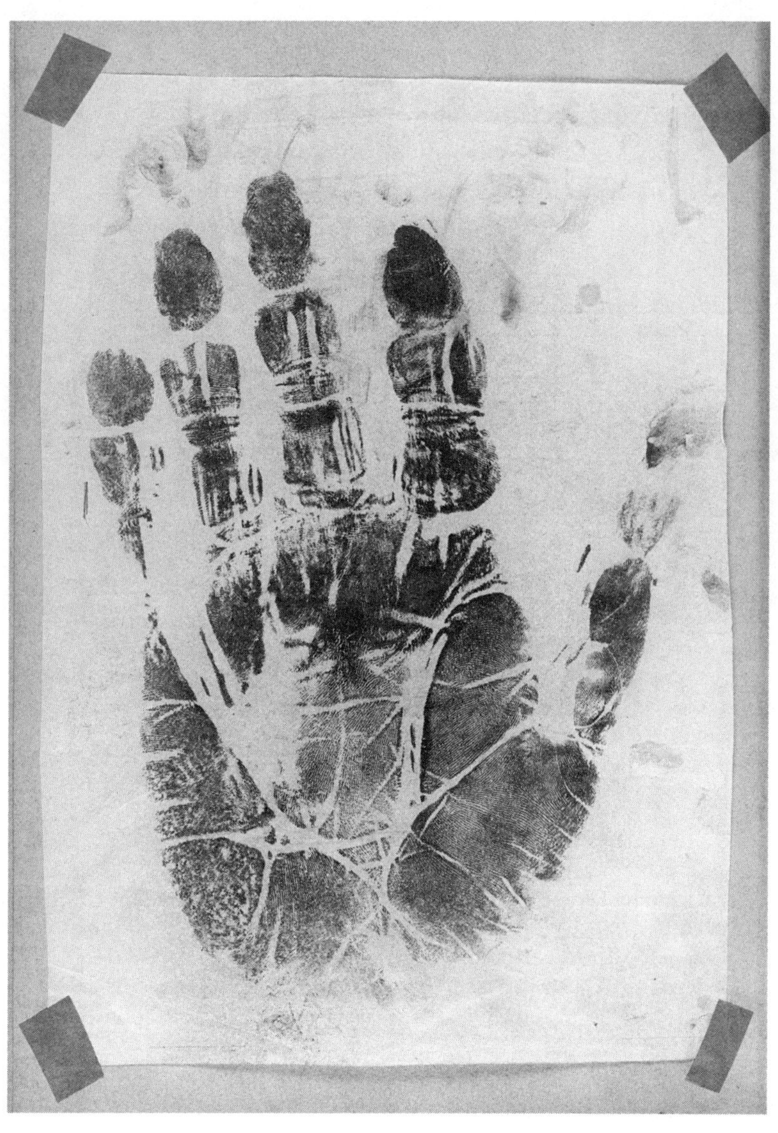

1.4 The handprint of Mok, a lowland gorilla, taken two days post-mortem in London Zoo, 1938. British Psychological Society Archive, Charlotte Wolff Papers, the Wellcome Collection, London PSY/WOL/3/7. By kind permission, British Psychological Society.

This is a history of the mysterious, curious, and often complex codes by which signs of the hand have been interpreted. It is a journey over many centuries and registers of knowledge, from elite and arcane to street-level exchange of the most commercial kind. We shall meet an idiosyncratic suite of hand readers, sometimes learned, sometimes outrageously deceptive, sometimes earnest, and more often than I ever expected, medically and scientifically trained: physician John Bulwer, for example, who in the seventeenth century considered hands able to speak in a silent universal language, "chirologia"; Charles Bell, one of the most distinguished anatomists of the early nineteenth century, who wrote a major natural theology treatise on *The Hand* and its "eloquence," divinely inspired and designed, he thought; a Brahmin-trained Irish palmist sought by prime ministers and shahs in the 1890s, who went by "Cheiro," and for whom the "language of the hand," as for Richard Saunders, was astrological, numerical, and geometric. We shall meet the Ellis family, proprietors and practitioners of vernacular seaside health, a brazenly entrepreneurial troupe of hand and face readers. They were pier dwellers who authorized their work with their own institutional invention, the too-grandiose British Institute for Mental Science, like magic conjured out of nothing. In the late nineteenth-century occult-revival boom, many practitioners gathered up the accoutrement of "science." Yet some aspired to scientific palmistry in more methodological terms, arguing for a modernized "medical palmistry" that embraced diagnostics: the dashing Edward Heron-Allen FRS (Fellow of the Royal Society) was one, and sensible Katharine St. Hill FCS (Fellow of the Chirological Society) was another. She explicitly eschewed chiromancy, and claimed that it was only in the sense of "prognosis" that she told the future. Lecturer in palmistry and director of a palmistry college, she was also author of a palmistry textbook, and examiner of medical palmistry students. But before our amusement becomes too superior, we should note that she was invited into London's hospitals and asylums to pursue research on the correlations between palm lines and particular illnesses.

The many prints of hands taken by these palmists, and their varying diagnoses of health, self, and character, were far more similar than is often recognized to the well-known work of another hand reader, the statistician polymath Francis Galton, second cousin of Charles Darwin. For years Galton took thousands upon thousands of fingerprints and verified statistically that each pattern was unique, a great new means of identification of the true self by emerging biometric states. It is little known, however, that the father of fingerprinting took palm prints too. We shall meet the early twentieth-century anatomists who named those Galtonian patterns "dermatoglyphs," the US physicians Harris Hawthorne Wilder and Harold Cummins. They wondered if hand marks correlated not to individuals but to different human types, to "race." Meanwhile, having fled Nazi Germany to climb into the chimpanzee cages of the London Zoological Gardens, Charlotte Wolff was, like those anatomists, looking for the so-called "simian line" in both human and nonhuman primates.

It all sounds like fringe bio-medicine, and in many cases it was. Yet in light of this accumulated observation, mid-twentieth-century clinicians, physicians, and geneticists began to understand that some palm lines *were* marks—signs—of particular syndromes. They were "phenotypes" that before long were linked to genotypes of certain chromosomal conditions, in the first instance Down syndrome. This was the work that Lionel Penrose did, as it so happens as Galton Professor of Human Genetics, previously Galton Professor of Eugenics, University College London. He and his team were correct. Neonatal specialists still look for that single palmar line, and it is still sometimes called "the simian crease."[4]

One might expect a book on palmistry to be primarily about the occult. To some considerable extent, this is. But at core, *Decoding the Hand* explains an unexpected history of health, medicine, and science, showing how the hand has been part of the history of lay therapeutics, diagnostics, and prognostics, and of more conventional medical sciences, human sciences, and biosciences. Not only for fortune-telling palmists were the future and past laid bare in the hand, but for all kinds of other experts in bodies and minds as well: physicians, psychologists, psychiatrists, embryologists, anatomists, endocrinologists, primatologists, geneticists, and physical anthropologists.

In all technical senses, the long history of chiromancy, palmistry, and medical hand reading is a history of decoding bodily signs. It was initially curious to me that one London-based Fellow of the Royal College of Physicians should research palm lines and then publish on the philosophy of language, in the 1920s lobbying for medicine's uptake of a "Theory of Signs." Yet on reflection, it is clear that signs are everyday medical business. Every physician, then as now, is taught the difference between medical signs, to be expertly interpreted, and symptoms, subjectively felt by the patient. Every physician would recognize Richard Saunders's routine seventeenth-century examination of the hand, the fingers, and the nails, their shape and proportion, the hand's moisture, dryness, color, and texture. The interpretation of meaning from those signs is radically different, however. He was thinking about humors and their balance, as well as about the relative position of Jupiter, the planet, and the finger. A current physician might be thinking about thyrotoxicosis if the hands are warm and moist, or a septic emboli if a macular lesion is observable on the palm, or perhaps lung cancer or heart disease if the fingers are "clubbed," that is, if the nail drops greater than 180 degrees. The modern-day physician may well need Saunders's measuring compass. The point is: these were, and are, all signs to be read.[5]

And so, *Decoding the Hand* is both a history of science and medicine, *and* a history of bodily semiotics. This sign has that meaning regarding health or ill health, character, fortune, or the future, an individual or a familial past, or even a species' deep past. This is so both in a clinical tradition in which a physician is taught to interpret a bodily sign and in a linguistic tradition through which the meanings of signs and symbols are deciphered. Well might medical students today be taught "The Body as Text." From medieval instruction to look for a triangle or a quadrangle in the palm, to Saunders's advice to note the hand's shape, color, and texture as well as the number of "rascettes" in the wrist, to Francis Galton's manual on how to identify this whorl or that double loop on fingertips, this is one long history of reading signs for meaning.[6]

In these chapters, I reproduce many exquisite drawings of hands through which these systems of signs and symbols can be explored. They might be strict re-presentations of the lines, angles, crosses, hatchings, loops, and whorls on the palms and fingers. They might show astrological symbols, the symbol of Saturn for the finger of Saturn, of Mars for the mount of Mars. Less familiar, perhaps, is the Sanskrit tradition in which figurative images are depicted—an elephant, a set of scales, a trident—each with their ascribed meaning. In other traditions, they might show Hebrew letters indicating the name of one's personal archangel, or Latin script that would more mundanely reveal the first letter of the name of one's future spouse. They might show Roman numerals or Arabic numerals: significantly, we count those ten Arabic numbers on our "digits." We will see how an arithmetical and geometric set of signs, a linguistic and cultural set of signs, and a medical and health set of signs have been interwoven for centuries. These many systems for decoding made the hidden (the occult) visible: a doctrine of signatures, as the physician Paracelsus put it in his sixteenth-century treatise on chiromancy: "Nothing is so secret in a human being that it does not have an outward sign."[7]

Two purposes for hand reading recur over time and place: chiromantic divining of the future, and physiognomic reading of internal character, health, and self. The latter is especially important: the classical, medieval, and early modern tradition of physiognomy that was to become highly popular in the eighteenth century; how the exterior surface of the body reveals an interior. For all manner of practitioners, it was not just the classic physiognomic eye and face, but the hands as well, that were windows to souls, to the future, and to fate, as well as to character, to temperament, and to diagnosis and prognosis. And just like a window, like glass, the body sign was inscribed on a surface that was the interface between inside and outside, between a human interior and an exterior world.

With the hand as our constant, we can see how comprehensions of the human body have changed over time with regard to its inside and outside. In Saunders's world, the microcosm of the human body was directly related to the macrocosm of the heavenly bodies. By contrast, in Lionel Penrose's world, the most profound secret of the human body, and of individual identity, lay inside, not outside. When he and his team looked at the human palm, they searched for deep interiority, eventually of the chromosome, the structure inside the nucleus of a cell. Yet reading signs of the hand turns out to be a history of an even more inscrutable and complex interior: the mind. The hand reveals a remarkable history of changing comprehensions of the human self, from the "soul" of classical physiognomy, to eighteenth-century "temperament," nineteenth-century "character," and twentieth-century "personality." Even the psychoanalytic "id" and "ego" were sometimes read in the hand. In this way, *Decoding the Hand* tracks not just chiromancy and physiognomy, an anatomy and physiology of the hand, but in them an unlikely antecedent of modern "psy" disciplines and practices. Hand readers were "seers," initially those who could apparently divine the future, and later those who could see inside the self of another, right into the secrets of their soul.[8]

This is also a book, then, that tracks ways of discovering and diagnosing identity, in registers high and low. Often this is fun. In so many libraries and archives across the world, I've laughed out loud as I read *You and Your Hand*, and *Your Luck's in Your Hand*, and *Palmistry Self-Taught*. The sometimes gaudy history of self-improvement, and the impossibly bold commercial supply of cheap instruction about how to "know thyself"—typically authorized by Aristotle, Aeschylus, Anaxagoras, Socrates, or Plato—is irresistible. Yet when I read the best and worst of these catchpenny pamphlets, amusement turned into something else. Despite myself, I became enchanted by the eloquence of some palm-reading instructors and the beguiling lists of words they conjured for a thousand human types, characters, and temperaments. One "Table of Qualities," drawn from physiognomy and phrenology, reads like an A-to-Z of Darwin's expression of emotions (and that, indeed, is another connection traced herein): "Acquisitiveness, Adaptability, Ambition, Anger, Benevolence, Calculation, Caprice, Conventionality, Courage, Courtesy, Deceit, Economy,

Egotism, Eloquence, Enmity, Fatalism, Impulse, Justice, Melancholy, Perception, Pride, Prudence, Reverence, Tact, Timidity, Vanity, Wit." Palmistry, it turns out, was poetry as well; another language of the hand.

Over and over again, practitioners and clinicians, palmists and chiromancers conceptualized just that: the language of the hand. In these chapters, we follow not only the long history of how body signs are read, but also of how the body speaks, and was understood to do so sometimes in entirely literal ways. As one South Indian palmist put it in 1918: "By signs and gestures, the hand is able to express at least approximately what the mouth communicates by words."[9] Communicating silent meaning to be apprehended at a distance, gesture can be so codified as to be a language itself. In 1644, physician John Bulwer set out the origins of sign language in his enchanting book, *Chirologia: or the naturall language of the hand. Composed of the speaking motions, and discoursing gestures thereof. Whereunto is added Chironomia: or, the art of manuall rhetoricke. Consisting of the natural expressions, digested by art in the hand, as the chiefest instrument of eloquence* (fig. I.5). Two hundred years later, anatomist Charles Bell nominated the "eloquent" hand. Three hundred years later, psychoanalytically trained doctor Charlotte Wolff wrote *A Psychology of Gesture* (1945). The eloquent hand had turned into the psychoanalytic body that speaks.

Like a *fin-de-siècle* conjurer, in this book I make unlikely historical figures shake hands across the centuries, across registers, and across disciplines—Richard Saunders and Lionel Penrose, John Bulwer and Charlotte Wolff, Charles Bell and Cheiro, even Aristotle and Blackpool's outrageous Ida Ellis. In doing so, I consider how the body's secrets were revealed both as a kind of wonder, and as plain anatomy. Reading signs of the hand turns out to be an unexpectedly useful case for ongoing historical assessment of modern "disenchantment": the undoing of magic over the centuries, the rationalization of belief.[10] The most common recent

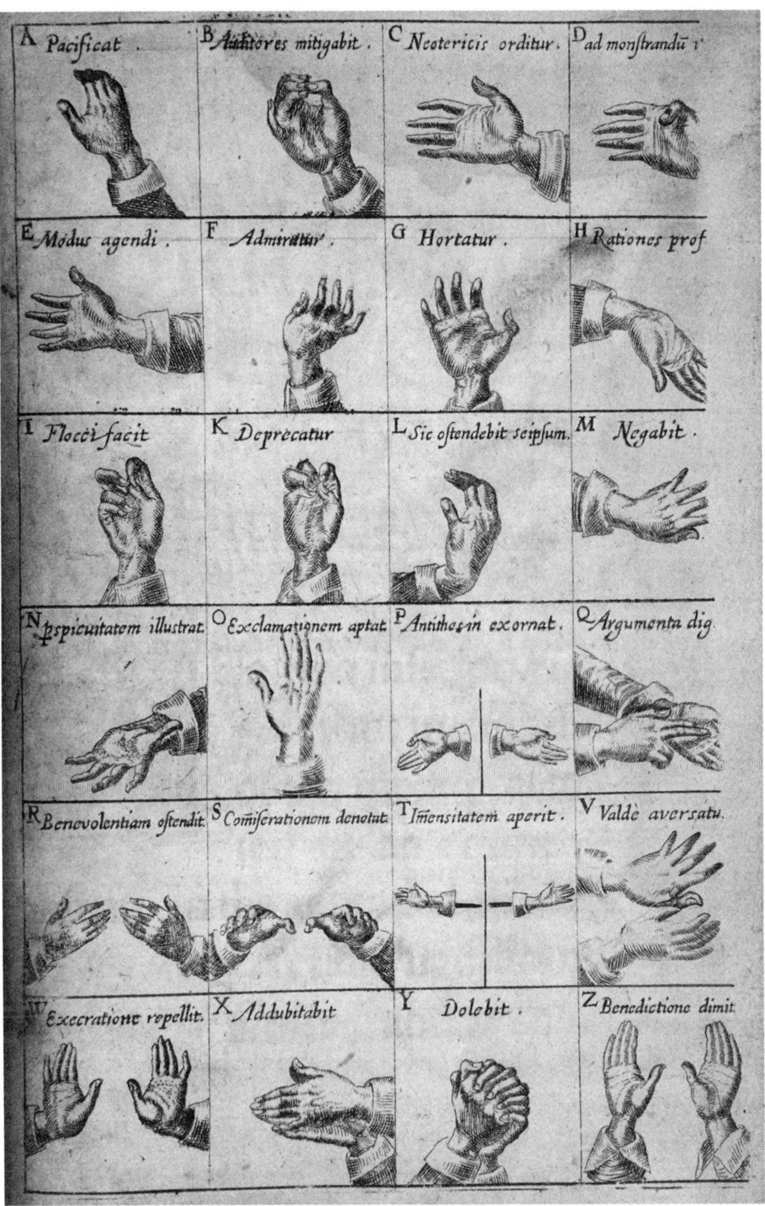

1.5 The eloquent hand. From John Bulwer, "The Canons of Rhetoricians Touching: The artificial managing of the Hand in Speaking," in *Chirologia: Or the Naturall Language of the Hand* (London, 1644), 65. Courtesy the Wellcome Collection, London.

consensus among historians is to perceive magic enduring into late modernity—the nineteenth and twentieth centuries—or to argue for an active "re-enchantment." The history of an occult revival has been important here. Of course, palm-based fortune-telling on piers and in parlors continued, and we might well read this as "secular magic in a rational age," as it has been put. For one scholar of modern enchantment, the desire for magic and wonder is best understood as a practice "that delights but does not delude."[11]

And yet, so many hand readers I have met in libraries and archives across the world were not working in cultural economies of delusion or delight at all. More of them were working as physicians and scientists. Put another way, while it is entirely possible to tell a long history from Richard Saunders's astrology to the leisure and pleasure of palmistry at the seaside, the story I tell discloses another modern trajectory: how occult hand reading and anatomical and medical hand reading have a linked history from the beginning, and how the *claims* to scientific practice of so many occult palmists in some instances actually *became* sciences: medical, psychological, zoological, and more. And so, after decades of revisions toward a now popular "re-enchantment" thesis, scientific and medical hand reading proves to be an instance of the original: modern disenchantment.

Past editors of *The Lancet* agree with me. In 1964, they proclaimed that "interpreting the palms of the hands (and the soles of the feet) is now a respectable science."[12] And yet for all my interest in exploring a medical history of hand reading, even of palmistry, this is not simply a history "from magic to science," as the *Lancet* editorial implies. Reading signs of the hand was part of diagnostic, therapeutic, and prognostic work all along, from physician-chiromancers in the late medieval and early modern period to the endocrinologists in the 1920s who diagnosed by the shape of the hand, to twenty-first-century medical students instructed how to read the body as text.

I won't be adjudicating whether various claims—either from palmists or from orthodox physicians—were true or not, to put it simply. This is neither a history of the occult that was nonsense all along, nor, in tracking forward to the surprising uptake of hand signs in medical genetics, is it a story of palmistry having been "true" all along, folk practice persecuted then vindicated. We might very well be amused or shocked by palmistry, just as one often is by early surgery or bloodletting or hypnosis or mesmerism. Yet it ends up being far more challenging as well as interesting to think hard back into a time and place when bodies, senses, minds, souls, health, disease, diagnostics, therapeutics, and prognostics were comprehended and experienced in completely unfamiliar ways. To dismiss Richard Saunders's astrological chiromancy, for example, is, in the end, to dissemble, to pretend that we moderns are too clever for it. Yet I can barely count on one hand the scholars who can fully understand the physics of the universe *circa* 1663, still less the theologies and philosophies behind it, drawn with great historical complexity from Persian, Greek, Arabic, and Hebrew sources into the Latin and then vernacular West. Still fewer can fully appreciate Vedic palmistry, or even have access to ancient Sanskrit texts that make up *Hast Samudrika Shastra*, perhaps an original palmistry, as I discuss in the first chapter on the early languages of the hand.

From ancient, stenciled hands in caves to fingerprint wax seals, the hand and its meaning are always and everywhere. My colleagues in Chinese history have shown me how hands have long been instruments for systems of counting, of measuring, and for memory—mnemonics—as well as for future-telling and fortune-telling. "The history of Cheirology is the history of the world," one practitioner claimed in 1927, from her well-heeled address in London. Not dissimilarly, with modern London at its geographic and historical center, *Decoding the Hand* reaches outward and backward, to understand where this knowledge came from. In the first section, we look back to the expansive classical, medieval, and early modern Indo-European world. In Greek, Sanskrit, Hebrew, Arabic, and Latin, physiognomies, chiromancies, prognostics, and diagnostics were recorded in precious manuscripts and as early printed books, versions, and remnants of which the moderns—Charles Bell, Edward Heron-Allen, Francis Galton, and Katharine St. Hill—read and appreciated. Next, the particular Indo-European history of Roma hand reading is investigated, along with the fascination of eighteenth- and nineteenth-century antiquarians with the so-called Gipsy diaspora. Did they, their language, and their palmistry originate in (hermetic) Egypt or in (Vedic) India? This is a hand-based history of the Occident, the Orient, and the occult, as well as a personal history of nineteenth-century orientalists, often fluent in Roma dialects, many of whom had their palms read, and a few of whom were palmists themselves.[13]

At least one, Edward Heron-Allen, was also a Fellow of the Royal Society. And indeed, the hand was hardly immaterial to many of his distinguished scientist-fellows. It was key evidence in comparative anatomy and for evolutionary biology. In the nineteenth century, humanity's deep past was being retold through the primate hand. One day, standing upright, the hand of the primate called *Homo sapiens* was free to grasp and to evolve differently, uniquely. It is no coincidence that it was in the 1860s that the term "haptic" arose, from the Greek *haptikós*, not just to touch, but to grasp or perceive. The hand that is free for matters beyond locomotion "gives him [Man] universal dominion," Charles Darwin wrote in *The Descent of Man*. What a revolutionary idea! Friedrich Engels thought so, borrowing directly from Darwin to wrangle the human hand into his dialectics of nature, his search for the material drivers of history.[14]

In the second section, we shall explore the specific ways in which palmistry was part of the English occult revival in the decades either side of 1900, and how physiognomy, phrenology, and palmistry came together in highly popular vernacular health practice. And yet, by 1900, this very booming success meant that hundreds of palmists or "chirologists" were subject to policing and to prosecution. For centuries, English law had interrupted the work of Romani fortune tellers, usually through vagrancy laws—already old in Isaac Newton's time—that criminalized the profiting from "subtle crafts" of deceit. The telling of the future was the specific offense, and so signs of the hand became *evidence*, not just in medical domains, but in legal domains too. We shall also see how a witchcraft act was pulled out from the bottom legal drawer in an effort to contain this new phenomenon, which apparently crossed too many class, gender, and ethno-cultural lines.

It was in this context that scientific and medical palmists overtly eschewed fortune-telling—prognostication—and reverted to the diagnostics of health and character, rather more in the physiognomic tradition. Some wished to detach the practice from an occult tradition completely, and certainly from the Roma tradition of *dukkerin*. Palmistry was explicitly, if aspirationally, reimagined, restructured, and reorganized as rational, not intuitive, and as systematically knowable, teachable, and examinable. It was disenchanted. This was a practice and a profession to be regulated and self-regulated, the fraudulent identified and kept at bay.

In the third section, we examine how aspirations for scientific and medical palmistry aspirations actually *became* scientific. Charlotte Wolff, "chirologist" to Aldous and to Mok, is the great crossover figure. There is no doubt that Huxley, Woolf, Ravel, and countless others perceived her, even celebrated her, as a kind of analyst-seer, yet she herself never understood hand reading in a tradition of wonder, occult, or magic. In fact, she reserved that assessment for the work of some of her medical supervisors in various asylums, who routinely deployed hypnotic therapies. Wolff, rather, comprehended her own hand-reading expertise within an anatomical tradition, beginning with Charles Bell. Twentieth-century anatomists took these accumulating ideas and gathered hand-based data from groups of humans across the world, subpopulations both proximate and remote, and applied the new tools of population genetics to read race, identity, the past, and the future. It was a doctrine of signatures for the biometric age.

That part of the primate body called the hand—*cheiro*—holds great historical and cultural weight. When sixteenth-century God stretches out his hand to Adam on the Sistine Chapel ceiling, it is life-giving; this is near-touch between Man and the Divine. When eighteenth-century Englishmen offered their handshake to people on the beaches of other worlds, it was a gesture laden with exchange possibilities or with potentially lethal misunderstanding. When twentieth-century Jane Goodall and Dian Fossey reached out to the wild chimpanzee and gorilla, iconic photographs caught first contact between Woman and Simian. Hands are symbols of seeking, reaching, connecting. And touch is the deepest sense, they say.

1.1 Johannes Praetorius, *Ludicrum chiromanticum Praetorii: seu thesaurus chiromantiae* (Jena, 1661). Reproduced by permission of the Master and Fellows of Trinity College, Cambridge, Wren Library, L.12.34.

PART I

THE EARLY LANGUAGES *of the* HAND

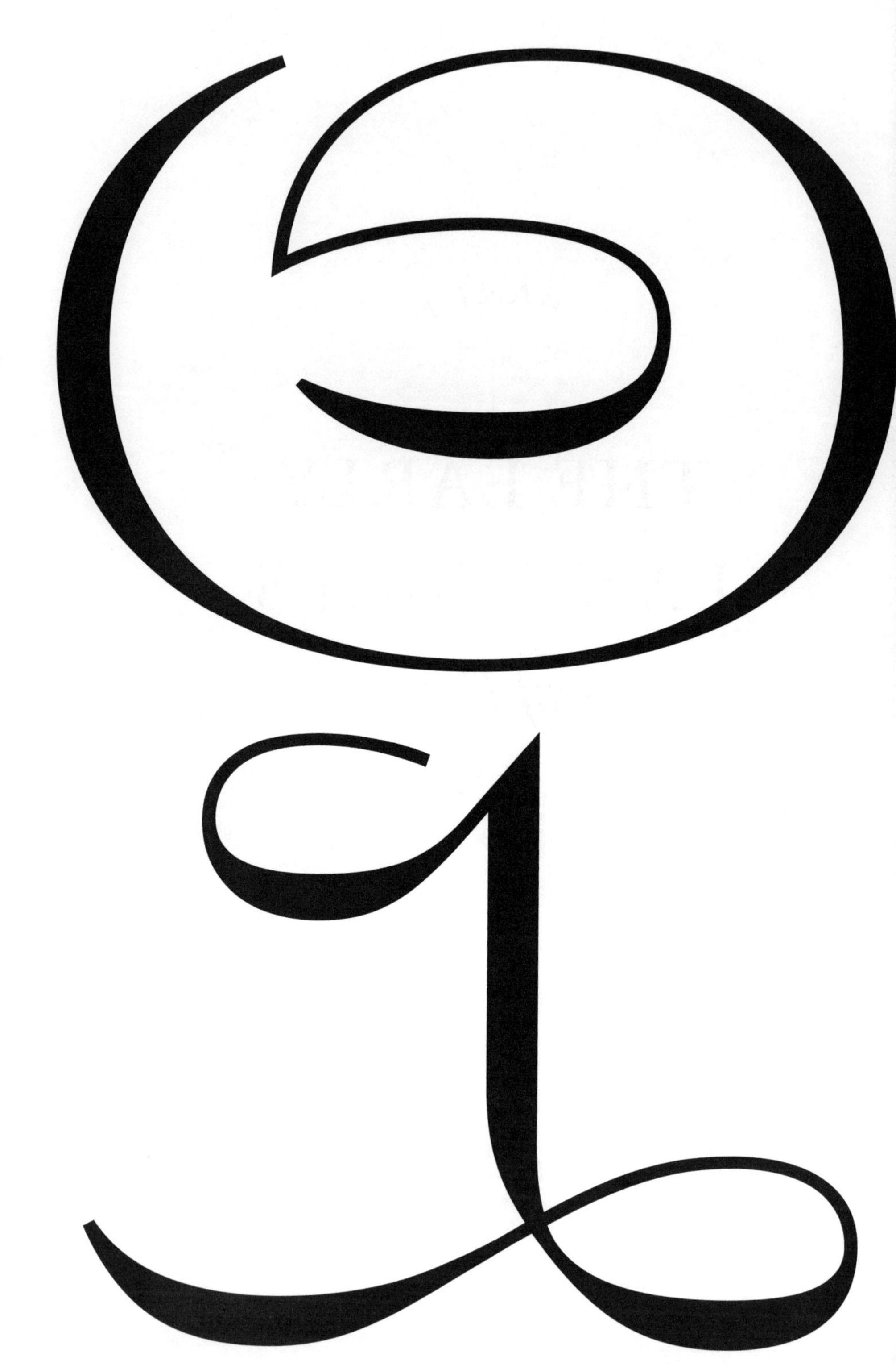

ONE

Origins

PHYSIOGNOMY AND CHIROMANCY

There is a seventeenth-century compendium of chiromancy that is as thick as a large hand turned sideways. It opens with a fine woodcut packed with verses, figures, scenes, and symbols. A reader in 1661 could decipher it more readily than most of us now. Even so, this condensed riot of early modern signs reveals the great reach of the hand across knowledges, across beliefs, across orders of men and women. It tells of theology and astrology, of Antiquity and the Scriptures, of God and false prophets, of the learned and elevated, as well as of commerce and chaos on the street. The eclectic Leipzig historian Johannes Praetorius pulled together major treatises on chiromancy, and in his woodcut lays open many hands for our consideration (figs. 1.1 and 1.2).

First there is God's hand, designed into the top right corner, proclaiming from beyond the firmament. We read Job 27, verse 1, or at least its first clause: *Docebo vos per manum Dei*: I will teach you by the hand of God. But that's not all from Job: "He sealeth up the hand of every man; that all men may know his work." It's a mysterious verse, one through which all manner of chiromancers, Christian and Jewish, were for centuries to authorize their practice.

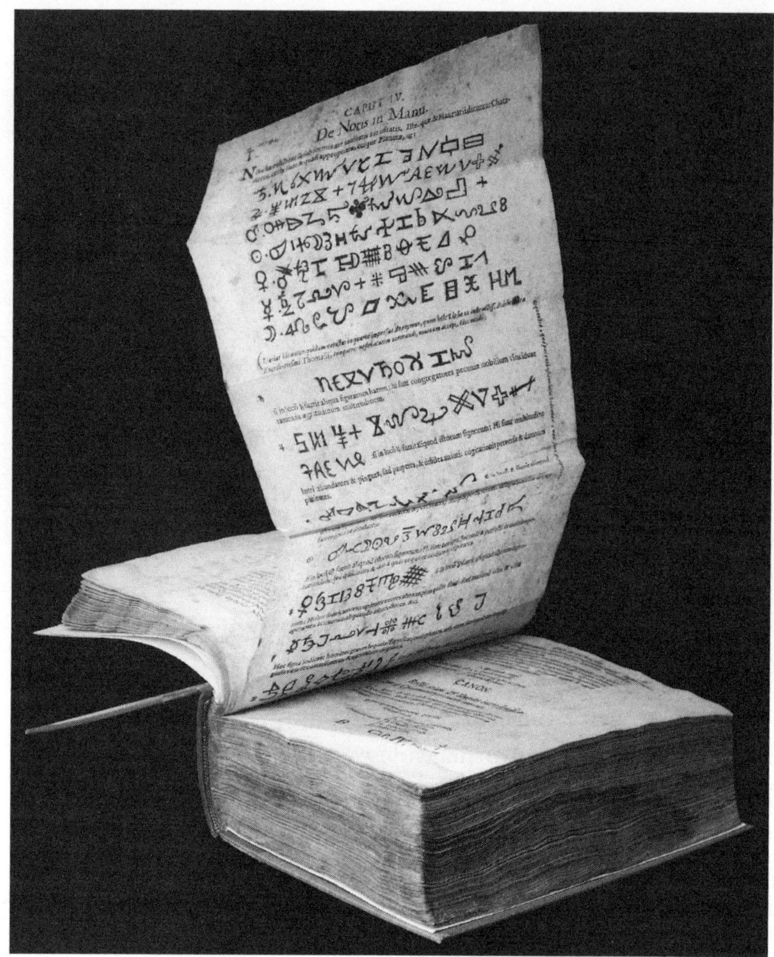

1.2 "De notis in manu." A 1661 compendium of chiromancy shows the signs of the hand to be decoded. Johannes Praetorius, *Ludicrum chiromanticum Praetorii: seu thesaurus chiromantiae* (Jena, 1661), 914. Collection/Alamy Stock Photo.

The busy woodcut then descends from the Scriptures to the Classics. The Roman lyric poet whom the English called Horace tells us something far more ordinary from his famed Epistles. A simple movement of one's thumb could show support or disfavor: a sign language. Lower again are Praetorius's seventeenth-century contemporaries, more or less, three figures who need decoding. The first is Hermes Trismegistus, half Egyptian, half

Greek, half mythic, half real, and foundational to the hermetic tradition that by Praetorius's time linked magic and science, the divine and the material. Next to him, two learned men are depicted reading palms: the reformist astrologer-priest Johannes ab Indagine ("John Indagine," as translators of the *Book of Palmestry* "englished" him), and the Paracelsian professor and physician, Rodolphus Goclenius, author of *Physiognomica et Chiromantica Specialia* (1621). But this compendium was not just about the lofty and the learned. At the base of the woodcut are scenes of everyday chaotic sociability. On the street, everyone is deceiving everyone else, and only the pickpockets are gaining. There is a scriptural warning about *Pseudo-prophetis*: "Beware of false prophets!" By 1661, commercial and trickster palmistry was already linked to the *Zigeuner* of the German lands, or those whom the English called "Egyptians." Bringing it all together in the center are two large hands and a face, each marked with signs and symbols from the astrological alphabet, as well as stars, circles, lines, and crosses.

In medieval and Renaissance cosmologies, the physical world contained signatures of the Divine. And so, lines and shapes on the hand could be conceptualized as "engraven by God and Nature," as one astrologer put it. The planet Jupiter and the finger of Jupiter were sympathetically related, for example, and communicated "by reason of the inseparable connexion and cognition of the superior bodies with the inferior, the Macrocosme with the Microcosme." Physician Richard Saunders likened signs on the hand to a kind of Divinely inspired, celestially linked somatic heraldry, by which the planets would "blazen" the hand with signs. Such marks were literally a code, a language of the hand, and marks of a cosmology through which human interiors, fleshly exteriors, and various Divines were connected.[1]

ORIGINS AND SECRETS

Claims about the origins of chiromancy form a major part of its own long history. It is a constant in the sources: in medieval manuscripts, in early modern compilations of revived ancient knowledge, in medical palmistry "grammars" from the 1880s, and even in a geneticist's history of palmistry from the 1970s. This tradition of narrating the origins of chiromancy was, from the earliest times, multilingual and multicultural: Hebrew manuscripts tell stories of Sanskrit origins; Arabic manuscripts tell stories of Persian and Greek origins; Latin manuscripts tell stories of Divine origin. These ancient and medieval worlds were often far more connected than we might imagine. Scholars of Judaic mysticism, for example, study an early translation into Hebrew of an Arabic text, *Re'iyyat ha-Yadayim le-Eḥad me-Ḥakhmei Hodu* (Reading the Hands by an Indian Sage). The palm-reading sage is even named in thirteenth-century Hebrew versions as "Nidarnar." Across each of these connected traditions, physiognomy and chiromancy were typically linked. Chiromancy—reading character or the future from the hand—was understood to be a more precise practice of physiognomy, the larger system of reading signs from the whole body, and especially the face.[2]

The English word "physiognomy" is Greek-derived from a set of ancient texts ascribed to Aristotle. In them, physiognomic signs were widely drawn and corporeally general. They could be "movements, gestures of the body, color, characteristic facial expression, the growth of the hair, the smoothness of the skin, the voice, condition of the flesh, the parts of the body, and the build of the body as a whole." Physiognomy entailed reading flesh as hard, firm, or smooth; movements as lethargic or rapid; the voice as deep (signifying courage) or as high-pitched (signifying cowardice). Sometimes classical physiognomy interpreted the expression of emotion, a keen and learned observer being able to discern "impudence," for example, in small, bright, and wide-open eyes that also showed "heavy blood-shot lids slightly bulging." Physiognomic signs might be a general bearing or carriage: men who hold themselves straight, who move with an easy gait, or who have great beards are fierce. Those are gentle who have moist flesh, carry themselves with their head thrown back, and have hair that starts high up on the head.[3]

Countless such aphorisms were repeated by classical, medieval, and early modern readers of bodies, and in books and manuscripts authored by those who inherited multiple versions of Aristotle's physiognomy. Soft hair signals cowardice, and coarse hair courage. Thin lips signify pride of soul. A nose that is broad at the tip means laziness. But how was this claimed to be knowable and known? Physiognomic methods were based on syllogistic argument and on correlations that were observable across the entire animal kingdom. That soft hair signals timidity and coarse hair courage drew from the observation that timid animals (deer, hares) have the softest coats, while the bravest animals (lions, boars) have the coarsest coats. This animal kingdom included variations in humans too: the inhabitants of the north, we learn, are "brave and coarse-haired," while those in the south have soft hair and are "cowardly." The face and especially the eyes held the clearest signs, "derived from those parts in which intelligence is most manifest." But what about the extremities in Aristotle's physiognomy? We learn that a strong character is signified by a large and shapely foot, one that is well articulated and sinewy. The hand, by contrast, is primarily an expressive organ, aligned with carriage and gesture: the "pathic," for example, carries his hands upturned "and flabbily," a sign of his suffering. This sign has that meaning: hence, we have bodily semiotics. Indeed, the Aristotelian tract was translated in the early twentieth century as just that: "the semeiotics [*sic*] of human character."[4]

In other Aristotelian texts, *The History of Animals* and *The Parts of Animals*, palm creases are specifically described. And from there, Aristotle set out a philosophy of embodiment. Human hands are the instrument of instruments, or the tool of tools, he famously claimed, reaching back to an earlier, Persian philosopher. Anaxagoras had written that "the possession of these hands is the cause of man being of all animals the most intelligent." But Aristotle thought this "endowment with hands" was a consequence rather than a cause of this intelligence. Manual and intellectual skills were combined, the physician Galen later agreed, as would, later again, the Renaissance anatomist Vesalius: "Man has the power to learn workmanship, to handle an instrument with the hands, the instrument of all instruments, to enquire with his reason into everything and to govern it." A philosophy of hands, reason, nature, and human nature was born and was long-lived.[5]

A second corpus of antique texts was also devoted to the interpretation of bodily signs, and had a far more intricate system for hand reading. Ancient Sanskrit manuscripts on astronomy and astrology, on divination, omens, and mathematics, included instruction on the meaning of body features, *Samudrika*, and signs of the hand, in particular, in *Hast Samudrika Shastra*, "the Indian Science of Hand Reading," as one practitioner translated in 1960.[6] Fragments survive from around 600 CE regarding this knowledge, some attributed to gods, some to sages, others that are authored by ordinary, if learned, humans. This is far more a somatic cosmology than Aristotle's physiognomy, linking a heavenly macrocosm to the surface of the human body and to inner being. And, unlike the classical Aristotelian physiognomic corpus, this was a method of divination, or prognostication, in which the hand and the palm might tell of death, birth, marriage, health, and fortunes. Marks are sometimes auspicious, sometimes they signal character, and sometimes the future. All this was (and is) linked to astrology, through *jyotiḥśāstra*, texts themselves originating in ancient Vedic scriptures, written in archaic Sanskrit.[7]

The teaching of this knowledge endured in both oral and written traditions. Manuscripts are rare, and even now many are kept secret. In the nineteenth century, a few books printed in India set out *Samudrika* in English. One translated from Marathi in 1888 explains, for example: "He on whose palm there are marks like the form of a fish, will succeed in all his undertakings, and will be blessed with wealth and many sons. There is no doubt in this." And: "If at the roots of the index and middle fingers there be a mark like the barley corn, he will be happy, own houses, and be blessed with many sons and a dutiful wife." It instructs on the proper reading of the right hand for men, and the left hand for women, and then expands the knowledge to other parts of the human body. Like Aristotle's physiognomy, one can read signs of the head, the chest, the lip, the brow, the foot, the armpits, the cheeks, the navel. And: "A man with a long penis is wretched, and with a thick one very poor, but a short one is always happy." Fingers, hair, and eyes were read especially closely for women, where the learned text reads more like a beauty imperative: "A woman whose eye-brows look joined or touch one another, whose breasts are not of the same size, that is, one small and the other large; whose speech is cruel, who quarrels with her husband, whose feet not becoming the body, such a woman is a cause of poverty to her husband." The consumer of this physiognomy is unmistakably male. Complex tables of lucky hours, of omens, and of "Time as Written by Shiva" are included in this knowledge-translation from ancient Sanskrit to Marathi to English, and packaged in 1888 as "palmistry, chirognomancy, and other prognostics."[8]

In South Asian traditions, *Hast Samudrika Shastra* represents not just a regional chiromancy, but is sometimes cast as an original knowledge that disseminated across the ancient world—Persia, Egypt, Tibet, and even China. "In the dawn of the world's history," wrote one Indian palmist-historian, "Aryan sages in India discovered that it was possible to know the latent abilities and motive powers, the faults and virtues of man from the marks and signs to be found all over his body. To this science they gave the name Samudrika Shastra." Its ancient Sanskrit texts became not just the responsibility of "a few leading Brahmin families," but also their secrets. He thought they guarded this ancient lore *too* closely, still refusing access even to those, like himself, with expertise to read and utilize the precious manuscripts.[9]

A third early language of the hand was medieval Hebrew, as well as Aramaic. Jewish physiognomic treatises are extant from the first century CE. A beautiful medieval hand is held in the British Library, from c. 1280 (fig. 1.3). There it lies in a North French Hebrew Miscellany bound with the five books of the Pentateuch (Genesis, Exodus, Leviticus, Numbers, and Deuteronomy), the Haftarot (readings from the Prophets), the Five Songs (Song of Songs, Book of Ruth, Book of Lamentations, Ecclesiastes, and Book of Esther), as well as legal texts, poetry, calendars, and the book of Tobit in Hebrew. Esoteric methods of reading the body were refined in Jewish mysticism that emerged in the Iberian twelfth century, known as the Kabbalah. Highly complex (and, like *Hast*

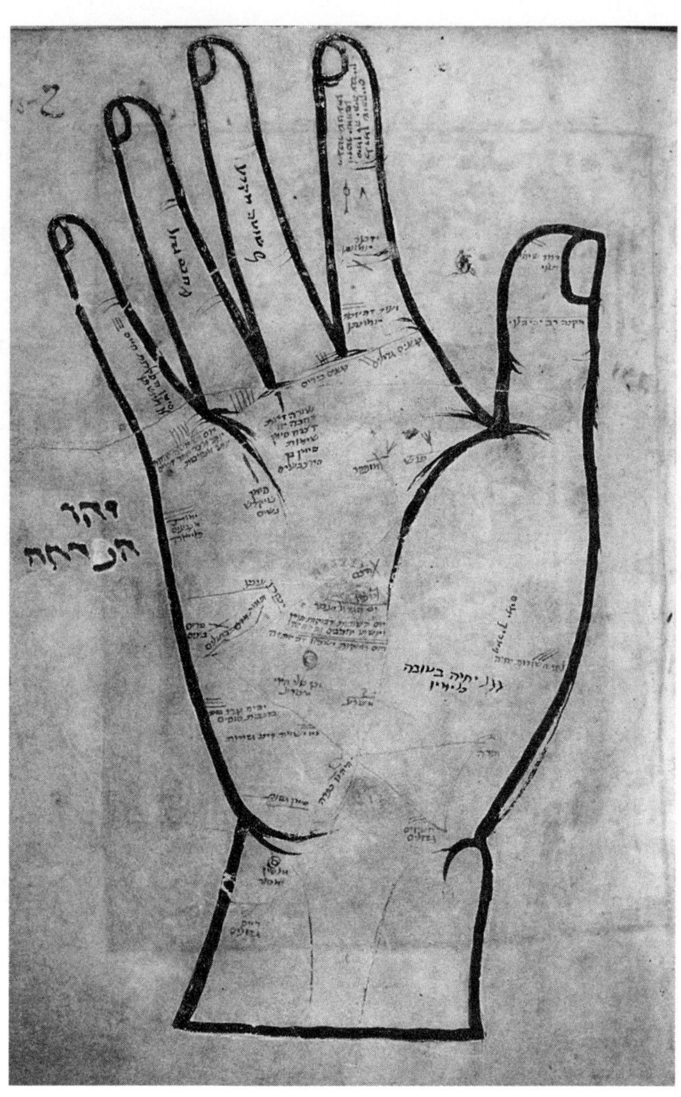

1.3 A medieval Jewish hand, c. 1280. From North French Hebrew Miscellany, British Library, Add MS 11639 f. 115r. British Library Hebrew Manuscripts Digitisation Project, Creative Commons Public Domain.

Samudrika Shastra, closely guarded and continuing to the present), the Kabbalah searches for hidden knowledge of the universe that can be discerned in and on the body, where marks of souls are engraved. Or, put the other way around, explains one scholar of medieval Jewish physiognomy, "the organs of the human body are understood to be the attire of the soul."[10] The foundational kabbalistic text, the *Zohar*, includes passages on the hand, sometimes interpreted as correspondence between the fingers and letters of the Hebrew alphabet, each signifying character, and sometimes structured explicitly through astrological correspondences between parts of the human body and celestial bodies. The idea of a macrocosm manifesting in the microcosm of the human body was to become familiar in early modern chiromancy, but few renditions are as rich as one passage from the *Zohar*, which draws the cosmological relation between the heavens and the marks of the skin: "What then is the human if not skin and flesh, and bones and sinews," it begins. "Skin, covering all, corresponds to those heavens covering all."[11] And so, celestial impressions and traces are left. "Those are stars and constellations of this skin, a covering heaven, through which concealed matters and secrets are shown and revealed." Over subsequent centuries, the Kabbalah, its core idea of ten *Sephirot* or levels of emanations of the Divine, and three levels or planes of the soul were adopted, and rewritten. Constituting a complex understanding of the inner and the outer, from the soul through the body to a Divine, many versions came to be reimagined and reincarnated. Continuous in Hebrew texts and as Jewish practice, Kabbalah was revived and Christianized in Renaissance Florence in the 1400s, then in English chiromancy and natural magic of the 1600s, and yet again in French esotericism from the mid-1800s, leading to the well-known revival of occult practices. This occult knowledge is perhaps the strongest, if strangest, manifestation of a connection between ancient and modern hands.[12]

A final set of ancient secrets of the hand lay within the Arabic *Kitab sirr al-asrar* (The Secret Book of Secrets), fragments of which in Latin became *Secretum secretorum*. This was the basis of perhaps the most copied manuscript of the entire Middle Ages, with versions multiplying not just in Latin but in Hebrew, Catalonian, English, Welsh, German, Icelandic, Croatian, and more. These translations were foundational for the wide medieval uptake of Aristotle from the twelfth century, including the classical physiognomic treatises. And indeed, a physiognomic section of the *Secretum secretorum* was often carved out, and copied as manuscripts and later in print. A 1528 English abstract of the "secrete of secretes of Arystotle prince of philosophers," for example, included detail on "the physonomy [*sic*] of people," setting out the body from head to toe. Midway are the hands, with the classic aphorisms: "Longe palmed handes with longe fingers, is ordeined to lerne many sciences, and artes, and special handy craftes, and to be of good governaunce. Fingers short and thicke, betoken foly." The *Secretum secretorum* had a long medieval and early modern life, its wide dissemination part of the reason that Aristotle, physiognomy, and chiromancy came to be so connected, even into the nineteenth and twentieth centuries.[13]

Together, Ancient Greek physiognomy, Sanskrit *Hast Samudrika*, Jewish Kabbalah, and Arabic and then Latin books of secrets signal two enduring dimensions and purposes for hand reading, which over time and place both separated and overlapped. First lies a more straightforward reading of the external body for character, for humors, and for health, clearest in the Aristotelian tradition, including its interpretation within early Arabic medical scholarship. In a manner, this was diagnostics. Within these traditions, somatic signs were generally not read for fortunes, rather for health or (what later came to be translated into English as) "character." In the Sanskrit and Hebrew traditions, however, we see a second purpose for hand reading: divination, reading the body for signs of the future or of fortune. This was not quite "prognostics," but instead "prognostication." Prognostication was part of medieval Arabic worlds too, although geomancy was more common than chiromancy: lines drawn in the sand, or sand sprinkled into shapes by which omens of the future might be read. This was *'ilm al-raml*, or a "science of sand," more mathematical than mystical or supernatural. Sixteen lines are drawn that generate four tetragrams to be read through a binary code, which is itself decoded through set procedures that capture the patterns of the four elements: fire, air, water, earth. Certainly divining, or prognostication, or fortune-telling was widespread across medieval Islamic, Christian, and Judaic worlds, evident also in Chinese (Ming) physiognomies of fortune, all as differentiated as they were complex.[14]

In medieval Christendom, divining of all kinds was theologically fraught. It did not accord with the doctrine of human free will, on the one hand, and with the belief that foreknowledge was a Divine privilege, on the other. As Thomas Aquinas put it, attempts to foresee usurp what belongs to God. Yet Aquinas also distinguished carefully between different divinatory acts and arts, between natural and artificial prognostication, between those that appeared unmediated and those involving human intervention. Geomancy in this view was deemed to be an artificial process—the sprinkling of sand—and so was not just sinful, but diabolically inspired. Chiromancy, by contrast, was not entirely dismissed, because it arguably had a natural basis; that is, the signs on the hand were present without human intervention. Similarly, dreams were natural, not artificial, and so could be interpreted with some theological legitimacy, or so some exponents argued. Other theologians, however, as well as everyday men and women of the Christian God, were less convinced, and some understood that divining from the hand necessarily involved the Devil's assistance, intruding on knowledge that was God's privilege alone, secrets that were never to be sought, let alone disclosed.[15]

The mixed medieval message of Christian chiromancy is caught tellingly in a manuscript in Krakow that shows one scribe's effort to set out chiromantic knowledge, including an image of the marked hand, and a later scribe's desperate effort to strike through, erase, and undo the record of this forbidden art. In another instance from the late medieval period, Bavarian physician Johann Hartlieb wrote *Buch aller verboten Kunst* (Book of All Forbidden Arts), setting out the full suite of divining practices—geomancy (reading lines in the soil or sand), hydromancy (reading water), aeromancy (the interpretation of atmospheres), pyromancy (the observation of flames), scapulimancy (the reading of a sheep's shoulder bone), and necromancy (summoning the spirits of the dead)—in addition to chiromancy. And a decade or so later, Hartlieb produced

Die Kunst Ciromantia, an elegant and highly illustrated text, now communicating through woodblock prints this apparently forbidden art of chiromancy, for wide consumption and in magnificent detail. In different medieval moments and places across Christian Europe, then, divining from the hand—chiromancy—was at times theologically forbidden, and at other times, and more mildly, illicit but interesting, perhaps fraudulent but not sinful, and not diabolical. Over time, chiromancy was wrangled into doctrinal toleration in England and across the rest of Christian Europe, slowly turning into a licit art, not a heretical one; part of scholastic natural philosophy.[16]

THE SLOW EMERGENCE OF AN ASTROLOGICAL HAND

One means through which chiromancy shifted from a forbidden art toward a *natural* magic, and thereby a natural science, was its gradual attachment to astrology. While this is counterintuitive to the modern mind, seeming to render hand reading more, rather than less "magical," in fact the reverse was the case. Great knowledge traditions on the heavens, calendars, and time in Persian, in Sanskrit, and in Arabic had been actively integrated in late Antiquity and in early medieval times and places, drawing astrology into Christian ecclesiastical philosophies, making it part of the natural order. In this unfamiliar cosmology, at least for a time, astrological signs could make hand reading *more* scientifically and ecclesiastically acceptable, rather than less.[17]

By the late medieval period in Christian Europe, astrological conceptions of the human body were widely appreciated, becoming orthodox, scientifically and theologically speaking. Certainly, astrological medicine was increasingly familiar, even as debates between "judicial astrology" (celestial influence on *human* life) and "natural astrology" (celestial influence on natural earthly phenomena) continued. For judicial astrologers and physicians, the whole human body came to be understood as connected to celestial structures, a complex doctrine of correspondences between parts of the body and fixed stars, constellations, planets, the twelve zodiac signs, the sun, and the moon. "Zodiac man" was a representative system that showed how parts of the body were linked to particular heavenly structures: Aries to the head, Virgo to the belly, Scorpio to the genitalia, and so forth, corresponding to conditions of the eyes or the liver or the bladder. On occasion late medieval medical practitioners considered these correspondences when scheduling procedures and interventions, for surgery or bloodletting, for example. This is why astrological almanacs—annual calendars that plotted the relative movement of celestial bodies—were important. Almanacs were even created for physicians in a portable folded or rolled form, to be unfurled at the bedside.[18]

And yet astrological *hands*—now so familiar—were slow to emerge. In many late medieval manuscripts and early modern printed books on chiromancy, astrological signs, symbols, and names are not evident. And so, while there were *medical* manuscripts built from astrological principles, many contemporary *palmistry* manuscripts were not. The earliest surviving Latin chiromancy manuscript in England, from the twelfth century, has no astrological trace. What would later become the fingers of Jupiter, Saturn, Sol, and Mercury are simply numbered. The hand's parts are described plainly as "tabula," for example, the flat part, or the table of the hand. The three main lines are simply the higher line, the middle line, and the oblique line. These form a triangle, we are told, the different shapes and angles of which mean things: evidence of theft, or of an honorable death, or death in battle. Likewise, a beautiful pair of illuminated hands from the late thirteenth century shows no astrological symbols (fig. 1.4). Nor does a unique rolled manuscript from the 1400s. The latter was written in Middle English for a female patron, suggesting that chiromancy had become both lay and vernacular, operating well beyond the worlds of men and women religious, and beyond the medieval universities. Like the medical almanacs, it was also made to be transported and unfurled, as the wealthy and literate patron read her own hands and those of others. What futures might she have foretold? Predictions might be made about marriage and about children. She might foresee routes toward wealth or ill fortune,

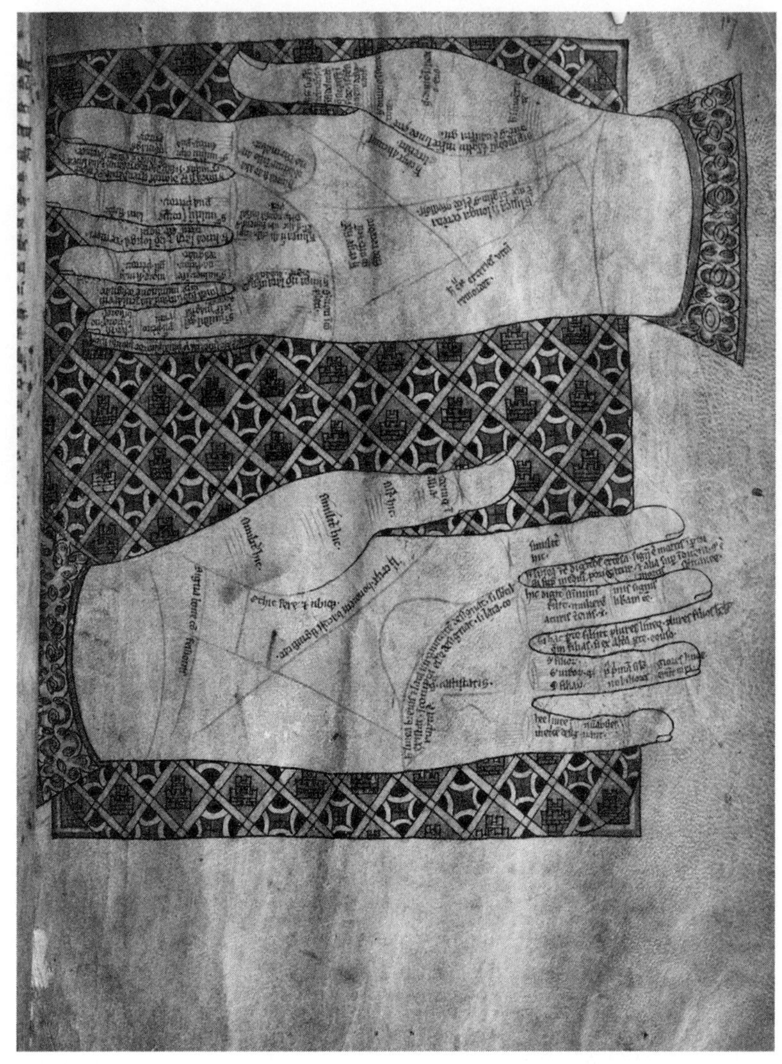

1.4 Chiromancy compiled in a medical miscellany, c. 1292. Bodleian Library, Oxford, MS. Ashmole 399, f. 17r. Photo: © Bodleian Libraries, University of Oxford.

or modes and places of death. She might show how and where to recover lost or stolen property. But in this late medieval case, she was still not interpreting signs on the hand that were astrological. Her description of the hand is simple and bodily: the line of liver and stomach, the line of life, the line of the head. The fingers are simply nominated as the little, ring, middle, and index fingers. The transfer of astrological systems and signs to the fingers, lines, and mounts of the hand was a gradual business up until the Renaissance and the Florentine hermeticism of the 1400s.[19]

Across the fifteenth and sixteenth centuries, new printing techniques meant that the occasional illuminated medieval hand turned into a whole new genre of body art. Quickly, a style of illustrated hands emerged in printed books in which astrological signs became more common, but not universal even then. Hartlieb's *Die Kunst Ciromantia* was a book entirely made up of illustrations of the hand, with symbols and text within. Forty-four woodcut blocks depict a male right hand on the recto (the right-hand page) and the left hand of a woman on the verso (the left-hand page), following from the idea—apparently well established by this time—that those are the hands on which fate is written, for men and for women. A 1490 book attributed to Aristotle, *Cyromancia Aristotilis cum figuris* (a rare copy was owned by the biostatistician Karl Pearson, Galton Professor of Eugenics at University College London in the early twentieth century), shows the female and male hand, the soft and hard hand, with text that still signals a physiognomic, not an astrological body: the line of liver, the line of the stomach, the line of life (fig. 1.5).[20]

Printed not long after is perhaps the most extensive and elaborate example of the new genre of depicted hands, chiromancy manuals in every sense. Bartolommeo della Rocca, known as Cocles, presented his *Physiognomiae & Chiromantiae Compendium* almost entirely through visual representation, not text. No fewer than 169 woodcuts set out individual hand marks: lines, stars, crosses, hatchings, circles, angles. Its purpose was to teach not just the meaning of each mark, but how to recognize it in the first place. One by one, each distinctive sign is depicted inside otherwise unadorned hands that are decoratively framed, possibly with Paracelsian botanicals. One shows clearly a single transverse crease, what nineteenth-century physical anthropologists, psychiatrists, and evolutionary biologists came to call "the simian line" (fig. 1.6). All of the illustrations in this particular chiromancy compendium are of this order except one. This shows a composite astrological

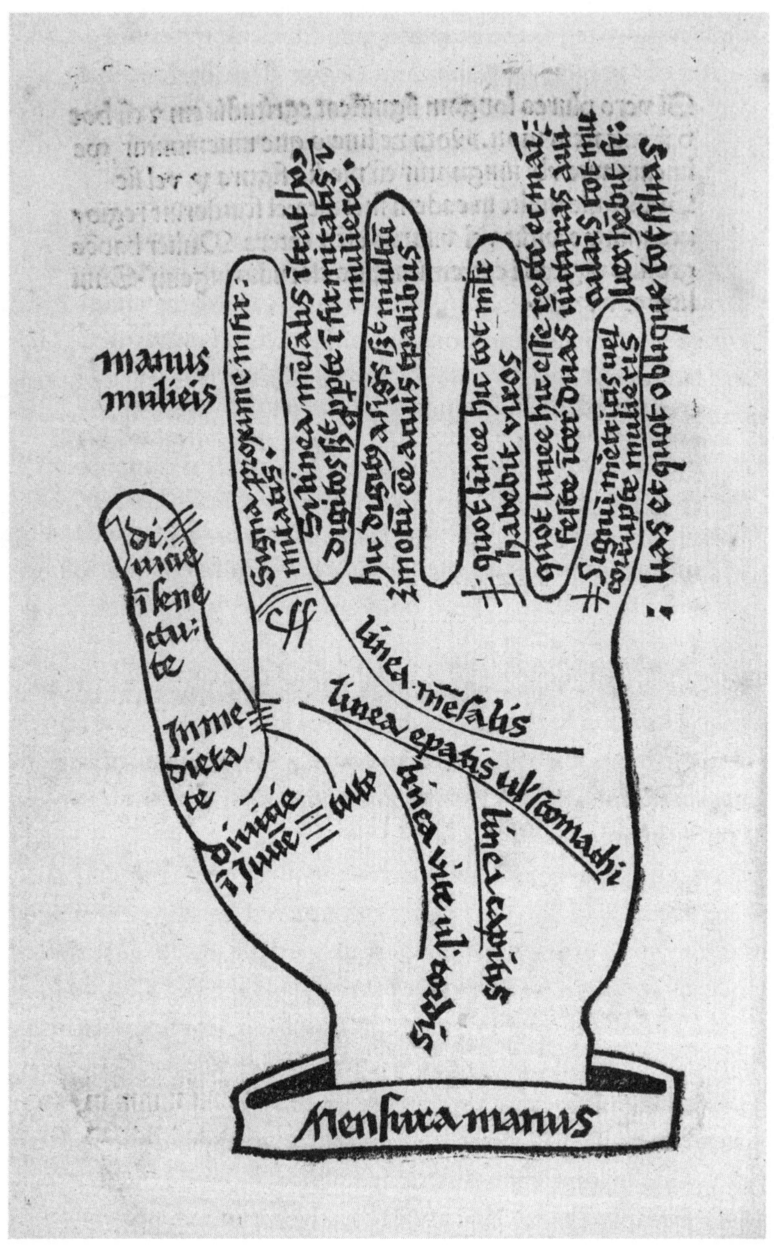

1.5 The hand of woman, "Aristotle," *Cyromancia Aristotilis cum figuris* (Ulm, 1490). This copy was owned by Karl Pearson of University College London, and bequeathed to the Cambridge University Library in 1936. By kind permission, Cambridge University Library.

1.6 In the late nineteenth century, this came to be known as the simian line. From Bartolemmeo della Rocca [Cocles], *Physiognomiae et chiromantiae compendium* (Strasbourg, 1536), plate 41. Courtesy the Wellcome Collection, London.

hand, the fingers named in middle German: *der Daum*, *Zeiger*, *Mittelfinger*, *Goldtfinger*, and *Orrfinger* (the thumb, pointer, middle finger, gold [or ring] finger, and ear finger). They were marked, but not yet named, by celestial symbols: the finger of Mercury, of the sun, of Saturn, and of Jupiter, the fields and mounts of Venus, of Mars, and of the moon. Enchantingly, other early modern chiromancers nominated "the milky way" in the hands.[21]

Cocles was "englished" in 1556 by Londoner Thomas Hill, who was in the habit of popularizing all manner of boutique knowledges. In Hill's translation, fingers were still simply numbered with some denoted for their function, or their "office": the first finger (the thumb), the second (the pointing finger), the third (the middle finger), the fourth (the ring finger), and the fifth, with which "we daily empty and clense the eares." The creases of the hand were carefully named, some descriptively, "the table line," some anatomically, "the liver line," and some astrologically, "Saturnia." In various editions, Thomas Hill used the word "palmestry," not "chiromancy." So did English translators of the German reformist theologian Johannes ab Indagine. A series of sixteenth- and seventeenth-century editions are titled as (some version of) *The Book of Palmestry and Physiognomy*. This text starts routinely with the ancient Greeks and their (imagined) "Chyromancia," but it continues far more astrologically than Hill's translation of Cocles. It explains how the names of the planets are appointed to parts of the hand "whereby judgement maye be gathered." Knowing the fingers, the mounts, the lines, and their bodily and cosmological correspondence enables "manual divinations, according to the places of the seven planets." Indeed, chiromancy is the primary knowledge of the three parts of Indagine's book, organized sequentially as Chiromancy, Physiognomy, Astrology. It was increasingly routine to explain the various parts and marks of the hand through the signs of the planets and the signs of the zodiac, common by the seventeenth century.[22]

It was physicians, in the main, who produced so many glorious compilations of hand knowledge. They were the self-declared experts in bodily semiotics, reading all manner of somatic signs. John Indagine, for example, instructed on the effect of the heavenly bodies on the four temperaments: Cholericus, Phlegmaticus, Melancholicus, and Sanguineus, and how this could be read in the hand. We shall see how this was retrieved and revived in the nineteenth century. Likewise, an early modern Spanish physician-chiromancer, Torreblanca, explained that certain lines on the hand are connected to the heart, the liver, the stomach, the brain, and the genitals. But reading the hand extended well beyond decoding simply the shape or patterns thereon: "In judging these lines you must pay attention to their substance, colour, and continuance, together with the disposition of the correspondent member." By means of the hepatic line, for example, "it is easy to form an accurate judgment as to the state of a person's liver, and of his powers of digestion." Here, medical chiromancy was a method of diagnosis. As a physiognomic reading, "complexion" also indicated a person's nature or temperament, diagnosis of a humoral disposition. In one English chiromancy from 1543, for example, readers were instructed to read the line of life according to whether the subject was choleric or sanguine. Far more than the future might be told. We might learn also about "the government of our selves."[23]

THE KABBALISTIC CHIROMANCY OF RICHARD SAUNDERS

Increasingly familiar in Anglophone chiromancy, the astrological hand was one expression of the so-called hermetic tradition. This was refined in Renaissance Florence in a highly influential revival and rethinking of early Kabbalah, Islamic astrology, and especially of a set of Greek and (purportedly) ancient Egyptian writings that came to be known as the *Hermetica*. It was so called after Hermes Trismegistus, a Hellenistic figure, part human, part god, who authored—allegedly—this hermetic tradition. Consolidating alchemy, magic, and astrology, this was "occult" or hidden knowledge brought into the Renaissance light under the patronage of the Medicis. Hermes Trismegistus appeared in Arabic, Persian, and Greek texts, and Florentine philosophers and theologians worked hard, and successfully, to synthesize the hermetic corpus with the Christian Scriptures: Hermes Trismegistus entered Latin texts as a kind of pagan prophet who foresaw the One God of the Abrahamic religions, and whose philosophies could, or were made to, align with Christian doctrine. This was the hermetic context out of which the physician, astrologer, alchemist, religious reformer, and chiromancer Paracelsus (1493–1541) acquired and practiced his arts and sciences. In England, esoteric hermetic ideas were extended influentially by mathematician, physician, and kabbalist-inspired Oxford-educated occultist Robert Fludd (1574–1637). And by the time Praetorius gathered together his riotous chiromantic compendium in the 1660s, which included Fludd's work, the figure of Hermes Trismegistus on its title page was familiar to scholars right across Europe, including young Isaac Newton in Cambridge. Reconceived via Christianized versions of Kabbalah and hermeticism, then, chiromancy had become a cosmology. From the lines and signatures "engraven in the hand," not only might our own selves and souls be discerned, we might also get to "the soul of the world," as it was put in one kabbalistic chiromancy of 1652.[24]

This is the tradition in which seventeenth-century physician Richard Saunders practiced and comprehended his chiromancy and physiognomy. Born in Warwickshire in 1613, Saunders was perhaps best known for his almanacs: tables that set out the movement of the celestial bodies so that medical procedures could be ideally timed. At one level, observing the hands was part of any physician's diagnostic routine: "in their visitation of the sick that by their face and hands they may discover their condition." At another level, this diagnosis was possible because of a profound connection—correspondence—between the face, the hands, the stars, and the planets, an astrological medicine that by the mid-seventeenth century had reached, and perhaps passed, its high moment, and that Saunders and others were keen to sustain, or perhaps revive.[25]

Richard Saunders was a friend of master astrologer William Lilly, whose endorsement helped authorize Saunders's standing and his several books, especially his lavish, expensive, and highly illustrated *Physiognomie and Chiromancie*. In it, Lilly ostentatiously congratulated Saunders for making public "in the English Tongue" an apparently new science of chiromancy. It was far superior, Lilly applauded, to the work of long-winded Cocles, rustic Indagine, or abbreviated Goclenius. Indeed, "not a man of all Europe comes near him." Little wonder that Saunders published this praise at the beginning of his book, alongside a dedication to the influential Elias Ashmole, founding Fellow of the Royal Society of London for Improving Natural Knowledge. Ashmole was also an antiquarian and collector, including of ancient and medieval chiromancy manuscripts. His collection is now in the Bodleian Library in Oxford and forms the foundation of the Ashmolean Museum. As we shall see, any number of nineteenth- and twentieth-century palmists and antiquarians—some scientist-fellows of that very Royal Society—sat with this Ashmolean collection of chiromancy manuscripts, recapitulating the search for origins and secrets. I have done the same myself.[26]

In addition to being a founding member of the Royal Society, Elias Ashmole was founding Fellow of the London-based Society of Astrologers, along with William Lilly and Richard Saunders, meeting from 1647 until the 1680s. They were brought together by mathematics as well as the stars, and by a strong sense that astrology was under attack, worse still that it was occasionally mocked, and so needed reauthorizing theologically and scientifically, and explaining to a much wider audience. It was in this context that Saunders's *Physiognomie and Chiromancie* was written, read, published, and republished not just in new editions "much enlarged," but in pocket editions as well, including *Palmistry, the Secrets thereof Disclosed*, acquired by Newton.[27]

Saunders set down his compilation of prognostications from the whole body. It included metoposcopy, through which lines of the forehead were read, as well as methods for the reading of moles: "the symmetrical proportions and signal moles of the body: fully and accurately handled, with their natural-predictive-significations." His book was also about "the art of memory," and the interpretation of dreams, already well known from medieval divination. Saunders set out a chiromancy that began with the fundamentals: "There are seven planets." They each have power over "inferior bodies," and they each govern parts of the body "and especially have their Material significant position in the Hands." He noted, immediately, the character or temperament of the marks conventionally attached to Saturn, Jupiter, Mars, Sol, Venus, Mercury, and Luna. Then, the second fact: there are twelve signs of the zodiac. But what is the zodiac? "It is nothing else but an imaginary Circle in the Heavens, regulating the Year, Months and Seasons." That circle is divided into 12 and then into 360, and the sun moves through the middle, but obliquely. There are four principal parts of the Zodiac linked to the Seasons, he explained, and as the sun, moon, and planets move, the seasons are created and from this "proceeds the generation and corruption of all things sensible and insensible." Saunders set out the "antipathies" and "sympathies" between what is inside the body, what is on the surface of the human body (hand creases, moles, forehead lines), and what is far beyond the body in the celestial spheres (fig. 1.7).[28]

More or less contemporaneous with Praetorius's compilation in 1661, Richard Saunders's book also represented layers of accumulated knowledge, from which he borrowed freely. Yet if Praetorius pulled it all together as a codex, with himself as a kind of editor, Saunders plagiarized liberally and made himself the seeming author. Still, at least he took the time to list all his sources, from Antiquity to his own present. Alphabetical rather than chronological, this bibliography reveals a wild and wonderful knowledge tradition, including ancient Greek sources via Latin translations of Arabic translations, Old Testament sources, Persian and Arabic sources, the work of medieval ecclesiastical scholars, and of hermetic Kabbalists of his own time: Albertus Magnus, Aristotle, Cicero, Fludd, Galen, Hippocrates, Johannes de Indagine, Michael Scotus, Moses, Pythagoras, Rhazes, Savonarola, Solomon, and Xenophon. By the time we come to "Z," his final three informants show the great knowledge realm out of which seventeenth-century English chiromancy arose: Zabarella, the Aristotelian philosopher from Renaissance Padua; Zoroaster, the ancient Persian prophet; and Zopyrus, the Persian trickster who cut off his own nose and ears in the Babylonian wars (so Saunders learned from Herodotus in his *Histories*).[29]

For all those well-known or little-known, real and mythical scholars, the most important chiromantic authority for seventeenth-century Richard Saunders was not ancient, but contemporary, wrote not in Latin but French, was not a Protestant chiromancer but a Catholic priest, and was not Aristotelian but kabbalistic. In 1620, the French curé Jean Belot published *Instruction familière et très facile pour apprendre les sciences de chiromance et physiognomie*. Saunders was probably working from a larger compilation, first published in Rouen in 1640, which added metoposcopy and an account of "Lullian" sciences, a twelfth-century logic and philosophy of truth designed to prove statements about God through

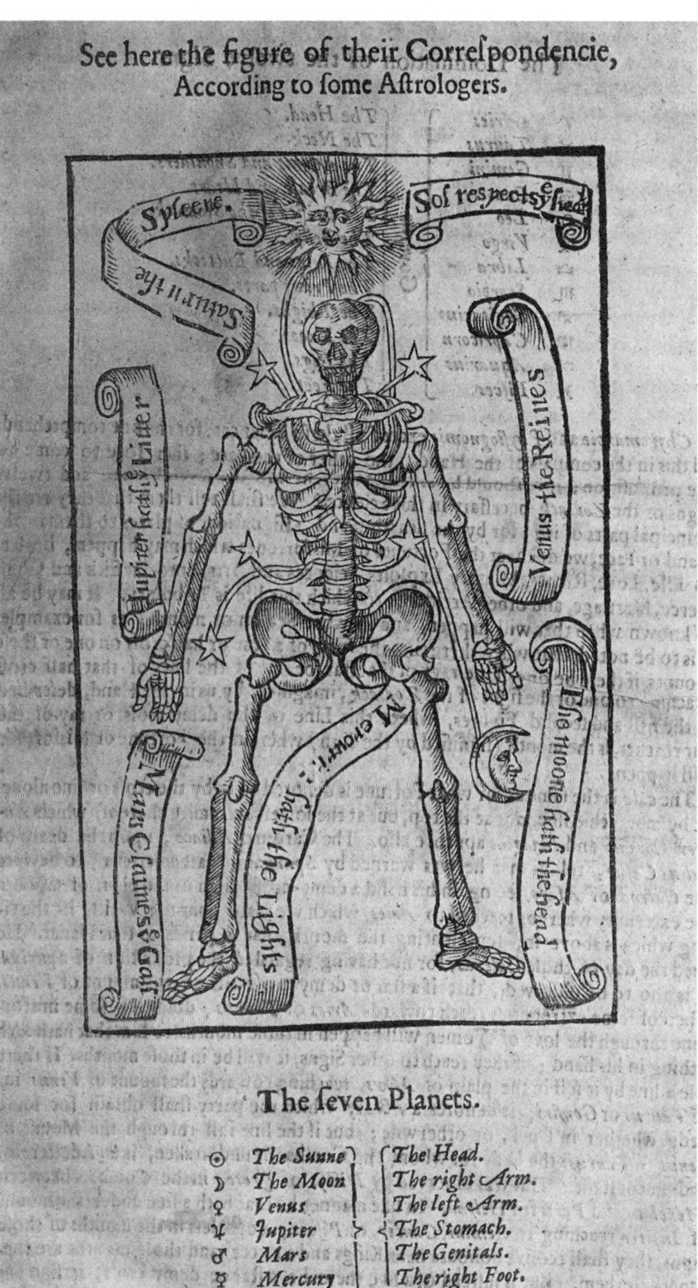

1.7 Zodiac, the man of signs. Richard Saunders, *Physiognomie and Chiromancie* (London, 1653), 23. Courtesy the Wellcome Collection, London.

complex memory diagrams. Large parts of Saunders's book were simply Belot translated and restated. We know that Isaac Newton owned a slightly later edition of Belot's work published in Rouen in 1662, and would have studied the original French version of Saunders's anglicized astrological hand, showing *oriens* and *occidens* (fig. 1.8). Via Jean Belot, the mysticism and hidden (occult) knowledge of the Kabbalah is explained by Saunders in great detail, knowledge drawn, he tells us, from Hebrews and Chaldeans, as well as Greeks and Latins. The French curé himself had been strongly influenced by Hebrew scholarship—citing astrological philosopher Abraham ben Ezra, for example—and was immersed in the Jewish texts that made up the Kabbalah specifically. The Kabbalah was being unveiled and turned into Latin, French, and English knowledge. One immensely important set of volumes was titled just that: *Kabbala Denudata*, Kabbala unveiled (1677).[30]

This was not just astrology, but an entire cosmology. A Supreme or Eternal, even a "world-soul" is the One who generates and emanates a corporeal world. Saunders set out the ten *Sephirot*, the "emanations" of this God-Infinite. There was, and is, an anthropomorphism in Kabbalah philosophy that lent itself to correspondences between bodies, meaning, spirit, soul, and world. Levels of consciousness or will are ordered hierarchically in a tree-of-life image and sometimes in an image/metaphor that corresponded to the upright human body. The first or topmost *Sephirah* is the Keter or crown, corresponding to the skull or head, for example. The left and right arms, including hands, are Chesed (kindness) and Gevurah (strength). Versions of Kabbalah describe three, four, or five "worlds," respectively of emanation, of creation, of formation, of action. Belot and Saunders denoted three worlds and indicated how they manifested on the hand. As we shall see in the next chapter, it was these three worlds that French chiromancers held to in their nineteenth-century revival of kabbalistic chiromancy.

Saunders's method of divination was so directly kabbalistic that an early chapter was devoted to explaining its hierarchies of angels and archangels, clearly stating that in doing so he disclosed natural magic secrets. Various orders of angels are apparent throughout Hebrew Scriptures, and are especially detailed and significant in

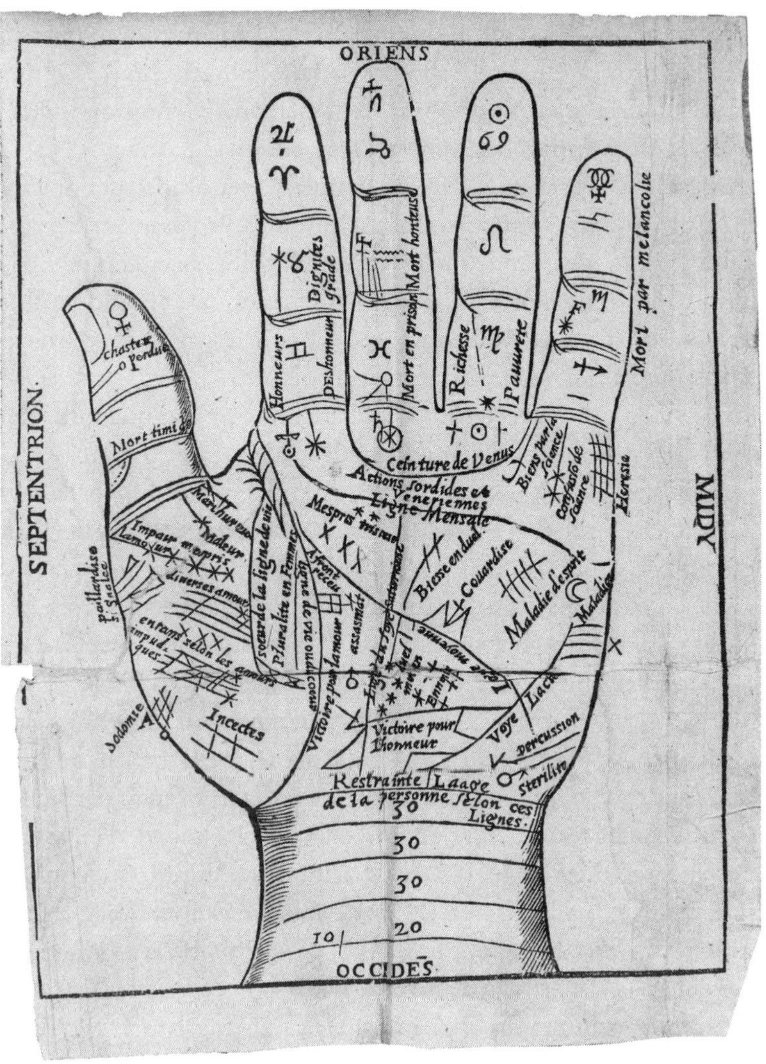

1.8 Jean Belot's astrological and kabbalistic hand, which Richard Saunders anglicized. Isaac Newton's copy. Jean Belot, *Les oeuvres . . . contenant la chiromence, physionomie, l'art de memoire de Raymond Lulle* (Rouen, 1662). Reproduced by permission of the Master and Fellows of Trinity College, Cambridge, Isaac Newton Library, NQ.16.69.

the Kabbalah. But Saunders considered his own intervention to be an improvement on the methods of the "ancient *Magi*," setting out a "shorter way" to discover one's personal Genius or Guardian Angel. We should all know, after all, "what Angel it is that Rules and Governs us." This is a manual, a set of instructions for the practitioner. "Supposing thyself to be a Professor of Chyromancy," take the hand, and observe the color, size, and breadth of the lines and veins. According to the color, one should be able to judge the humor. Red lines, he sets out in a straightforward example, signal a person with the nature of Fire, who is "Cholerick." So far, so good: this is basic Galenic reading of complexions. But then it all gets far more complicated.[31]

Red lines in the palm would signal a Fiery Genius, probably "of the Hierarchy of *Gargatel* Emperor of the fiery Region," or possibly of some princes under him, Taviel, Tubiel, or Gaviel. This, he explains, can be confirmed through geomancy techniques, and through attention to the sacred significance of Hebrew names and letters, especially *aleph*, *mem*, and *schin*, which represent three worlds, three elements, air, water, and fire. Chiromantically speaking, in a particular area of a person's hand (determined by their nativity), the (Hebrew) letter of their particular Genius can be found. But that is not all. Special knowledge, previously secret knowledge, is needed to interpret this, first according to that angel being Oriental, Occidental, Meridional, or Septentrional (fig. 1.9). Second, this is all to be calculated according to a table that sets out the 30 degrees in every sign, corresponding with different combinations of Hebrew letters, themselves drawn from intricate readings of time and date of birth (fig. 1.10).

This, Saunders proclaims, is his *simplified* method, although its complexity makes it almost incomprehensible to most of us. His apparently quick, easy, efficient, and for him modern method ends up like this example:

> According to our latter figure, if it be of Aries, and that thy Nativity happen in the first degree, which is vulgarly called the Head, but mark what Letter happens on the Degree of thy Ascendant from one to 30 Degrees; take the Letter thou findest on

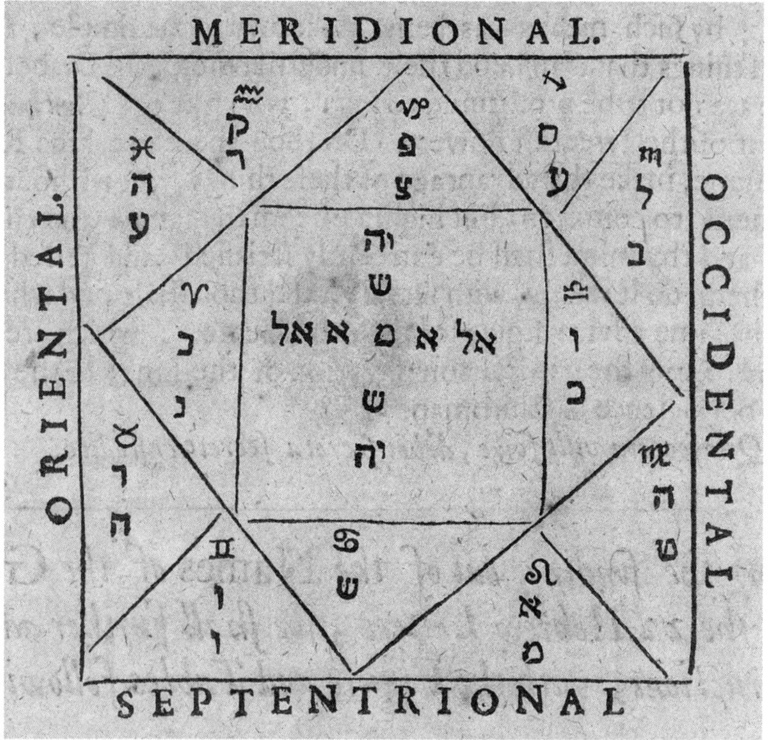

1.9 Richard Saunders's rendition of Kabbalah astrology, chiromancy, and geomancy, in *Physiognomie and Chiromancie* (London, 1671), 48. Copy in author's possession.

> that Degree, being the Ascendant of thy Nativity, and let that serve for the second which is in the following degree, and so the third if occasion serve, and then the Name of God according to the Climate, whether it be Oriental, Occidental, Meridional, or Septentrional.

Moreover, attention to the day and time of this reading offers "a more full discovery." Choose a day when the sun enters the first, seventh, fourteenth, or twenty-first degree, and when Mercury is in a good aspect. Then, on that particular day and before dawn, look toward the place of the sign, "whether it be East, West, North or South, and that with Prayeres to God." There you will find him, which is to say, one's personal Angel, or Genius. Once found, and

ש א

	♂	♀	☿	☽	☉	☿	♀	♂	♃	♄	♄	♃		
	♈	♉	♊	♋	♌	♍	♎	♏	♐	♑	♒	♓	D	
1	כ	ל	ל	ש	מ	כ	י	ט	ט	ח	ח	ל	♈	
2	ב	כ	כ	מ	ש	ל	ב	י	י	ט	ט	כ	♏	
3	ה	מ	מ	א	א	ב	ל	כ	כ	י	י	ס		
4	ו	ע	ע	ש	מ	ס	כ	ל	ל	כ	ס	ע		♂
5	ז	פ	פ	מ	ש	ע	ס	נ	נ	ל	ל	פ	♉	
6	ח	צ	צ	א	א	פ	ע	ס	ס	נ	ר	צ	♎	♀
7	ט	ק	ק	ש	מ	צ	פ	ע	ע	ס	ס	ק		
8	י	ר	ר	מ	ש	ק	צ	פ	פ	ע	ע	ר		
9	ב	ה	ע	א	א	ר	ס	צ	צ	פ	פ	ח		
10	ל	ע	כ	ש	מ	ח	ר	ק	ק	צ	צ	ע		
11	נ	כ	ב	מ	ש	ע	ת	ר	ר	ק	ק	ב	♊	
12	ה	ב	ר	א	א	ב	ע	ה	ה	ר	ר	נ		☿
13	ע	ד	ח	ש	מ	נ	ב	ע	ע	ח	ח	ר	♍	
14	פ	ו	ו	מ	ש	ד	ר	ב	ב	ע	ע	ה		
15	צ	ז	ז	א	א	ח	ד	ב	ב	כ	כ	ו		
16	ק	ת	ח	ש	מ	ו	ח	ד	ד	ב	ב	ז		
17	ל	ט	ט	מ	ש	ז	ו	ה	ה	ד	ד	ח		
18	ח	ר	ר	א	א	ח	ז	ו	ו	ה	ח	ט	♋	☽
19	ע	כ	ב	ש	מ	ט	ח	ז	ז	ו	ו	י		
20	ב	ב	כ	מ	ש	י	ט	ה	ה	ז	ז	כ		
21	נ	נ	נ	א	א	ב	ו	ט	ט	ח	ח	ב		
22	ר	ר	ד	ש	מ	כ	כ	י	י	ט	ט	ג		
23	ה	ת	ת	מ	ש	נ	ב	ב	ב	י	י	ד	♌	☉
24	ו	ו	ו	א	א	ד	נ	כ	כ	כ	כ	ח		
25	ז	ז	ז	ש	מ	ה	ר	נ	נ	ב	ב	ו		
26	ה	ת	ה	מ	ש	ו	ה	ד	ד	נ	נ	ז	♐	
27	ט	ט	ט	א	א	ז	ו	ה	ה	ד	ד	ה	♓	♃
28	ר	י	ר	ש	מ	ח	ז	ו	ו	ז	ח	ט		
29	ב	כ	ח	מ	ש	ט	ח	ז	ז	ו	ו	י	♑	
30	ה	מ	ו	א	א	י	ט	ח	ח	ז	ו	כ	♒	♄

1.10 The calculus of chiromancy. "As concerning the finding out of the Names of the Genius according to the 22 Hebrew Letters, we shall further add these instructions, with the Figure and Tables following." Richard Saunders, *Physiognomie and Chiromancie* (London, 1671), 49. Copy in author's possession.

dutifully approached and known, all sciences "Divine and Sacred" will be revealed to you, "and that without much study otherwise."[32] In this kabbalistic chiromancy, horoscope, and geomancy, signs of the hand were not just read or discerned, they were calculated.

SIGNS AND SIGNATURES

It is clear that chiromancy in the early modern period, and indeed throughout its strange journey into modern medicine and science, entailed a system of reading signs and symbols. Seventeenth-century experts were explicit: "a way of knowing the internal affections of natural bodies through signs." For Richard Saunders, the Zodiac was best described as the Sign Carrier. And another chiromantic scholar changed his name, his very identity, accordingly: Johannes Rosenbach became Johannes *ab Indagine*, "investigation or sign." The physician who put this together as a system and a whole-of-nature philosophy was the astrologer, diviner, and chiromancer Theophrastus von Hohenheim, better known as Paracelsus.[33]

Paracelsus had all manner of symbols at his disposal, for all kinds of matter that were his day-to-day business: symbols for his three primes, sulfur, mercury, and salt; for the four elements, air, earth, fire, and water; for metals that corresponded with the planets, copper with Venus, silver with the moon, iron with Mars, quicksilver with Mercury; for various alchemical compounds and alloys, sulphates, or nitric acid, for example. Signs and symbols were not just inventions of human art, however, but were also, and more importantly, evident in natural matter, through which the invisible may be discerned from the visible. "Nothing is without a sign, that is, nature does not let anything proceed from it without marking it with a sign of what it entails," Paracelsus wrote. "It follows from this that the physiognomy and chiromancy of natural things should be understood to the highest degree by every physician." This was known as his "doctrine of signatures," often applied to plants, but for Paracelsus applicable across all nature, including the human body, including hands.[34]

Paracelsus set out methods for reading these signatures, and chiromancy was foremost when it came to the interpretation of signs on human bodies. This he understood as the study of the extremities, of feet as well as hands. Chiromancy as a first method was followed by physiognomy, reading signs of the face and the head, and then the shape of the whole body, as well as manners and gestures. All these "signatures" together, read correctly, would reveal the nature of the inner or hidden individual. These were the arts—uncertain arts, he called them—that physicians needed to acquire, and with them, access to a hidden world, an invisible realm, was enabled. Knowledge that had been sequestered in various books of secrets was now knowledge from which all could learn. A manual for interpreting hands, the Paracelsian *Chiromantia* offered a theory or model, or even a metaphor for reading underlying signs throughout nature. It was an art for decoding messages that were hidden: in a mountain, in a plant, in a stone, in a hand. These were *signata*, "signatures of things." This was not just a language *of* the hand, but writing *by* a Divine Hand. "The Finger of God hath left an Inscription upon all his works," wrote the French-, Dutch-, and Italian-educated physician Thomas Browne in his *Religio Medici*: "By these Letters God calls the Stars by their names; and by this Alphabet Adam assigned to every creature a name peculiar to its nature." Browne described what is written by the finger and the pen of God into human faces and hands, readable through physiognomy and chiromancy: "There are mystically in our faces certain Characters which carry in them the motto of our Souls, wherein he that can read A. B. C. may read our natures." These figures in the hands, he continued, are "delineated by a Pencil that never works in vain."[35]

Readable in and on actual flesh, facsimiles of this Divine inscription were also transferred, transcribed, and re-presented on vellum and paper, in the many hand illustrations that thrived in the new print culture. Different semiotic systems unfolded across the several traditions of chiromancy, and we shall follow these languages of the hand into the nineteenth and twentieth centuries, even to a London physician's calls in the 1920s for a new philosophy of language to inform medical sign reading (chapter 9). In the early modern period, one system was strictly symbolic, the codified astrological alphabet, written onto the representation of the hand. The planet Jupiter and the finger of Jupiter corresponded, for example, and in countless prints and woodcuts that finger was indicated by Jupiter's symbol, originally a *Z* for zeta, denoting Zeus, with a stroke through it to indicate abbreviation. This morphed into Jupiter's symbol, ♃. But "Jupiter" might also be written out in Roman, Arabic, or Hebrew script, another set of signs, or technically "glyphs," that required particular expertise to be properly understood: literacy (fig. 1.11).

A second kind of sign was more strictly representational, copying the form and shape of the mark in the palm itself: the lines, the crosses, the stars, the grids, the circles that one might literally see. The text in the earliest medieval chiromancy manuscript in England reproduced such visible shapes, and ascribed meaning to them, including an acute triangle that denoted a stipend furnished by the Church, or a line with crosses through it that signified as many captivities as there were crosses. And "if around the foot of the first natural line (a mark) like a 'c' should occur . . . he will be a bishop." The foldout page of Praetorius's compendium (fig. 1.2) similarly instructed on patterns to be read directly from the hand. And later again, a 1675 *Polygraphice* on the techniques and arts of drawing, engraving, etching, limning, painting, washing, varnishing, gilding, coloring, and dyeing instructed also on the drawing of hands, explaining good and evil lines, marks, and characters. This was also written by a medical doctor (fig. 1.12).[36]

In South Asian traditions, signs were frequently figurative. They were not abstract shapes, but marks that looked like a known object, animal, or plant: "He on whose palm is a mark resembling a mountain, a bracelet, a female's womb, a human head, will become a king's minister." And "He on whose palm there are marks like a pair of scales, a village, a thunder-bolt, will be a successful trader." Such marks might include a spear or an earring or a trident. But symbols as well as figurative signs might also be discerned; "the mark of a *Swastik*," for example, on the palm of a woman's hand and on the sole of her foot, means she is destined to become a queen.[37]

2. Chyromancy, or

2 The touchins & cuttings thereof. 3 The placeing of the same.
Also, unto the Hand are appropriated the 7. Planets, & 12. Signes.

The 7. Planets. are

1.	2.	3.	4.	5.	6.	7.
♄ Saturne.	♃ Iupiter.	♂ Mars.	☉ Sol.	♀ Venus.	☿ Mercury.	☽ Luna.

The 12. Signes, are

1.	2.	3.	4.	5.	6.
♈ Aries.	♉ Taurus.	♊ Gemini.	♋ Cancer.	♌ Leo.	♍ Virgo.

7.	8.	9.	10.	11.	12.
♎ Libra.	♏ Scorpio.	♐ Sagittari9	♑ Capricorn9	♒ Aquarius.	♓ Pisces.

Now, to the Thumbe belongeth ♀ Venus; to the fore finger ♃ Jupiter, to the midle finger ♄ Saturne

1.11 A physician details the planets and twelve signs. "Chyromancy or Palmestrye. Also Physiognomy and Metoposcopie." 1648, MS 8727. Courtesy the Wellcome Collection, London.

The lines and angles of the palm also invited a more straightforward geometry. A Middle English manuscript (c. 1440) held the triangle in the hand to be the most consistent shape to be read. If its lines were of equal length—an equilateral triangle—this signaled steadfastness, amiability, and "by nature a perfect constitution." If the angle made by the line of life and the line of the

A note concerning the Good and Evil lines, marks or Characters.

XX. *The good lines marks or Characters are* parallels as = or || double or treble and the like, Croſſes as + or ×: double Croſſes and the like: Stars as the Sextile Aſpect ✱ or the like: Ladders-ſteps and Quadrangles as □ or ▭: the trine aſpect as Δ: Angles as the right or acute, or a mult-angle, &c. the Characters of *Jupiter* and *Venus*, as ♃ ♀, and other the like a kin to theſe.

1.12 Seventeenth-century signifiers. "Good and evil lines, marks or characters." William Salmon, *Polygraphice: or, The arts of drawing, engraving, etching, limning, painting, washing, varnishing, gilding, colouring, dying, beautifying and perfuming . . . To which is added, a discourse of perspective and chiromancy* (London, 1675), 406. Courtesy the Wellcome Collection, London.

stomach and liver was small and sharp, it betokened avarice and covetousness. Indagine's printed books included chapters "Of the Triangle" and "Of the Quadrangle." Much later, in the transfer of learned knowledge into a vernacular eighteenth century, a simple geometry of the palm continued to be explained. A tiny pamphlet titled *The Conjurer's Guide!* set out that "the line of the brain usually called the liver line, reaches to the table line, making a triangle."[38]

In his abridged *Palmistry*, Richard Saunders included another set of signs: not those engraved into the hand, but those made by the hand, a sign language that communicated characters of the English alphabet. Inner writing became an outward reading. Saunders was borrowing again, however, this time from physician John Bulwer, for whom the gestural significance inherent in physiognomy was reworked into codes of silent language. Yet this was not intended for mute humans. Bulwer speculated upon gesture as a pure and unmediated communication, direct from the mind to the hands, "the chiefest instruments of eloquence." The meaning of gesture was understood by humans everywhere, he thought: "For, after one manner almost we clappe our Hands in joy, wring them in sorrow, advance them in prayer and admiration; shake our Head in disdaine, wrinkle our Forehead in dislike." He was presenting gesture as a universal language, spoken with the hands. This was another extension of physiognomy, in which a natural language of the Hand and the natural language of the Head were aligned; "Chirologia" and "Cephalelogia," he called them, respectively. The hand was an extension of the mind, an idea we shall see developed by nineteenth-century physiologists (chapter 3) (fig. I.5).[39]

Bulwer's *Chirologia* was a study of the senses: "For as the Tongue speaketh to the Eare, so Gesture speaketh to the Eye." But here, "tongue" was not just a metaphor for language; rather, Bulwer presented the hand as itself speaking as a kind of second mouth: "The Hand, that busie instrumente, is most talkative, whose *language* is as easily perceived and understood, as if Man had another mouth or fountaine of discourse in his *Hand*." And this second mouth—the hand—in its gestures communicated the interior in an unmediated and pure way: "the signifying faculties of the soule." The hand could do so with such authority that we might think of it as the "*Spokesman* of the Body it speakes for all the members thereof." On the title page of Bulwer's *Chirologia* is the hand of God, just like Praetorius's compendium. There also is the figure of Hermes Trismegistus, identifiable by the wings of Mercury, and by his symbol, the caduceus, the two serpents wound around the staff (not coincidentally the modern symbol for medicine). But unlike Praetorius's title page, we are explicitly shown here everything that the hand can signify and communicate on behalf of the mind: intellect, memory, knowledge, will, arithmetic, elocution, oratory, logic, and reason (fig. 1.13).[40]

EIGHTEENTH-CENTURY HANDS

Bulwer's seventeenth-century *Chirologia* and Saunders's hermetic and kabbalistic *Physiognomie and Chiromancie* were the high-water mark of learned Anglophone languages of the hand. By the eighteenth century, English chiromancy had crossed over into a lay

1.13 John Bulwer, *Chirologia: Or, the Naturall Language of the Hand* (London, 1644), frontispiece. Courtesy the Folger Shakespeare Library, Call #: STC B5462. Image: 51106.

register, and in that dissemination its high learning was diminished. Like physiognomy more broadly, chiromancy became for many an attenuated curiosity, part of fortune-based leisure and pleasure, even a game. Cheap and simple how-to-tell-fortunes chapbooks began to be printed over the eighteenth century, not a few in Grub Street. There was certainly a market for them, bolstered by an already longstanding association between "Gipsy" lore and palmistry. Yet these kinds of fortune-telling booklets are better understood as simplified renditions of early modern chiromancy and physiognomy, rather than as documentation of the oral *dukkerin* of Romani women across the British Isles.

At the same time, the slippage between ancient "Egyptian" hermetic chiromancy and folk palmistry of the Egyptians or Gipsies was useful for those increasingly aware that Gipsy lore itself was marketable in a new and cheap print culture. *An Old Egyptian Fortune-Teller's Last Legacy*, printed at the Flying Horse, Grub Street, set out in 1775 a range of simple methods: the wheel of fortune told by pricking with a pin, fortunes told by the roll of dice, by the reading of moles, and by the interpretation of dreams: "To dream you fly, signifies hasty news of strange things . . . To dream you swim in a tempestuous water, denotes trouble . . . To dream of the cackling of geese, signifies troublesome visitants . . . To dream you are at a feast and eat greedily, denotes sickness."[41]

The art of physiognomy was summarized too, detailing the significance of lines in the face, as well as marks of the hand (fig. 1.14). In 1785, a printer in Bow Church Yard produced a similar set of thin and delicate (which is to say cheap) few pages, also explaining throwing the die, the wheel of fortune, and how to read lines in the face and hand. The latter art "is of High Esteem in most Nations," but this little pamphlet of distractions was a simplified version. *The Conjurer's Guide!* (1800) is another tiny publication for parlor or drawing-room entertainment, made to sit inside the hand of the reader. It explicitly addressed women as well as men. As if drawn straight from Richard Saunders's *Physiognomie and Chiromancie*—and each of these booklets was likely a précis of just that—the hand book explains, again, the art of palmistry, the reading of moles, and the interpretation of dreams. Character

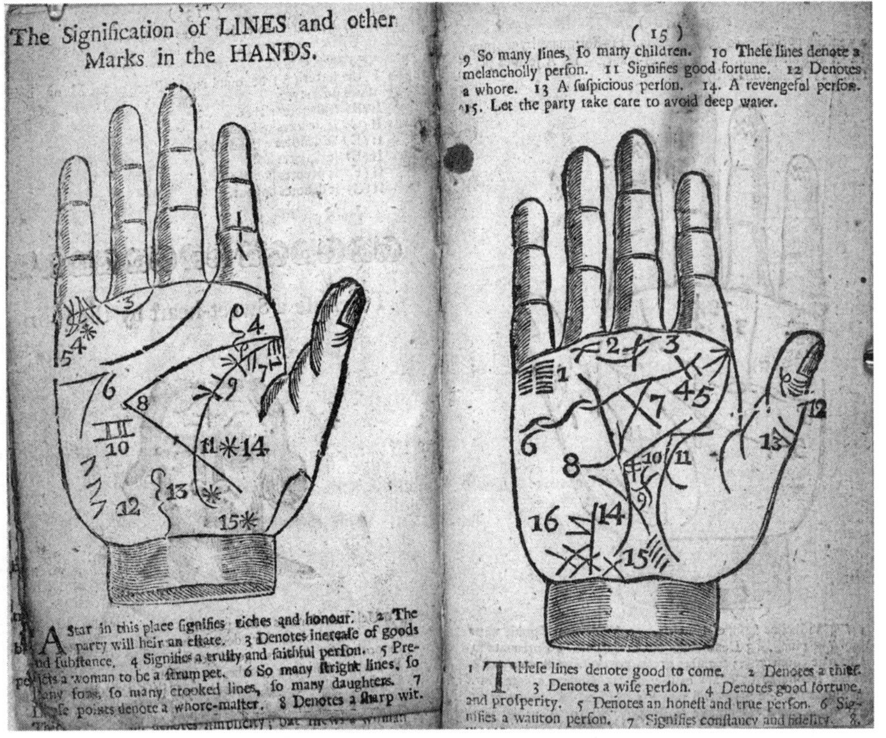

The Signification of LINES and other Marks in the HANDS.

(15)

9 So many lines, ſo many children. 10 Theſe lines denote a melancholly perſon. 11 Signifies good fortune. 12 Denotes a whore. 13 A ſuſpicious perſon. 14. A revengeful perſon. 15. Let the party take care to avoid deep water.

A Star in this place ſignifies riches and honour. 2 The party will heir an eſtate. 3 Denotes increaſe of goods and ſubſtance. 4 Signifies a trulty and faithful perſon. 5 Prejects a woman to be a ſtrumpet. 6 So many ſtright lines, ſo many ſons, ſo many crooked lines, ſo many daughters. 7 Theſe points denote a whore-maſter. 8 Denotes a ſharp wit.

1 THeſe lines denote good to come. 2 Denotes a thief. 3 Denotes a wiſe perſon. 4 Denotes good fortune, and proſperity. 5 Denotes an honeſt and true perſon. 6 Signifies a wanton perſon. 7 Signifies conſtancy and fidelity. 8.

1.14 The signification of lines and other marks in the hand. [Anon.], *An Old Egyptian Fortune-Teller's Last Legacy* (London: Printed at the Flying-Horse. Grubtree, 1775). Courtesy the British Library.

might be revealed, but simple aphorisms repeated into the eighteenth century were shorn of Aristotelian philosophy, medieval theology, or kabbalistic cosmology.[42]

On the whole, eighteenth-century lay chiromancies and physiognomies in English offered the mere surface of the astrological and hermetic tradition. It was the nineteenth-century occult revivalists who deepened that knowledge again. And yet, this secularized chiromancy that gathered up early modern learning and turned it into a curiosity was still a system of sign reading. *The Conjurer's Guide* (1800), for example, disclosed how "secret characters, marks, or letters" were sometimes inscribed on the palm. They were even called "sacred characters," but for this eighteenth-century audience, it had to be simple: "An *A* found between the mount of the moon and hollow of the hand, denotes sickness and losses . . . If a *T* be on the mount of Venus, it denotes success in love affairs; but if it comes so low as to cut the line of life, then it denotes crosses and misfortunes in love, and much mischief." This writing on the hand was certainly palmistry for the literate—those who had their letters—and yet it was semiotically immature compared to what had come before. To take another example, the *High German Fortune Teller* instructs:

> There are many certain Letters in the Hand, engraved by Nature, as G, H, L, M, and the like; these constantly, or for the most Part, stand for the Name of the Party you shall marry, that is, the first Letter of the Christian Name, and sometimes the Lines in the Hands make these. The Veins in the Wrist likewise make the Figure of Letters that answer the same Purpose.

Everything here was attenuated and secularized. If Richard Saunders had explained the meanings, at once precise and cosmological, of the Hebrew letters *aleph*, *mem*, and *schin*, the *High German Fortune-Teller* of 1785 conveyed easy letters in Latin script: "G, H, L, M." And if Kabbalist chiromancy, explained by Saunders just a century earlier, served to determine the name of one's Genius or Angel, thereby accessing secrets of all Nature and all *Scientia*, here the name to be conjured was simply that of a future spouse. Eighteenth-century laypeople might thinly appreciate that the letters were "engraved by Nature," but they could not be expected to interpret complex lines, crosses, and astrological symbols that for centuries had crowded the images of hands in Latin, English, Hebrew, French, German, Persian, or Arabic texts. They might be literate, but they were not necessarily learned. Any connection between the signs of the hand and the soul of the world was momentarily suspended, at least for these Anglophone readers of hands.[43]

Over these centuries, English and Hebrew chiromancy traveled in different directions: the former became popular, secularized, and light; the latter remained learned, complex, and cosmological. In 1700, for example, *Wit's Cabinet* was a companion for gentlemen and ladies, containing information on the interpretation of

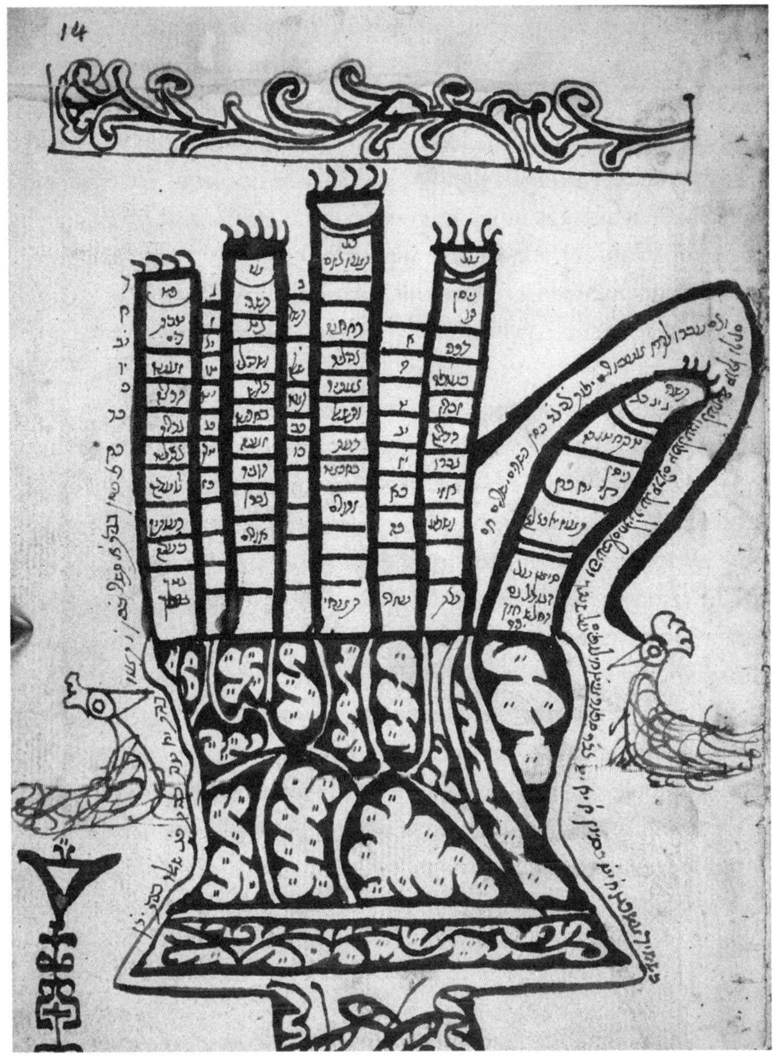

1.15 Hand diagram in Josef ben Shem Tob ben Yeshu'ah Hai, She'erit Yosef, Algeria, 1804. BL Or 9782, f. 14r. Photo from the collections of the British Library, "Ktiv" Project, the National Library of Israel.

dreams, physiognomy and palmistry, the right preparation for cosmetics, and a "choice collection of songs." At the same time, in Amsterdam, Jacob ben Mordecai's book of wisdom was published, ascribing palmistry, physiognomy, and astrology again to Aristotle (who, he claimed, had converted to Judaism). Likewise, in 1775, those who appreciated *An Old Egyptian Fortune-Teller's Last Legacy*, printed at the Flying Horse, Grub Street, would be unlikely to appreciate another 1775 book of hands and fortunes, the Kabbalist treatise by rabbi Isaac Ḥayyim Cantarini printed in Tunisia. And at the beginning of the next century in London, *The Conjurer's Guide!*—with an exclamation point—was printed in 1800 "for the benefit of young men, maids, wives," while in Algeria at the same time, a highly complex hand diagram was drawn from the work of Joseph ben Shem Ṭov ben Yeshu'ah Ḥai, to accompany his metrical account of the calendar (fig. 1.15).[44]

In eighteenth-century London, such learned books were already becoming orientalized, manuscripts collected by antiquarians, perhaps for the new British Museum Library established in 1753, perhaps for the India Museum of the East India Company. One precious copy of Mordecai's book of chiromancy was held in the library of the "White Jews of Cochin in India," and collected by the Rev. Buchanan in 1806, less as a book of wisdom than as antiquarian curiosity, sequestered as a new kind of secret to enjoy. It was now the low-end Grub Street printers who packaged and sold simple fortune books for use on the street, at the inn, or in parlors. *This* language of the hand had been not so much disenchanted, as re-enchanted into matter that "delights but does not delude," moving toward a secular magic for a rational age.[45]

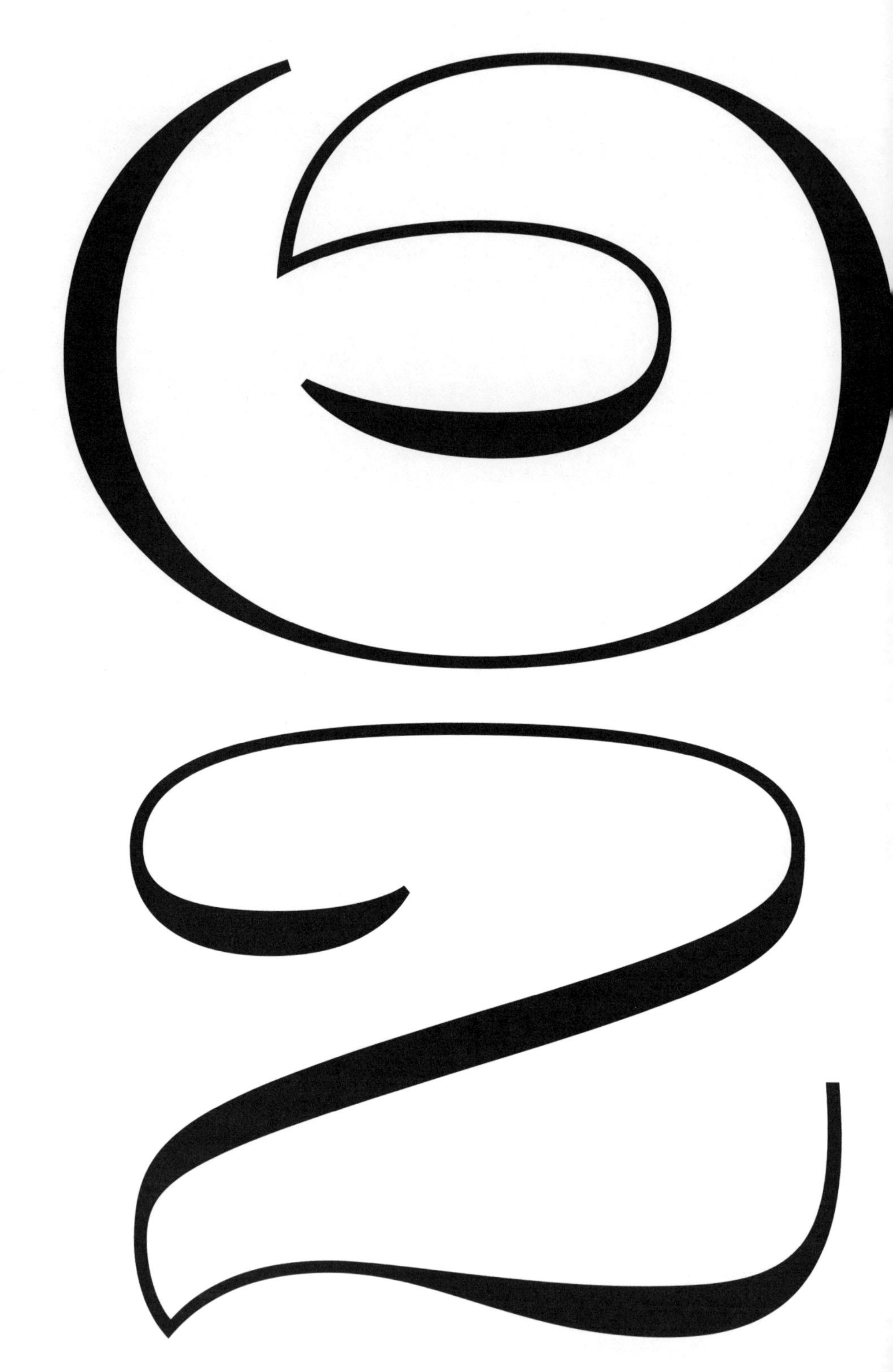

TWO

Reading the Palm Across Indo-European Worlds

Vast is the Romani word for hand, and *dukkerin* for fortune-telling. So Roma people across Europe, the Middle East, and North Africa informed a suite of nineteenth-century linguists and folklorists. One philologist, fluent in Romani as well as Arabic, Persian, Hindustani, Turkish, and Sanskrit, translated the hand in a sacred Persian text as "*dast*. Hand. God's attribute of Power." He was less interested in the hand itself, and more in the words for it, as he joined a dozen other learned orientalists who were mapping a newfound phenomenon that was as old as the ages: an Indo-European language family, sometimes called Aryan or Indo-Aryan. The origin of European Gipsies, Zincali, Çingeneler, Gitanos, Romanichal, Bohémiens, and Sinti was a clue in this mapping of connections between the Orient and the Occident.[1]

In the sixteenth and seventeenth centuries, Travellers in Britain were known as "Egyptians"—hence "Gipsies"—an association that linked them on occasion to the hermetic occult tradition, with its own imagined antique Egyptian past. In his *Religio Medici* (The Religion of a Physician, 1643), Thomas Browne put it this way: the actual ancient Egyptians had a close knowledge of mystical sciences, including chiromancy, "to which those vagabond and counterfeit Egyptians did after pretend." Perhaps they retained "a few corrupted principles," he conceded, and with these might occasionally "verifie their prognosticks," but for him the most notable skill of these modern, traveling Egyptians was fraudulence. By the nineteenth century, however, the apparent link with Egypt had fallen away, and it became far more common to assert Roma origins to be Indian or Persian, and insofar as hand reading was concerned, this linked them more to the widely practiced Sanskrit-based *Hast Samudrika Shastra* than to a Greek-Arabic-Hebrew-Latin hermetic occult.[2]

If learned men set down knowledge of the signs and language of the hand in medieval and early modern times, it was Romani women who undertook folk hand reading, a practice that endured in lanes and markets, tents and fairs, fields and town squares. Across time and place, a thousand palms were held in Romani hands, and fortunes told. This fortune-telling was also represented—overrepresented—in orientalist paintings and etchings, in linguistic and folkloric studies, in poetry and nineteenth-century novels. The "Gipsy" fortune teller was perhaps the most iconic hand reader of the nineteenth century. And yet she was not the only one. Put another way, it was not only via the figure of the Gipsy that palmistry was orientalized. A mid-century revival of kabbalistic knowledge in France fed into a wider middle-class enthusiasm for occult practices, apparently "hidden" or "secret" knowledges. This animated a new chiromancy. While it is tempting to understand this as a bourgeois incorporation of Roma folk practices, in fact it was far more directly a revival of text-based astrological palmistry of the seventeenth century.

At the same time, another manifestation of the occult revival *was* linked with South Asia, but again not via Roma origins and connections, and even in opposition to them. From the 1880s, the massively successful theosophical movement actively retrieved a South Asian cosmology of the body, explicitly linking Eastern and Western esotericism. Theosophists brought anglicized versions of *Hast Samudrika Shastra* to hungry consumers of palmistry books and journals across the English-speaking world, and even to Anglophone Indians. At the same time, South Asian palmists themselves were modernizing ancient practices in a thriving new trade in palmistry books written in Tamil or Urdu, Hindustani or English. Perhaps more money could be made selling books of the hand than reading the palm itself. Hands were reaching across the Indo-European world in unexpected ways, forging rich and strange links between the Orient, the Occident, and the occult.

EGYPTIANS IN ENGLAND

Almost as soon as the people known as Egyptians arrived in England, laws were passed, initially to remove them from the realm as aliens, and later to regulate and punish various practices, including fortune-telling by the reading of palms. In 1530, Henry VIII passed an act nominating "Egipcions," people who were "outlandysshe" or foreign. A second act was passed in 1554, in the reign of Mary I, "for the punishement of certayne Persons calling themselves Egyptians." These Egyptian Acts were not strictly or simply vagabond or vagrancy laws. Henry, for example, had introduced an entirely separate Vagabonds Act in 1530, an early poor law that established both punishment for vagrants and itinerants able to work, and to some extent relief for those unable. Yet over time, laws governing so-called Gipsies and so-called vagabonds gradually converged.[3]

Palmistry was specifically mentioned from the beginning, both in the Egyptian Acts and in subsequent vagabond and vagrancy laws. In the earliest Tudor act, Egyptians "used greate subtyll and crafty meanes to deceyve the People, beryng them in Hande [persuading them] that they by Palmestre could telle Menne and Womens Fortunes." Palmistry and fortune-telling were proscribed also in Mary's Egyptian Act (1554) and Elizabeth's Vagabond Act (1597), described again as "subtle crafts." Over the seventeenth and eighteenth centuries, even as vagrancy laws were extended to include more kinds of people—"lunatics," for example, in 1714—both Gipsies and palmistry were still specified. The 1744 Vagrancy Act under George II rendered fraudulent not just palmistry but also physiognomy, criminalizing those "pretending to have skill in physiognomy, palmistry, or the like crafty science, or pretending to tell fortunes, or using any subtle craft to deceive and impose on any of his Majesty's subjects." From these earliest years in the British Isles, as across Europe, imposture and Egyptian identity were conflated.[4]

In the next century, too, palmistry was outlawed in the longstanding Vagrancy Act (1824), which served to manage an ever-wider range of people deemed indigent. This particular law was introduced in large part to address the homelessness of thousands of decommissioned soldiers and sailors after a generation of war with the French, but it nonetheless retained a version of the clause on deceit by fortune-telling, bringing Tudor law into the modern world in the strangest way. This was the Act that was to trouble a growing number of urban palmists and clairvoyants in the age of occult revival, men and women who were definitely not indigent, rather living in fine rooms in London's affluent West End. Any number of them were charged and convicted, even into the twentieth century, as we shall see (chapter 5).[5]

Notwithstanding all this lawmaking on subtle means of deceit, the tradition of selling and telling fortunes via the palm came to be recognized everywhere. Yet who were these "Egyptians" so connected to palmistry? There were any number of stories alive across Europe about their origins: that they were descendants of the ancient Moors in Spain, or possibly of the Tartars, or of the Persians. Entering Europe in the fifteenth century, "unknown and uninvited," according to an eighteenth-century German historian, they were generically "an eastern people, and have eastern notions." But their Egyptian origin was the most frequent early modern claim. A common account was that when Roma first came into the German lands, learned men reached for the Scriptures to understand the sudden appearance of foreigners who were skilled in divination. They saw before them the manifestation of Ezekiel 30:26: "I will scatter the Egyptians among the nations, and will disperse them through the countries."[6]

This old story of how Roma came to be *imagined* as Egyptian itself fascinated nineteenth-century folklorists, antiquarians, and linguists. They decided, as a rule, that the Egyptian origin story was a mythical but useful one, expediently claimed by Gipsies across Europe. "The fame of Egypt in astrology, magic, and soothsaying, was universal," one chronicler wrote in the early nineteenth century, "and they could not have devised a more artful expedient than the profession of this knowledge." And yet, by the mid-nineteenth century, the Egyptian origin thesis was held to be untenable, after all manner of folklorists had themselves traveled to Egypt in search of Gipsies in their supposed place of origin. They proved hard to locate, and as one claimed, those he could find spoke neither Coptic nor Arabic, but Romani dialects. They were not, in other words, native to Egypt, but were foreigners there too.[7] Another commentator, writing in 1856, was unequivocal: "The Gipsies are strangers in the land of Egypt." A consensus started to emerge that Hindustan was the mother country of the so-called Egyptians, and a replacement expulsion story was attached to this idea. They were "the relics of a wandering race expelled from Hindustan," descended from a low caste who had been persecuted during the violent early fifteenth-century war of Timur Beg.[8]

The Indian origin of European, Middle Eastern, and North African Gipsies was increasingly a linguistically based claim. The German historian Heinrich Grellmann disseminated this tradition in the 1780s, declaring a striking similarity between Roma languages and the dialects of northern Hindustan. English enthusiasts in the Romantic era quickly translated his work, delighting in expanding the vocabularies that were the key evidence of an Indo-European connection. In long tables and glossaries, people could read the translations of "Gypsey" words into Hindustani, Turkish, and English, and see for themselves the connection. Dialects were increasingly refined, distinguishing between "English Gypsey," "Turkish Gypsey," "Cingari," and "Hindustani," the latter divided into "Bengalese" and "Mahratta." It was in this early search for linguistic connections that the art of palmistry was also perceived to be evidence of an Indo-European link. In *Die Zigeuner* (1783), Grellmann wrote that "Fortune-telling is practised all over the East; but the peculiar kind professed by Gypsies, viz: chiromancy, constantly referring to whether the parties shall be rich or poor, happy or unhappy in marriage, &c is nowhere met with but in India."[9]

Palmistry itself, in other words, was a cultural clue to Roma origins, and to a wider thesis on Indo-European connections. Grellmann even exclaimed that *Indian* chiromancy is "perfectly Gypseish." Over the next few generations, some of Europe's greatest orientalists and philologists pursued the claim, many of them acquiring fluency in Roma dialects. They all abound in Hindu words, it was increasingly asserted by linguists, travelers, folklorists, orientalists, and missionaries, some of whom were nothing short of obsessed with Roma society and culture: "folk" whose customs, language, economies, and rituals made for fascinating study, all amplified by the seeming mysteries of a nonliterate and secretive culture.[10]

George Borrow was one such, a Norfolk-born enthusiast, good with languages, who took long walking and riding tours over the 1830s and 1840s, sometimes independently, at other times as a representative of the British and Foreign Bible Society. By then Romani communities were visible and thriving, one source suggesting 18,000 people across the British Isles. Yet young Borrow was more interested in how the Gipsy question warranted an adventure in more exotic landscapes across Europe and Russia. After a five-year journey through the Iberian Peninsula, he wrote *The Zincali: An Account of the Gypsies in Spain* (1841), including a long appendix in which Zincali words were translated into English, Spanish, Persian, and Sanskrit. Later in life and living in London, he undertook extensive language studies of, and with, Romanichal in Wandsworth and Battersea, publishing *Romano Lavo-Lil, a word-book of the Anglo-Romany dialect* in 1874.[11]

Out of such fascinations came the Gypsy Lore Society, established in Edinburgh in 1888 and later moving to Liverpool, a small group of "folklorists"—we might say ethnographers—who investigated the language and culture, rites and ways (including palmistry) of "folk," implying nonliterate cultures. This Society's foundational research problem regarded origins, keeping on with the favored Indian thesis, including whether Romani was an early or a late descendant of Sanskrit. It was in such studies that Gipsy folklorists, and occasionally Romanichal themselves, came to be connected with scholarly orientalists who were just then deeply animated by debates on linguistic genealogies, including of Indo-European languages.[12]

Charles Godfrey Leland was another Gipsy scholar in this tradition, and the first president of the Gypsy Lore Society. A Princeton-educated American folklorist, he was a highly capable linguist, as well as devotee of magic, not because he believed in it, but because its powers and wonders enchanted him. Leland traveled widely across Europe and the Middle East, acquiring as many languages as friends. He was fluent in Romani dialects, and as fascinated as any by the idea of a North Indian origin of the Roma diaspora in Europe: "I have many Anglo-Romany words—purely Hindi as to origin—which I have verified again and again." The "gypsy numerals" *yeck*, *dui*, *trin*, *shtor*, *panj* were the same in Hindustani or Persian, he claimed. Leland was a word hunter as well as a Gipsy hunter, so to say.[13]

One day Leland regarded a man on the Marylebone Road in London and read his physiognomy: "a very dark man, poorly clad, whom I took for a gypsy; and no wonder, as his eyes had the very expression of the purest blood of the oldest families." Leland opened their conversation:

> "*Rakessa tu Romanes*?" (Can you talk gypsy?)
>
> "I know what you mean," he answered in English. "You ask me if I can talk gypsy. I know what those people are. But I'm a Mahometan Hindu from Calcutta. I get my living by making curry powder. Here is my card." Saying this he handed me a piece of paper, with his name written on it: *John Nano*.

For Leland, it was far more exciting that this man identified as "Mahometan Hindu" than Romanichal, especially as John Nano could easily understand the Romani tongue. He asked again, checking: What does "*Rakessa tu Romanes*" mean? And the man replied with more information than Leland himself either knew or expected: "It means, 'Can you talk Rom?' But *rakessa* is not a Hindu word. It's Panjabī." The London curry maker had more languages than the folklorist and linguist from Princeton.[14]

Leland brought an expert friend in to verify this claim, the eminent linguist and Cambridge professor of Arabic Edward Henry Palmer. After a childhood precociously acquiring Romani in and around Cambridge, Palmer had become fluent in Hindustani and was readily able to talk with John Nano in any tongue he preferred. John Nano recounted that he belonged to a "tribe of wanderers" who were equivalent in India to the Gipsies of Europe. He declared his people to be "the *real* gypsies of India, and just like the gypsies here. People in India called them Trablūs, which means Syrians, but they were full-blood Hindus, and not Syrians." John Nano told Leland and Palmer that his own people in India called themselves and their language *Rom*: "Rom meant in India a real gypsy." While Nano insisted that his "tribe of wanderers" were Indian, not Syrian, he also told Palmer and Leland that their language was particular, consisting of words rarely intelligible in other Indian dialects or languages. One such unique word was *manro*, which meant bread. Leland and Palmer were ecstatic: *manro* was the Gipsy word for bread "all over Europe."[15]

For such mid-nineteenth-century orientalists, Gipsy origin, Roma language, and a fascination with mysticism often went together. Accordingly, they were not infrequently both students of chiromancy and of what was starting to be called the "occult" in English, as well as scholars of languages. The multilingual Palmer, for example, was fluent in Romani dialects, as well as Persian, Hindustani, and Arabic, and he was also interested in hermeticism, magic, and divination. This translator of the Qur'an into English also published *Oriental Mysticism* from a Persian Sufi text, including an early English use of the term "theosophy." One expert praised Palmer's deep learning and great expertise, declaring him to be not just an explorer, professor, and born linguist, but also a "conjurer and thought reader." Palmer could speak Persian with such fluency that those able to listen correctly could perceive "the pure Shirázi twang." That admirer was Richard Francis Burton, the explorer, linguist, and translator of *A Thousand Nights and a Night*, one of the few Englishmen then alive who could discern discrete Persian intonations.[16]

Burton was also a member of the Gypsy Lore Society, and author of *The Jew, The Gypsy and El Islam*, in which he claimed, again, that palmistry had an origin in India and spread with Gipsies into Europe and northern Africa, including into Egypt.

> Their special privilege is the practice of palmistry and divination. The Fehemi takes the inquirer's right hand by the finger tips, and bends them gently backward so as to render the lines more visible. She mutters a spell while with all gravity she reads the book of destiny, and then reads the result; of course her hand must be crossed with silver.[17]

If Gipsies in England, originally called Egyptians, were later linguistically and culturally determined to be of Indian origin, here the diaspora in Egypt itself was linked to India by palmistry. Over many centuries, Roma *dukkerin* sat at the crossroads of oriental, occidental, and orientalist cultural geographies.[18]

DUKKERIN AND DECEIT

Early modern commentary on Egyptians in England was often explicitly about deceit, and those who should be protected from it, as from their own credulity. The early laws compounded this connection by nominating "counterfeit Egyptians," people who were passing as Egyptian, and later sometimes meaning people passing as Gipsies, and thereby performing a double pretense in the deceit of fortune-telling from the palm. Inheriting this connection, nineteenth-century folklorists also typically assessed Gipsy mysticism, including chiromancy, as clever and effective pretense.[19]

George Borrow thought it likely that Roma people had brought *dukkerin* with them "from the East," but he never considered it genuine, rather from the beginning "a means of fraud and robbery." Gipsy fortune tellers in his early nineteenth-century view were simply tricksters who would play on the wonders of the Egyptian—or hermetic—connection. His concern was to differentiate *dukkerin* from the learned and literate traditions of chiromancy. In doing so—and typically for nineteenth-century folklorists and antiquarians—he gathered information not just from the many Travellers he met on his long journeys in the 1830s and 1840s, but also from early modern authorities. In this instance, George Borrow drew from a Spanish book on magic by Torreblanca (1678), who described a "true and catholic chiromancy": the capacity to read a geometry of the hand, the lines and shapes of which connect with the brain, the liver, or the stomach. But this was entirely different from Zincali *pretense* of knowledge drawn from the hand. Torreblanca himself had declared that this was a "scandalous practice," even "a pact with the devil," and Borrow took some trouble to translate this dismissal of the Zincali capacity to tell the future—to divine—repeating and underscoring its fraudulence for nineteenth-century readers.

> A practice turned to profit by the wives of that rabble of abandoned miscreants whom the Italian call Cingari, the Latins Egyptians, and we Gitanos, who, notwithstanding that they are sent by the Turks into Spain for the purpose of acting as spies upon the Christian religion, pretend that they are wandering over the world in fulfilment of a penance enjoined upon them, part of which penance seems to be the living by fraud.[20]

Borrow agreed that these "witch-wives" had for four hundred years derived profit from fraudulent chiromancy.[21]

In a not dissimilar way, folklorist Charles Leland deployed the seventeenth-century Praetorius to comprehend a distinction between learned chiromancy and Gipsy "pretended chiromancy." Yet he was gentler and more genuinely inquiring than Borrow. Over his many walking tours, Leland collected his observations of customs and ceremonies of "fortune-telling, witch-doctoring, love-philtering, and other sorcery." Fortune-telling was interesting to him precisely because it was such an enduring folk practice found all over the Indo-European world, Gipsies "the wandering priests of that form of popular religion." Like his predecessor, Borrow, Leland maintained that a knowing deceit was at work, yet this pretense contained its own skill, even power, and was not to be simply dismissed as crude, let alone evil. Indeed, Leland was impressed. His Gipsy friends were as much skilled thought readers as palm readers, who had acquired the art of discerning character. Like classical physiognomy, this art was at least as much about eyes as hands. "The gypsy fortune-teller is accustomed for years to look keenly and earnestly into the eyes of those whom she *dukkers* or 'fortune-tells.' She is accustomed to make ignorant and credulous or imaginative girls feel that her mysterious insight penetrates 'with a power and with a sign' to their very souls."[22] Leland, like Borrow, and indeed like seventeenth-century Torreblanca, considered the learned astrological tradition of chiromancy to be altogether different. It was methodical, empirical, and masculine, while *dukkerin* was vernacular, happenstance, and feminine. The former were "learned and wise men," and even though Leland admired Gipsy fortune tellers, he did not grant them the equivalent status of "wise women." Pretense was the point: "Their women have all pretended to possess occult power since prehistoric times."[23]

In his travels throughout Europe, Central Asia, and the Middle East tracking magic and Gipsies and words, Leland constantly engaged with women about hands and palms, about *dukkerin*, and about how the future could be seen. And even though fortune-telling was a feminine practice, Leland himself was introduced into the skill, and on occasion read the palms of Gipsy women himself. He recounted a Russian journey in which he landed on the right dialect to converse with a woman who was curious about his own knowledge of Romani. But he was about to impress her with another Romani skill: *dukkerin*. The Russian woman, Liubasha, told Leland that her community had lost the art, but had heard that it was still known in England. "We hear that you Romanichals over the Black Water understand it. Oh, *rya*," she cried, eagerly, "you know so much,—you're such a deep Romany,—can't *you* tell fortunes?"

> "I should indeed know very little about Romany ways," I replied, gravely, "if I could not *pen dorriki*. But I tell you beforehand, *terni pen*, '*dorrikipen hi hokanipen*,' little sister, fortune-telling is deceiving. Yet what the lines say I can read."

Leland proceeded to read the palms of all the girls in the household, aware that this was the "chicken dressing the cook," as he put it. It was not in the natural order of things for the "Gorgio [to] tell fortunes to gypsies."[24]

Leland was an imposter, but an honest one. The most famous Gipsy in Victorian literature was far more fraudulent, we might say a highly successful "counterfeit Egyptian." Mr. Rochester in Charlotte Brontë's *Jane Eyre*, published in 1847, the same year as George Borrow's *Account of the Gipsies*, disguised himself twice over. First, he cross-dressed himself into a woman, and so, second, he passed as a "gypsy vagabond." Through feigned fortune-telling, he garnered information (for himself and for the reader) about how his betrothed and his yet-to-be-admitted real love (Jane Eyre) truthfully felt about him. It was a deceit that revealed a truth. Charlotte Brontë was able to use the device because the Gipsy encounter was an instantly recognizable tableau. Thousands of English etchings depicted comfortable pastoral scenes in which a Gipsy read (typically) a woman's palm, a social and cross-class encounter along a path or while picnicking in a field (fig. 2.1). Such etchings echoed a

2.1 Picnickers having their palms read at Burnham Beeches. Nineteenth century. Alamy Stock Photo.

"fortune teller" subgenre in higher orientalist art. Those paintings, however, typically depicted the fortune teller *in* "the Orient," as often representing masculine traditions of chiromancy and geomancy, as folk and feminine palmistry or card reading.[25]

For hundreds of nineteenth-century novelists, playwrights, and poets, the familiar "Gipsy" was a useful figure for signaling other dramatic possibilities—escape, transgression, desire, and danger—while fortune-telling from the hand was a device that might readily complicate a protagonist's self-awareness, enable a confessional reveal, or establish tensions between what might happen in the future and roads not taken. George Borrow turned his own extensive knowledge into novel form, after returning from his Spanish adventures, writing *Lavengro: The Scholar, the Gypsy, the Priest* (1851) and *The Romany Rye* (1857). Gipsies made regular appearances in Victorian poetry as well, a useful device, not least for the exceptionally clever. George Eliot's narrative poem *The Spanish Gypsy* (1868) was so learned that it sank into obscurity under its own weight, according to one critic. When Charles Leland visited her home, conversation turned immediately to Gipsies. She excitedly recounted her journeys to Spain, in which she visited the Zincali and learned from them, inspiring her grand novels and (too) learned verse.[26]

A far more popular manifestation of Romanichal in Victorian poetry was Matthew Arnold's "The Scholar-Gipsy" (1853). It was a story of another imposter, or at least an interloper, if a welcome one, another "counterfeit Egyptian." In the extended poem set long ago, an Oxford student abandons his studies to join a Gipsy community from whom he learns and appreciates new arts and powers of imagination. Arnold cast the Scholar-Gipsy as a kind of ghost figure who could still sometimes be glimpsed in the Oxfordshire countryside. He *was* a ghost, since the story was borrowed by Victorian Arnold from seventeenth-century Joseph Glanvill, an Oxford defender of the new experimental natural philosophies that were then negotiating hermeticism. Glanvill's original Scholar-Gipsy learned "that the people he went with were not such *impostours* as they were taken for, but that they had a traditional kind of learning among them, and could do wonders by the power of imagination." Seventeenth-century Glanvill was greatly interested in suggestibility and in thought transference, ideas that re-emerged in the Victorian era. He described, for example, maternal-fetal imprinting—"the wonderful signatures in the foetus caused by the imagination of the mother"—and wrote of thoughts transferring across minds, time, and space. One such process involved signs on the hand, a "secret conveyance" of meaning. It was a literal alphabetical code. Friends might agree on parts of the hand that would signal a letter of the alphabet, and if one were in, say, Paris, and the other in London, those parts would be "pricked" until words were spelled out. This would be felt—be sensible to, and thus readable by—the remote other. Glanvill called these "sympathized hands." He also called it "a new kind of chiromancy."[27]

By the time Matthew Arnold composed his Victorian rendition of the Scholar-Gipsy, by the time George Eliot wrote up her Gipsy poem that was too complex for anyone but herself to appreciate, and by the time Leland was admiring Romani powers of suggestion, this kind of thought transfer, and indeed the very idea of a new chiromancy, was on the cusp of a great second life within the nineteenth-century occult revival.

A NEW CHIROMANCY

The revival of occult practices in the nineteenth century is often understood to have arisen out of American-born "spiritualism," communication via a medium between the living and the dead. And yet a new chiromancy for the modern era was originally more French than it was American. Different from both *dukkerin* and physiognomy, modern palmistry is best understood as one of the novelties of a revived *occultisme*. A range of esoteric medieval and Renaissance practices—alchemy, astrology, natural magic, symbology, Kabbalah—resurfaced in the mid-nineteenth century, especially through the work of French Catholic clergyman turned occult magician Alphonse Louis Constant, who assumed the name Éliphas Lévi (1810–1875), what he constructed to be a

Hebrew rendition of his birth name. Perhaps the greatest modern *Magus*, Lévi was highly influential as an expositor of occult arts and sciences, if an unlikely one, the son of a shoemaker. His *Doctrine of Transcendental Magic* opened as a work that draws back the veil of knowledge from many ancient traditions, occult or hidden knowledge from India, Egypt, and Assyria. He retold, in other words, both origins and secrets. It was no ordinary story.

> Behind the veil of all the hieratic and mystical allegories of ancient doctrines, behind the darkness and strange ordeals of all initiations, under the seal of all sacred writings, in the ruins of Ninevah or Thebes, on the crumbling stones of old temples and on the blackened visage of the Assyrian or Egyptian sphinx, in the monstrous or marvellous paintings which interpret to the faithful of India the inspired pages of the Vedas, in the cryptic emblems of our old books on alchemy, in the ceremonies practised at receptions of all secret societies. There are found indications of a doctrine which is everywhere the same and everywhere carefully concealed.[28]

Who wouldn't be enchanted? Lévi's mission was to disclose all of these secrets, but especially "that of true magic which comprises the secrets of the Kabbalah." Perhaps his signature phenomenon was *le fluide astral*, matter through which the sun, the moon, and the planets influence each human individually from conception to death and beyond. Drawing on Paracelsus, he explained that the astral fluid or sometimes astral light imprints signs on human bodies, likely correspondences of the "astral kabbalistic alphabet."[29] It was an antique sensibility about the human body, actively transported into the modern world.

Charismatic Lévi gained as many esoteric followers as he lost clergymen-brothers in the Roman Catholic Church, into which he had been ordained in the 1830s. One admirer was his friend and proximate mystic the Parisian Adolphe Desbarrolles (1801–1886), "amiable and spiritual," the master Magus called him, "a veritable magician in chiromancy." Borrowing *le fluide astral*, and *la lumière astrale*, Desbarrolles posed an astral theory about hand signs. Made up of seven fluids, derived from the seven primary planets, a power is exerted on the body and mind by means of an Astral Light. His study, *Les mystères de la main révélés et expliqués* (1859), came to be widely referenced—almost canonically so—in Anglophone palmistry texts of the later nineteenth century, reintroducing the Kabbalah to an entirely new generation.[30]

Like seventeenth-century Saunders and Belot, Desbarrolles read three worlds within the human palm, corresponding to three spiritual realms, somewhat crudely rendered as the elemental, the celestial, and the intellectual. These "realms" were not just represented in drawings of the hand but were marked in the hand itself. They could be discerned in the letter *M*, inscribed by the intersections of the three main palmar lines. He marked each phalange of the fingers as *Monde divin*, *Monde intellectuel*, and *Monde matériel*, and indicated how the twelve months and four seasons corresponded to the four fingers, three phalanges on each. Alongside more familiar astrological symbols for Venus, Mars, or Luna, Desbarrolles located his hand cardinally, to the west, north, and east, as had Saunders and Belot in the seventeenth century: *Oest* signaled *Actif*, *Est* signaled *Passif*. Yet there are few classic palm lines here. Instead, there is a geometry: a straight line from *Femelle*, from *Midi* to *Nord*; a triangle in the center of the palm. Drawing explicitly on kabbalistic traditions, Desbarrolles placed the various *séphirotes* onto the hand, and named this triangle *Kether*, the Father, the Supreme. Surrounding *Kether* are the fields of other *séphirotes*, each with their meanings and their characters: *Binah* (freedom, driving power, initiative) and *Chocmah* (wisdom, sovereign reason) (fig. 2.2).

This new chiromancy did not just reveal character, health, or fortunes. It was part of a revived world religion, a cosmology. *Chiromancie nouvelle* reached toward the kind of "world-soul" that had enchanted seventeenth-century chiromancers. Desbarrolles nominated the "living scriptures" of the hand, which enabled vision to the past and to the future, and to God. The Divine could be perceived therein, thanks originally to the work of Éliphas Lévi, son of a shoemaker.[31]

For Lévi, Desbarrolles, and the occultists who came after them,

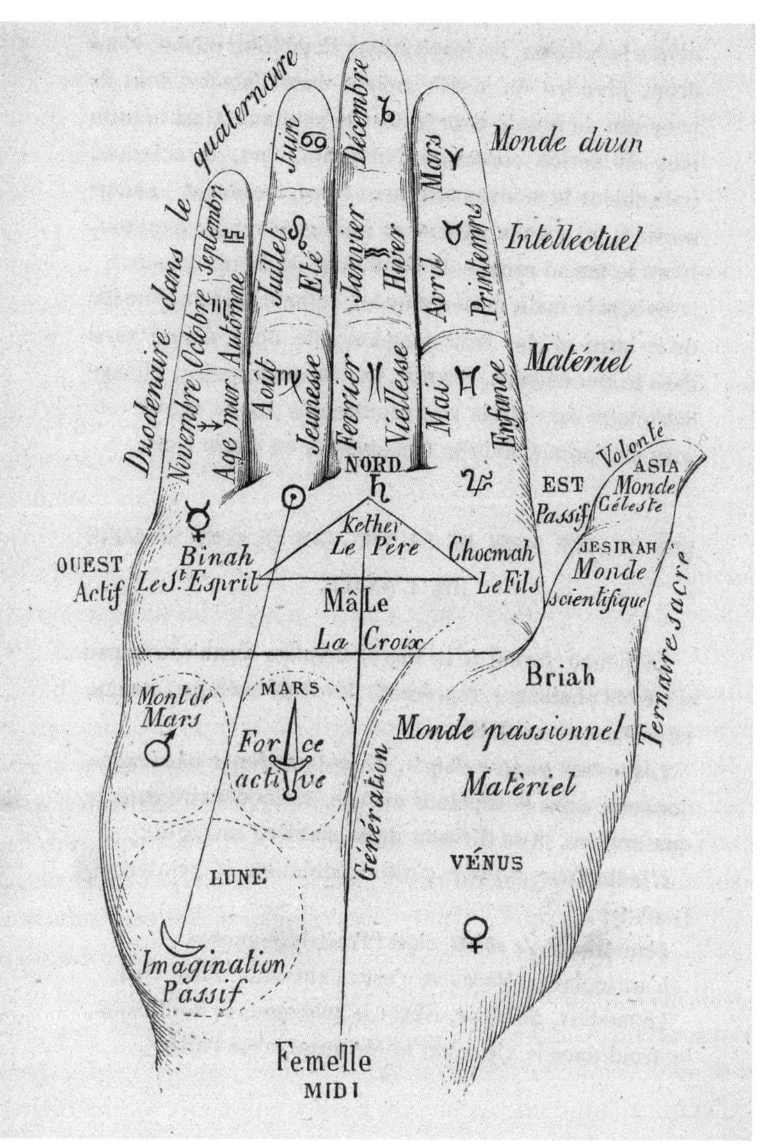

2.2 The nineteenth-century kabbalistic hand. Adolphe Desbarrolles, *Les mystères de la main révélés et expliqué* (Dentu, 1859), 292. Copy in author's possession.

the secrets to be disclosed about the hand, the Divine, and the soul of the world came from many traditions: Egyptian, hermetic, Hebrew, and Vedic. In truth, this was a mixed and messy Indo-European mélange, and it only became more so. For some, the common and original knowledge was always the Kabbalah, but as we have seen, the nineteenth-century revival was already a recapitulation of a seventeenth-century revival of an originally twelfth-century mystical theology. But even that origin was embellished with other antiquities by Victorian-era occultists. English palmist Rosa Baughan, for example, began her early 1880s manual with "the Holy Kabbala" known in ancient Chaldea, India, and Egypt at the very least, she claimed. For her, the Kabbalah was introduced to Greece by Pythagoras, acquired through his exchange with "the ancient Magi, during his travels in the East." This was knowledge of the secrets of nature, originally known only to the initiated. She explained a doctrine of silence, often represented by the figure of "Adda-Nari," whom the Egyptians called Isis. The Christianization of the Kabbalah was for her a process of democratization, the laudable disclosure of secrets, "a divine law in opening the way of light and life."[32] Baughan drew the image of Adda-Nari—"*l'Isis des Indiens*"—from Lévi and Desbarrolles, both of whom turned the Hindu male-female Ardhanarishvara into a kabbalistic deity: "The human figure is placed between a bridled bull and a tiger, thus forming the triangle of Kethar, Geburah, and Gedulah, or Chesed," explained Desbarrolles. "In the Indian symbol, the four magical signs of the Tarot are found in the four hands of Addha Nari." Even those hands were read in the new chiromancy. While the thumb and the first two fingers are held open (the chiromantic fingers of will, power, and fatality), the fingers representing light and science are held closed. Baughan, in turn, thought that the opening of those latter two fingers represented modern palmistry, in which ancient secrets were brought to light.[33]

This revived esotericism all got stranger, and if anything more hybrid, as the nineteenth century progressed. Lévi also influenced the founder of Western theosophy, Helena Blavatsky, who almost single-handedly created a new religion-science that spread like wildfire among European and American devotees. She definitely had a plan for the soul of the world. Clairvoyance was core to the phenomenon that was theosophy, but not even Blavatsky at its founding in 1875 could have foreseen the stratospheric success of her ideas. A Russian émigré to the United States, Blavatsky founded the religion-science there, later relocating its organizational center to Adyar, near Madras, India. From the beginning, it was a self-conscious union between Western esotericism and Eastern mysticism: part Indian, part Tibetan, part Persian, part Oriental, part Occidental. Theosophy effected a truly remarkable reenchantment of an Anglo-American nineteenth century that was increasingly subject to scientific materialism: it enabled a complete reimagining of Earth, of God and of gods, a vast new story of where humans came from and were going to. Thousands upon thousands of men and women became theosophists, the most successful "new religion" of the period, although Blavatsky always demanded its scientific status. Her magnum opus, *The Secret Doctrine*, was subtitled *The Synthesis of Science, Religion and Philosophy*. In it, she was seeking the soul of the world from its very origins. There was nothing modest about Madame Blavatsky: her first volume was *Cosmogenesis*, her second, *Anthropogenesis*.[34]

It is no exaggeration to consider leading theosophists to be philosophers of the body, presenting a new comprehension of the physical, the mental, and the sensual. The co-founder who led American theosophy, William Quan Judge, was an eloquent theorist in this regard, dismissing routine conceptualization of the material body as coterminous with the individual: "The body, as a mass of flesh, bones, muscles, nerves, brain matter, bile, mucous, blood, and skin is an object of exclusive care for too many people, who make it their god because they have come to identify themselves with it, meaning it [the body] only when they say 'I.'" He was convinced, rather, by the idea of an astral body, in which lie "the real organs of the outer sense organs" of sight, hearing, smell, and touch, and we see here Lévi's direct French influence. The astral body "has a complete system of nerves and arteries of its own for the conveyance of the astral fluid which is to that body as our blood is to the physical." Writing in the early 1890s, he used contemporary psychology about the mind (Freud's ideas about forces located in "the subconscious perception" and in "latent memory") to help explain the astral body too.[35]

Dozens of practices, including palmistry, were the means by which previously imperceptible planes and zones of thought could be brought into being, so it was claimed. This was a kind of clairvoyance in which ordinary physical senses were cast to one side in order to "see" in another dimension, with another sense, and in another temporality. "Clairvoyance" meant at base nothing more than clear-seeing, but also "the power to see what is hidden from ordinary physical sight." This was a higher faculty, latent in all, but developed by a few. Like Glanvill's sympathized hands, theosophists claimed that some held "clear sight" in spatial terms—the capacity to perceive scenes or events removed from the seer in space. Others held clairvoyant capacities in temporal terms, to see events removed in time, "or in other words the power of looking into the past or the future." Theosophists wrote of "etheric or aerial vibrations" to which some were more sensitive than others, intrigued also by the phenomena of mesmeric trances, of hypnosis, and of the dream state, each of which had long been approached and applied diagnostically and therapeutically by both conventional physicians and irregular health practitioners. William Quan Judge, too, thought that hypnosis touched the realms of the astral body, even though many hypnotizers themselves were baffled by their own powers. He was right. Physicists in this period were constantly inquiring experimentally into the nature of the matter at work in mesmerism, hypnotism, and thought reading. As we see in the next chapter, there was far less distance than might be imagined between these "occult" practices, including the new chiromancy, and scientifically trained experimentalists.[36]

INDO-EUROPEAN HANDS

Even as physicists and psychics came together to experiment in, and on, the nature of suggestion, and indeed of the unconscious, the core theosophist message was that "the real psychology is Oriental."[37] In such statements, they did not mean the orientalized Middle East, but the ancient philosophies of Central Asia, of Hindustan, and especially of Tibet. Blavatsky had declared herself Buddhist in 1880, and the provenance of her own "secret doctrine" was not in the genealogy of the Arabic and Latin *Secretum secretorum*, or the Kabbalah, but instead the "Book of Dzyan," sequestered in remote Tibetan libraries.[38]

Theosophy's Buddhist underpinning held a curious link to the reading of hands. Handprints and footprints were, and are, key in Buddhist art, and indeed the Buddha's handprints and footprints are sacred relics across South Asia and East Asia to this day. One theosophist explained that the wheel traditionally depicted on the Buddha's sole, the mark called the "chakraverti," *should* have been on the palm of the Buddha's hand. Sages at his birth were able to recognize and interpret that displacement, and this was taken to be somatic and textual evidence for the antiquity of chiromancy: since those sages were already able to read the signs of the hands and feet at Buddha's birth, palmistry must have predated the year 550, so it was argued.[39]

This was all explained for the Anglophone world in a study of *Indian Palmistry* by English theosophist Mrs. J. B. Dale. True to both theosophy and studies of chiromancy, she pulled information on antique origins from everywhere. Mixing her sources together, just like seventeenth-century Praetorius, she first cited the book of Job, less as a Scriptural or Divine text than an ancient one, from the world before Egypt. It was, she claimed, a Chaldean text from the origin of civilization, from the sacred Babylonian culture that first perceived astrology and numerology and chiromancy. This was truly ancient wisdom that the Egyptians, then the Greeks, inherited. But as a theosophist, palmistry was even more truly an "Indian" practice. Her version of the chiromantic hand was a busy conglomeration of animals, signs of the anchor, scales, arrows, shells, numbers, symbols of the sun and moon, but not from the astrological alphabet. The main lines she nominated as the Mother line (giver of life) and the Father line, the line of Head or Reason, of Fortune or Happiness, as well as the Sun line and the Moon line. Hatchings, triangles, and stars appear as well, as do spirals on the fingertips, some like Galton's contemporaneous whorls, others represented by the logarithmic spiral of a nautilus shell (fig. 2.3).

By the 1890s, when theosophy was booming, linguists had already clarified that the Roma diaspora in Europe had a North Indian origin, and that Gipsy fortune-telling from the palm was a remnant of *Hast Samudrika Shastra*, evidence of the distant connection. Yet as a rule theosophists could not have been less interested in folk practice, or more dismissive. William Quan Judge, for example, granted Gipsies a happenstance clairvoyant power, but not a learned one. Theirs was an instinctive naturalized capacity to read bodies, minds, and hands that derived from "being a strange and peculiar people living near to nature." It was genuine Vedic palmistry that was more interesting to theosophists, and it was they who most directly connected Anglophone palmistry to Indian practices. Writing on "Chirognomy" in 1884, Judge explained that in the Indian system, the fingers and the palm were separated out for different signs, the former relating to intellectual life and the latter to "animal life." "Animal instincts," for example, could be recognized in thick and hard palms, self-confidence in well-pronounced first joints. He also noted that the head and heart lines were joined "in many idiots." Hand-reading techniques in South Asia were explained for Anglophone practitioners. In men, the right hand was seen to be the "fate hand"—denoting events that would happen—while the left denoted that which had happened, or had been accomplished, or a current event "just passing away." The converse applied for female hands.[40]

Any palmist in Calcutta or Delhi, Chennai or Varanasi might well wonder why such basic instruction needed to be outlined. In many ways, chiromancy across South Asia was far more quotidian. Indian theosophists themselves conveyed its ordinariness. In 1882, Barad Kau Majumdar, for example, told his fellow theosophists of a journey he had recently made to Calcutta to meet a Tantrik mystic. Finding him, the mystic indicated that he had not yet "come to that state of Yoga which makes the Yogi a clairvoyant," but occasionally, in times of fasting, attained a lucidity that approached that state. For the moment, however, people sought him out for more mundane knowledge of astrology, palmistry, and medicine.[41]

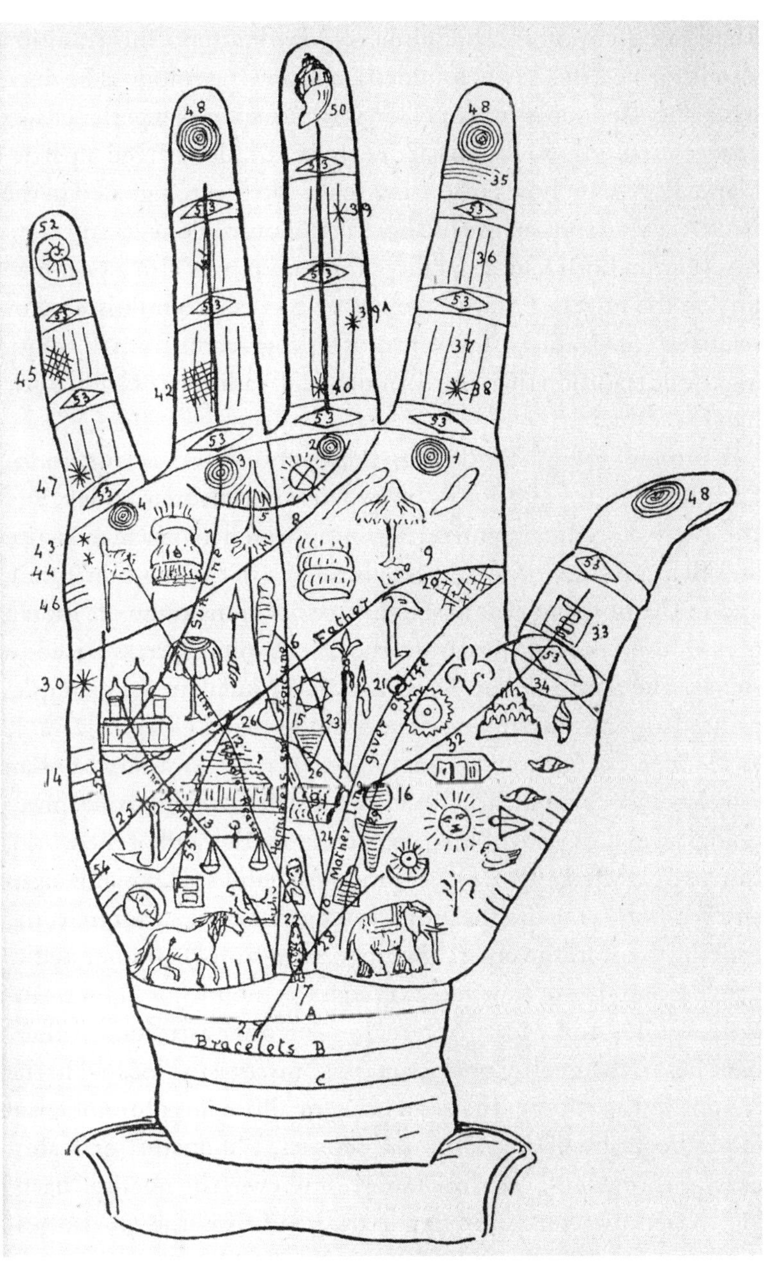

2.3 "The Reference Hand," from Mrs. J. B. Dale, *Indian Palmistry* (Madras and London: Theosophical Publishing Society, 1895), plate I. Copy in author's possession.

These were important, but also everyday matters. This tradition also informed the newly founded London Cheirological Society, which boasted some Indian fellows, whose special experience was appreciated. Mr. K. R. Chatterjee, for example, offered a particular expertise on how poison and snake bites were signaled in the hand—a blue spot on the life line, that if only in one hand meant recovery, in both hands, fatality. But if there was also a post on a line on the finger of Mars, "the poison had been administered by relatives." Such aphorisms were more recognizable from the physiognomic tradition than the cosmological traditions of kabbalistic hand reading.[42]

Palmistry of all kinds was being translated across Indo-European worlds and across many languages by new, quick, and cheap presses. *Hast Samudrika Shastra* appeared in new manuscript as well as printed forms across South Asia, in the south in Tamil, and in the north in Sindhi. English translations came off Indian presses too, and not just from the Theosophical Press based in Adyar. The South Indian Kumaraswami Mudaliar, for example, printed an entirely anglicized and modernized *Complete Treatise on the Science of Chirography and Chiromancy*. Here, we find not an astrological or a mystical hand at all, but a disenchanted one, phrenological and physiognomic: character, not the soul of the world, can be discerned (fig. 2.4). This modern hand still explains signs and symbols, as they appear on the hands: the bars, the circle, the trident, the star, the cross. The fingers are simply enumerated—first, second, third, fourth—although the mounts of Jupiter, Saturn, Apollo, and Mercury remain. The only astrological remnant lies in the quality or character of "mysticism" located in the Mount of the Moon. The remaining qualities are phrenological or physiognomic: "ambition" or "sadness," "industry" or "calmness," "shrewdness" or "brilliancy," and even the quality "dashing" has its own place and sign in the hand. The study is also gestural, even phenomenological, regarding bodily comportment in modernizing India. And yet, just like the theosophists and esoteric astral chiromancer Adolphe Desbarrolles, this South Indian also authorizes palmistry as ancient knowledge. It *was* part of a Hindu past, he insists: "The ancient monuments and temples of India

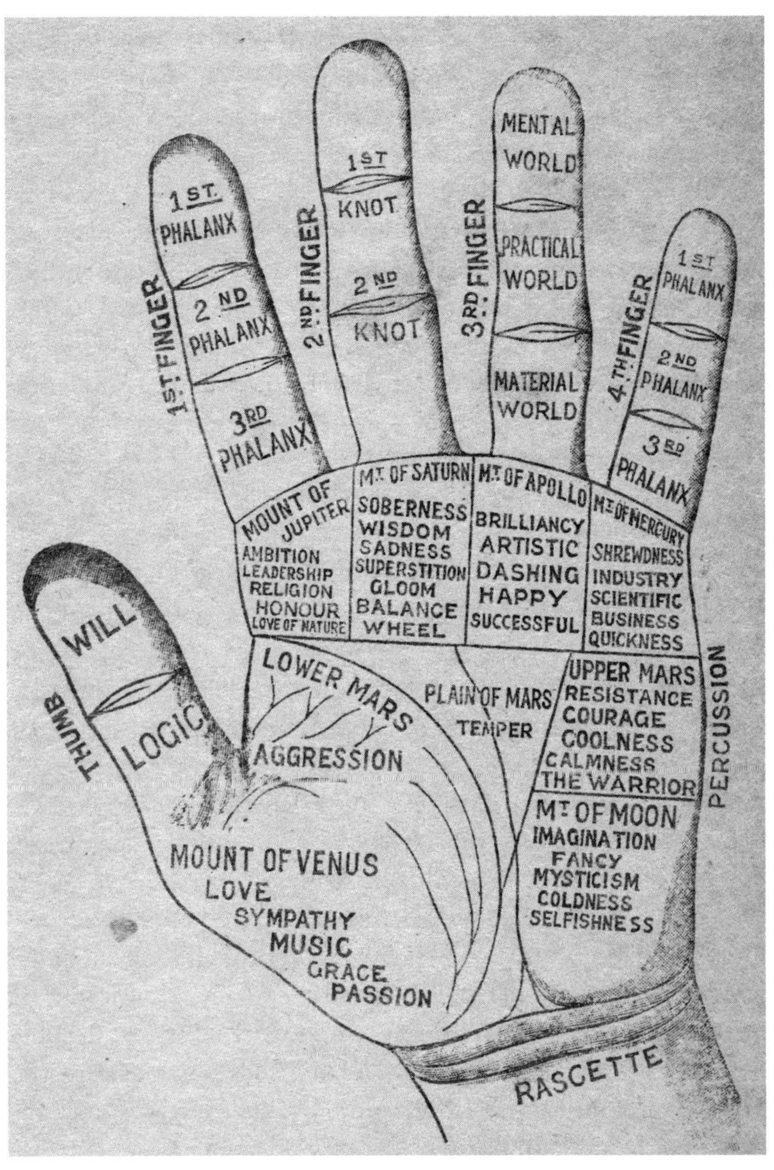

2.4 A modern Indian hand, phrenological more than astrological. V. S. Kumaraswami Mudaliar & Son, *A Complete Treatise on the Science of Chirography and Chiromancy* (Vellore: The Victoria Press, 1918), plate 27. Courtesy the British Library.

show clearly that the Hindus were experts in this science long before the birth of Christ." And this knowledge had a regional location too: "In the North-West Provinces of India, the Science of palmistry has been practised from time immemorial by the caste of astrologers."[43]

Out of those Northwestern Provinces, so the story goes, *Rom* had been expelled, journeying west, eventually into Eastern Europe, into the German lands, and into Tudor England, where ecclesiastical Christians imagined them as biblical "Egyptians." They brought with them knowledge of how to decode marks on the hand. Centuries later, one curious manifestation of the converging culture of modernity was the hybrid hand that folded Eastern and Western knowledge of the body together.

Palmistry of the late nineteenth-century occult revival is instantly recognizable and easily comprehended, or so it would seem. But precisely *what* was being revived? In fact, the multiple lines of knowledge and practice take some unraveling. *Dukkerin*, *Hast Samudrika Shastra*, *la chiromancie nouvelle*, theosophical palmistry, and a whole new Kabbalah together represented the strangest folding and refolding of oriental, occidental, and occult practices. This was an intense exchange of invention and tradition about the hand, its signs, and their meaning, and over many centuries. There were four broad traditions of palmistry, each of which in some manner "orientalized" the practice for occidental consumption. First, "Gipsy" fortune-telling from the palm. This was long understood as "outlandysshe"—foreign—but in fact for centuries was part of European culture. At the same time, this palmistry was newly interpreted by nineteenth-century orientalists as an attenuated remnant of Sanskrit *Hast Samudrika Shastra*. Second, and counter to the antiquarian interests of the orientalists, *Hast Samudrika Shastra* itself was actively modernized, not least by Indian practitioners themselves often publishing in English. Third, a new chiromancy revived Jewish Kabbalah, again, its secrets disclosed initially in fifteenth-century Florence and then again across seventeenth-century Europe. In its nineteenth-century revival, Kabbalah was once more mixed with "oriental" signs, symbols, myths, and figures, from the Egyptian Isis to Adda-Nari, "*Isis des Indiens*." Theosophists took inspiration from the French revivalists of the Kabbalah and turned it into a more Buddhist-focused philosophy of the body, a kind of South Asian *fluide astral*. And fourth, there was a mythic but enduring and discursively powerful link with antique "Egyptian" knowledge, explicitly through the learned hermetic tradition, and in a more muted way through the palmistry of "Egyptians" themselves, that is to say, "Gipsies."

In the deepest recesses of the Occident—in parlors in Paris, London, Chicago, New York—late nineteenth-century practitioners of the new chiromancy absorbed these revivals and reincarnated these hybridized traditions. As we shall see, they continued to claim or simply invent great antiquities for the signs of the hand, deep histories for the expertise to read them, and fantastic accoutrement, paraphernalia, and décor through which to signal "secrets." The inventions were almost always orientalized in some manner, via Gipsy association, via Sanskrit-based Indian cosmologies, or via new versions of hermeticism and Kabbalah that had so energized Richard Saunders and that had Isaac Newton thinking hard about geometry and other wonders in the seventeenth century. This was an Indo-European exchange of hands across worlds, that in a small corner of it settled into a recognizable late nineteenth-century occult revival. And yet, that was not all. With a different kind of accoutrement, tradition, and purpose altogether, the hand was also read by anatomists of the human body, some of whom also saw the Divine therein.

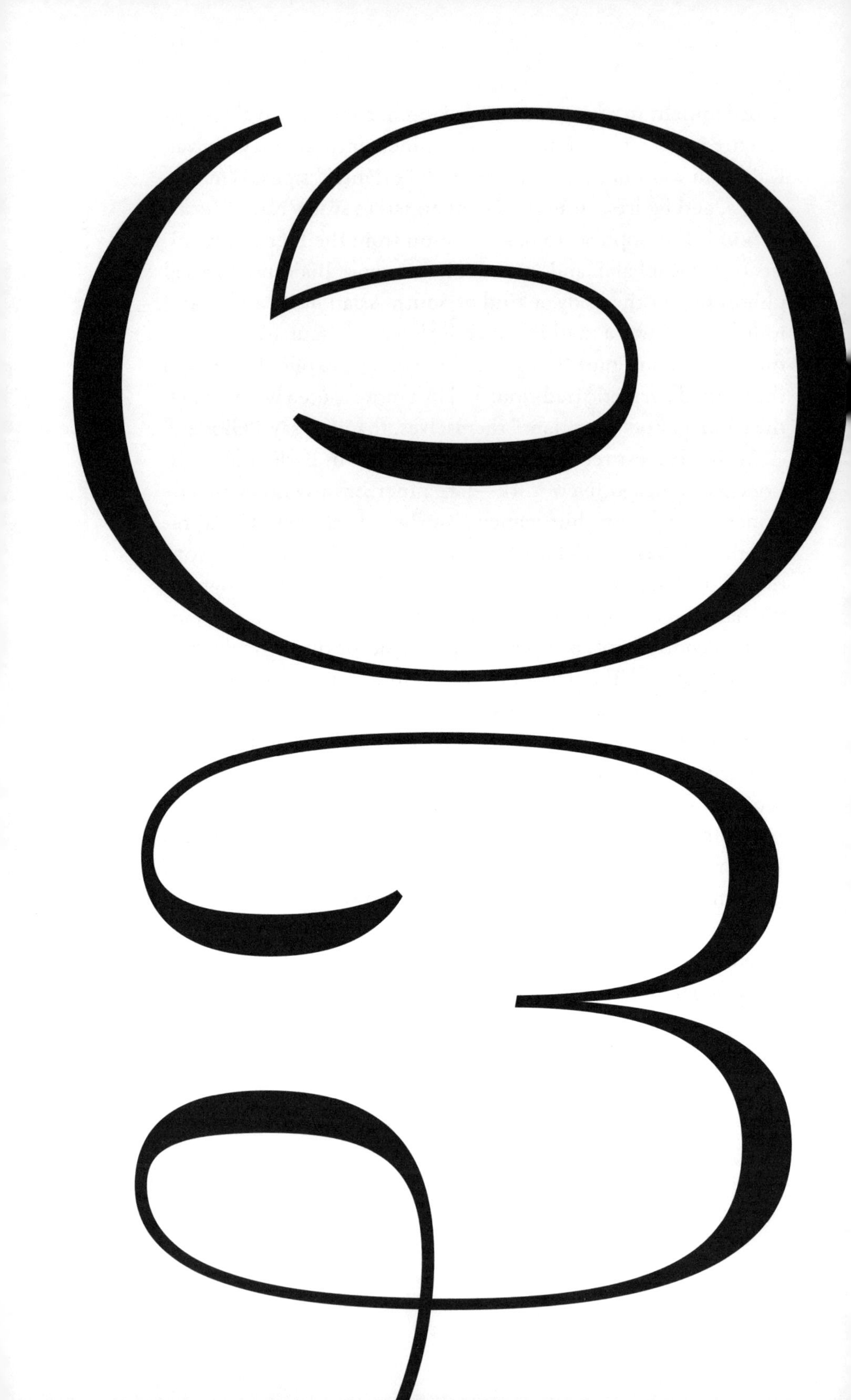

THREE

The Hand in Victorian Science and Medicine

While folklorists were traveling through Europe and the Middle East searching for *dukkerin*, while new chiromancers began to see in the palm the three worlds of Kabbalah, and while Indian palmists were modernizing *Hast Samudrika Shastra* for a new cheap press, nineteenth-century anatomists were setting down their own accounts of the human hand. A foundational medical discipline, anatomy was detailed and intricate, but also observationally basic, knowledge of bones, muscle, nerves, and skin drawn secondhand from old anatomical texts, and firsthand from human bodies, dead and alive. At first glance, these anatomical hands were pared-back nature, the very opposite to the wonders and secrets of chiromancy.

Yet was anatomy really that reduced? Was it just physical, not metaphysical? For many reasons, no. In the first instance, the senses were core anatomical business. The hand was routinely defined as an organ of touch, even *the* organ of touch, and so, of perception. And while this certainly required ever more refined understanding of the interaction of muscle, bone, skin, and especially nerves, nineteenth-century anatomists rarely stopped there. How was the hand connected to the brain that received its messages? And in what way was that brain in fact a mind, with a will that *sent* messages to the hand, to grasp or to write, to wave or to stroke? Sometimes those minds sent messages to hold another's hand in one's own in order to feel its shape and texture, and to observe its flexure lines. Both clinicians and chiromancers did that.

The hand, it turns out, was conceptualized by anatomists and physiologists not just as an organ of touch, but as an organ of the mind, and even of volition. It had a special place in new approaches to the nervous system, and in their derivative disciplines, neurology and psychology. Indeed, almost every variation of Victorian mind/body medicine had something to say about the hand. From touch, to the brain, to the mind, and even to the human will, the meaning of the hand was almost as expansive in nineteenth-century medicine and science as it was in occult knowledges of the cosmos.

This is not a conceit. Eminent Victorian anatomists saw the Divine in the human hand and said so in their anatomy lectures and in their textbooks. And even for those scientists driving a naturalist and materialist agenda in which God had nothing to do with the structure of nature or the human body, the hand still ended up a thing of wonder, holding significance far beyond metacarpal bones or ulnar nerves. For comparative anatomists inclined toward the transmutation of species over time—what came to be called "evolution"—the hand had long set bipedal humans apart from other primates and other vertebrates.[1]

Evolutionary biology and neurology were two domains in which great new horizons opened up over the nineteenth century. But old systems of thought endured too, and they were not necessarily at odds with the novelties, even for those at the frontiers of knowledge. Physiognomy resurfaced, and Darwin was one among many who thought hard about both faces and hands, and how an inner emotion "expressed" itself on a bodily surface. Phrenology too was the business of serious students of minds and bodies, with a specific application to the hand set out in the middle of the century. Even antique Galenic humors, and Hippocratic ideas about airs, waters, and places, remained plausible in the nineteenth century, linking latitudes and climates to different human constitutions and capacities, also understood to be encoded in the hand. From a combination of comparative anatomy, a medical geography of human difference, and emerging evolutionary biology of human variation, there came a strange set of ideas that correlated the human hand with human types, national, constitutional, and racial.

ANATOMISTS OF THE HAND

Jan Evangelista Purkyně—Purkinje—was a specialist in the senses, in the eye and vision, but he also studied the hand, fingers, and touch. Educated in Prague, in 1823 he wrote a treatise that set out nine patterns of fingerprints that were later key to Francis Galton's work. Yet Purkinje's treatise was also about the larger hand, and a lesson in the physiology of the senses. In any clinical examination, the hand should be considered first, since it is "the organ of touch itself." Purkinje's work was definitely tactile. Ascertain heat and cold, he instructed, and look at the appearance of the nails and the color of the flesh. Is the skin "hard or soft, moist, oily, clammy, or dry, warm or cold, elastic, rigid or spongy, smooth or rough, loose or taut"? With proper observation of these signs, differences might be revealed of "constitution, age, sex, or even family and nationality." The color of the skin varies as much as the human races vary, he noted, shades that are both innumerable and subtle and might only be reproduced properly "by the hand of a master painter." As we shall see, this was neither the first nor the last time that an anatomist considered art, just as artists themselves studied anatomies of the hand. Color also varies with sex, age, constitution, and temperament, as with states of bile or phlegm. Smell the skin too, he advised, the olfactory sense too often ignored. All these "properties of the sensibility of the skin" needed assessment before turning to the anatomical configuration of the hand.[2]

In making such claims in 1823, Purkinje felt compelled to distance his work from chiromancy, in the way that a physician in 1623 would not. "In my opinion the work of palmists ought to be considered equal to the prophecies of soothsayers and augurs," he dismissed, "who prophesy from the flights of birds and motions of the intestines." Richard Saunders would have wondered just what the problem was. Yet Purkinje was dissembling. In fact, he explicitly studied the crease patterns on human palms as well as on monkey palms and found himself reaching for traditional palmistry to explain his modern anatomy. Purkinje specified new terms: the lines of opposition, abduction, adduction, extension, and flexion. But in order to clarify, he then explained them as the life line, the line of Mars, the line of matrimony, the table line, the line of Venus, the line of Rascetta, and the lines of honor and fortune. In 1823, palmistry was one reference point for a modern anatomy of the hand.[3]

A decade after Purkinje's treatise appeared in Prague, a celebrated anatomical work on the hand was composed in London, at the Great Windmill Street School of Medicine. The Scottish anatomist Charles Bell's *The Hand: its mechanism and vital endowments as evincing design* (1833) was a foundational text for human anatomy, and it turns out for palmistry too. *The Hand* was certainly both point and counterpoint for nineteenth-century natural historians and evolutionary biologists, just then considering the similarities and differences between the human form and that of all other vertebrates. Bell (1774–1842) thought through this question in a particular tradition of natural theology, in which the designer of Nature is God.[4] It was a comparative anatomy that looked not just at the human hand, but also at its equivalent in bats, anteaters, birds, dolphins, elephants, camels, monkeys, and more. He examined the nails in humans, for example, alongside the hoof of a horse. And he compared the bones of a human hand to those of a "chimpanzee" from Borneo (as he had it). In fact, he called it a "paw," the whole point of his study the uniqueness of the human hand, God-given.[5]

When Bell focused on the bones of the hand, the human thumb emerged as particular: "On the length, strength, free lateral motion, and perfect mobility of the thumb, depends the power of the human hand." Like medieval and early modern chiromancers, nineteenth-century anatomists continued to name the thumb "pollex," derived from *pollere*, "to have power." But if *pollex* gave strength, for Bell it was nerves and what they enabled—touch, sense, sensibility, and perception—that gave hands their real capacity. Charles Bell was as eloquent about the hand as he thought the human hand itself was eloquent about the Divine.[6]

Every respectable Victorian doctor and natural scientist read and cited Charles Bell's *The Hand.* Some were partial to his natural theology: the great comparative anatomist Sir Richard Owen, for example, who knew the skeletal structures of other vertebrate species better than Bell, but who was equally taken with the significance of hands and feet. *On the Nature of Limbs* (1849) provided even more evidence for homologies across species that could be interpreted either as evidence for a divine "unity of plan" or for the *transmutation* of species. Others who read Charles Bell were set against his natural theology: the materialist agnostic Thomas Henry Huxley, for instance, who also obsessively researched the bones of primate extremities. Just what is a "hand" and what a "foot" in different species of primates?[7]

Charles Darwin read *The Hand* in 1839, just after a book on ethical philosophy and just before a book on Egyptian remains. For the field of evolutionary biology that Darwin was yet to revolutionize, the extremities of all kinds of vertebrates, including primates, including humans, were key. "What can be more curious," he wrote, "than that the hand of a man, formed for grasping, that of a mole for digging, the leg of the horse, the paddle of the porpoise, and the wing of the bat, should all be constructed on the same pattern, and should include the same bones, in the same relative positions?" Both he and Bell knew that human hands were similar to, but also different from, all these. Yet to what mechanism should that difference be ascribed? Darwin's evidence for natural selection began to suggest not so much unity of design by God, as natural and incremental evolution from one species to another. For him, human hands were not just homologically similar to a fin but had actually evolved from an ancient common ancestor *with* a fin. The hand was certainly not *given* to Man-the-Wise, as Bell and Anaxagoras would have it.[8]

Distinguished nineteenth-century anatomists and natural historians often analyzed the hand and the foot together. In 1861,

G. M. Humphry, lecturer in anatomy and physiology at Cambridge, gave two important lectures on just that. The humble and underestimated human foot came first, with its particular arrangement of bones that enabled man to stand upright. But the real significance was the flow-on effect. Relying on two, not four feet for locomotion, "the hand is set at liberty to minister to the will." Darwin drew on these insights when he came to think through human extremities in *The Descent of Man* (1871). "Man could not have attained his present dominant position in the world without the use of his hands," which, once liberated, could be put to increasingly complex uses. Darwin understood that because of the hand's connection with the mind and the intellect, humankind had acquired a species supremacy, dominion over the rest of nature, for better or worse. This was no small claim.[9]

All these anatomists and evolutionary biologists considered the human hand through basic body systems: the skeletal, muscular, nervous, and circulatory systems, and what came to be called the integumentary system of the skin. The skeletal system was foundational. Whether they were natural theologians after Charles Bell, or evolutionists after Charles Darwin, comparative anatomists agreed that humans stand alone because of the bones in their feet, not just freeing their hands, but also permitting a special mobility in their wrists and their thumbs. The thumb can "be moved to and fro," instructed Cambridge anatomist Humphry, "can be opposed to the other fingers, and to any part of them individually and collectively." Moreover, the human hand can fold the thumb beneath the fingers, a movement that Cocles had taken the time to illustrate in his 1536 *Physiognomiae et chiromantiae* (fig. 3.1). Late in the nineteenth century, this was linked to developmental psychology: "By a contraction of the flexor, and non-development of the extensor muscles the human infant hides its thumb in the palm of its hands until its will shall have developed itself." The metacarpal bones of the thumbs and of the ring and little fingers in a monkey, by contrast, have not the same amount of play upon the wrist, explained Humphry, which means the thumbs and the fingers of the animal cannot be opposed to one another. But it was not all about the celebrated thumb. The arrangement of bones

3.1 Both Cocles in 1536 and George Murray Humphry, Cambridge professor of anatomy and physiology in 1861, noted how the human hand can fold the thumb beneath the fingers. Bartolemmeo della Rocca [Cocles], *Physiognomiae et chiromantiae compendium* (Strasbourg, 1536), plate 159. Courtesy the Wellcome Collection, London.

in the human hand also means that fingers can be separated out, fanlike, something other primates cannot do. And the particular arrangement of bones in the wrist enables "pronation" and "supination," turning the palms downward and upward, while other primates "cannot completely supinate the hand." (We might say that this means they can't have their palms read, but we see in chapter 10 that this is precisely what happened to monkeys in the London Zoo in the 1930s.) Even in humans, Humphry explained, it's an awkward position, neither habitual nor natural. A thousand palmists were to find just that.[10]

This movement relied on the connection of muscles to bones, in this instance, what Humphry called a *pronator* muscle and a *supranator* muscle (fig. 3.2). Muscles move bones, but also guard them, even shape them, anatomists would say. Charles Bell knew this: "The inert and mechanical provisions of the bone always bear relation to the living muscular power of the limb; and exercise is as necessary to the perfect constitution and form of a bone as it is to the increase of the muscular power." Darwin wondered about the heritable effects of muscular use and disuse of the hands over generations. He noted that the hands of English laborers were at birth larger than those of the gentry—or so he thought. It all sounds like the Lamarckian inheritance of an acquired character, but Darwin was toying with natural selection in humans. And we shall see how Friedrich Engels deployed Bell and Darwin on the hand to integrate the human capacity to use tools in his dialectics of nature (fig. 3.3).[11]

A pathology of the hand ran parallel to this anatomy. Congenital disorders were detailed—extra fingers or united fingers—alongside diseases of bones, including osteosarcomas and chronic rheumatism, which affect joints. But how did anatomists see the creases in the hand? They were usually called "flexure lines," and considered within embryological anatomy or as anatomy of the skin. *Quain's Anatomy*—revised and relied upon throughout the century—detailed the "furrows" of the skin "as those so well known in the palm of the hand and at the joints of the fingers." Drawing from the many editions of *Quain's*, a 1920 anatomy of the hand pictured the relation of the palmar lines to the bones, though such explicit illustration was rare (fig. 3.4). Another account set out the crease lines for surgeons' use, like a grid or a map that could guide an incision. Flexure lines were "landmarks to the surgeon," a different semiotics of the hand altogether.[12]

The circulatory system—arteries and veins—fed those all-important muscles that created the flexure lines, and anatomists would sometimes claim that the hand was the most vascular member of the body that was subject to voluntary movement, with the exception of the tongue, according to one. The arrangements of arteries and veins vary widely across the body, and the particular

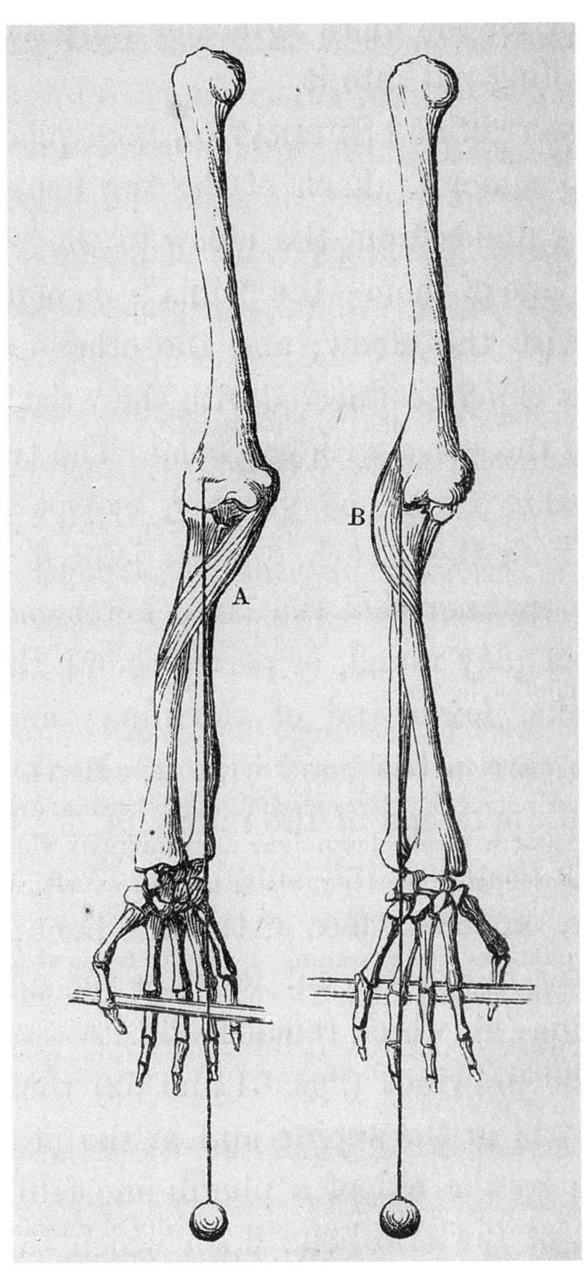

3.2 The hand supine and prone, movement enabled by the pronator and supranator muscles, essential for reading human hands. From George M. Humphry, *The Human Foot and the Human Hand* (Macmillan, 1861), 128. Courtesy the Wellcome Collection, London.

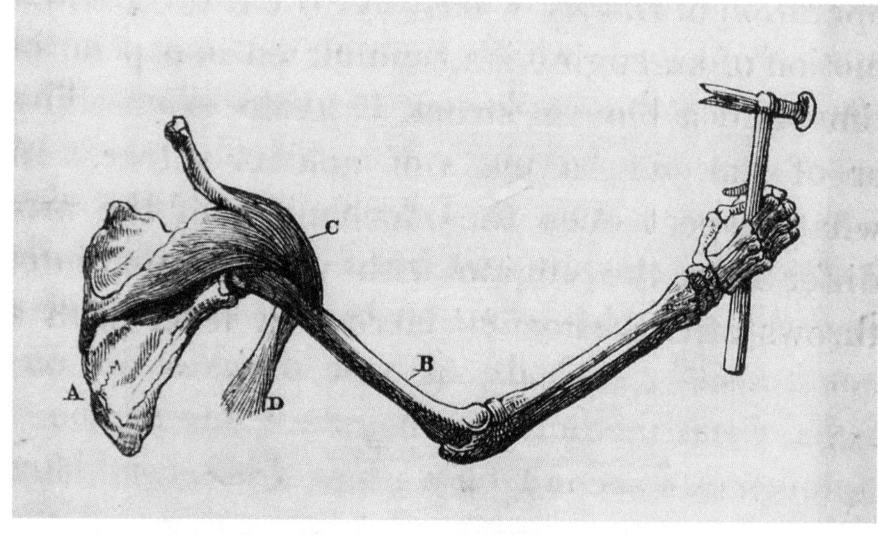

3.3 The muscles of the arm and hand. Charles Bell, *The Hand, Its Mechanism and Vital Endowment as Evincing Design*, 2nd ed. (William Pickering, 1833), 115. Courtesy the Wellcome Collection, London.

concentration in the hands meant that it regulated heat and cold, a vasomotor effect that takes place through the skin, the integumentary system. Quain, Humphry, and a hundred other anatomists represented the skin of the hand microscopically, its layers, papillae, oil glands, sweat glands, hair, and nails. They visualized a reduced biological understanding of what the human hand could be. But when anatomists turned from bones, muscles, arteries, and skin to *nerves*, this attenuated biology became truly expansive.[13]

THE HAND AND THE MIND

The nervous system was a particular specialty of the nineteenth century. The geography and topography of nerves around the face and the brain are still colonized by nineteenth-century neurological luminaries: Broca's area, Bell's palsy, Wernicke's area, Purkinje cells. In his time, Charles Bell was a leading anatomist of nerves

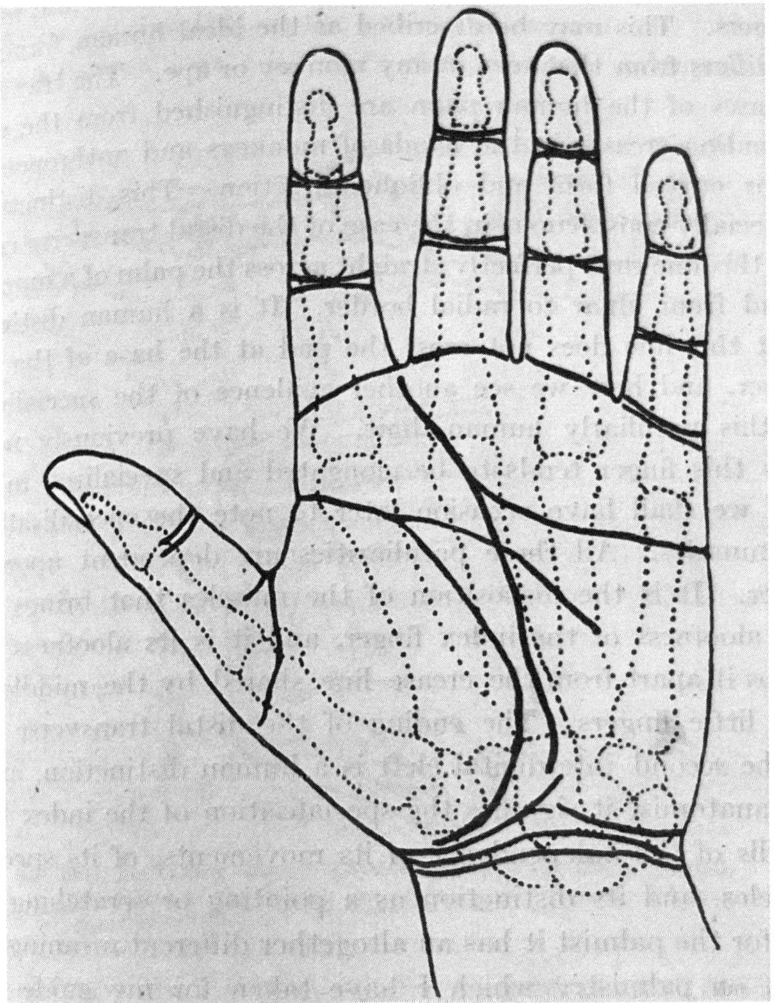

3.4 The palmar flexure lines drawn in relation to the bones of the hand, from *Quain's Anatomy*. In Frederic Wood Jones, *The Principles of Anatomy as Seen in the Hand* (J. & A. Churchill, 1920), 73. Copy in author's possession.

and the nervous system. He showed that they were compound: in one sheath lay two nerves from different roots, one conveying the influence of the brain to the muscle, the other from the muscle to the brain. Nerves thus connected the whole body in an elaborate system that Bell knew better than almost anyone then alive.

> The hand is not a thing appended, or put on, like an additional movement in a watch; but that a thousand intricate relations must be established throughout the body in connection with it; such as nerves of motion and nerves of sensation: there must be an original part of the composition of the brain which shall have relation to these new parts, before they can be put into activity.[14]

Bell's work on "the nervous circle" that connects voluntary muscles with the brain, sending messages to and fro, was well demonstrated by the responsiveness of the hand, asked by the brain to move in intricate and multiple ways, for a thousand purposes. But the messages the hand sent back, through the sense of touch, were also highly significant. While every other sense is superior in "brutes," the sense of touch is superior in humans, some claimed. Two chief nerves descend from the brain, along the arm and forearm, explained Humphry, and these "supply sensation to the palmar surface of the hand." But there, a distinction needed to be made between a sense of touch and what Humphry called "common feeling" or sensitivity to pain. For Bell, this was a double sense: "We must not only feel the contact of the object, but we must be sensible to the muscular effort which is made to reach it, or to grasp it with the fingers." This was a "sense of consciousness," a version of what was later called proprioception, a sixth sense.[15]

When early Victorians connected nerves of the hand to the brain, they found the most fascinating organ of all: the mind. Charles Bell was unequivocal, claiming that the hand develops "the endowments of the mind itself." In the next generation—for the mid-Victorians—the humble hand was finding its place at the center of a linked philosophy, physiology, and psychology of mind and body, and their intriguing and little-understood relations. "This perfection of the tactual apparatus has subserved the highest processes of the intellect," declared philosopher Herbert Spencer in his usual florid style, "by the complex and versatile adjustments of the human hands." Mysteries of consciousness, of perception and proprioception, as well as the neurology and psychology of five (or six or seven) senses, were thought through via the hand. Many were intrigued, for example, with how this organ of touch could sometimes substitute for other sense organs. Blind people can read through touch, and deaf people can talk through their hands. They did so through modern renditions of a sign language that originated with Bulwer's *Chirologia*.[16]

Like Bulwer, anatomists considered the hand an organ of expression, an instrument of the mind. Indeed, Bell and Humphry both claimed the hand to be *the* organ of expression, referring to its communicative power. "How effective an auxiliary to the orator is the wave of the hand, or, even, the movement of a finger," Humphry wrote. "Some men, indeed, seem to owe the efficiency of their declarations as much to the hand as to the tongue." Darwin also explored "gesture language" and "sign language," analyzing the placement of the hands in gestures of devotion, for example, the clapping of hands, the wringing of hands, and how the Anubis baboon in the London Zoological Gardens shook hands with his keeper. We shall see over the next few chapters how the language of the hand was common ground between the most materialist of biologists and the most esoteric of late Victorian chiromancers.[17]

We can also begin to perceive the crossover between a strict anatomy of nerves and the brain, and an emerging psychology of the mind, of the senses, and even of individual identity that by the century's end was manifesting as analytic psychology. While "the sense of feeling" enabled internal communication with one's own body, touch primarily enabled a sense of contact with external bodies, according to Humphry. And when Bell explained, far earlier, that the nervous system, via the hand, "enables us to distinguish what is external and what belongs to us," he was laying ground for the psychoanalytic significance of individuation of self from another. Self-identified "scientific cheiromant" Edward Heron-Allen (chapter 4) was inspired by Bell, like so many: "The act of touching our body with our hand calls forth a double sensation of touch, one through the hand, and the other through the part of the skin touched, whilst touching an external object causes only a single sensation of touch through the tactile organ." Freud wondered about just that from his base discipline of neurology.[18]

When anatomists, physiologists, and psychologists connected the hand to nerves, to the brain, and to the mind, they also connected it to the will. Bell had made this connection explicit, and anatomist Humphry followed through, stating uncategorically that "the hand is the organ of the will, more directly and completely under its influence than is any other part of the body." Darwin too considered hands to have admirably adapted, "to act in obedience to his [man's] will." Not just the mind, then, but human will itself had its own special organ. Yet, as every nineteenth-century reader would appreciate, the human will crossed over into another register: the doctrine of free will, a fundamental theological question. As it happens, free will and predestination were proximate to a philosophy of chiromancy, if we can call it that, as we shall see in future chapters. But a theology of will was also proximate to anatomy. Charles Bell would state that will categorically distinguished humans from all other animals: it exists in humans only, put there by God. Humphry also argued that free will is embedded in humans by God, as the "responsibility attaching to the selection between good and evil, and which is given to him to fit him to be the reasonable servant of his Maker."[19] Given that the hand was cast as the organ of the will, the instrument of voluntary and purposeful action directed by the individual mind, those fingers, palms, bones, nerves, and skin together carried great theological and psychological weight.

Humphry's understanding of the anatomy of the hand, just like Charles Bell's, was ultimately religious. He concluded his anatomy on the hand with a story about the Protestant martyr Thomas Cranmer burned at the stake on March 21, 1556, under the authority of Mary I. Holding out his right hand into the flames, Cranmer declared "this unworthy hand!" Humphry's nineteenth-century flourish was feigned incredulity. "Of whom or of what was that hand unworthy? Was it unworthy of Him who made it? Was it unworthy of him who bore it? Was it unworthy of the purposes for which it was made?" Cranmer the man of God, just like Bell and

Humphry the men of science, recognized Divine workmanship in the human hand, and argued that we should all strive to be worthy of such an eloquent instrument. In such a claim, anatomists in the natural theology tradition were of a piece with chiromancers. When one early twentieth-century palmist proclaimed that "God has, in his divine grace, written everything upon our hands in a language clear and easy of understanding," in a manner, Charles Bell and George Murray Humphry would have agreed. Even in nineteenth-century anatomy, then, the hand was an organ of Divine wonder.[20]

HANDS AND FACES: PHYSIOGNOMY AND PHRENOLOGY

Hands and faces were, and are, often twinned. Over years of research in countless old books, dusty papers, and institutional archives, I found as many faces staring back at me as hands reaching out, occasionally the visages of the famous. While scouring Isaac Newton's library for his books on the hand, I was shocked to meet him face-to-face. There, casually stored on the Wren Library shelf, was his death mask. It was at eye level, and although mine were definitely wide open, his were respectfully closed. Sculptors would use the death mask for faithful representation in marble, bringing him to life again. As we have seen, art and anatomy constantly cross, especially when it comes to faces and hands. Tellingly, I found scores of silhouettes as well, often in the collections of hand experts. Francis Galton's profile, for example, sits among his tens of thousands of fingerprints (fig. 8.2), another kind of signature. Silhouettes were both popular and purposeful. They were drawn and cut not just as sentimental likenesses, but also as profile studies of features to be decoded physiognomically.

Faces came to be read with a new intensity in the eighteenth century, in a revival of physiognomy led by Johann Kaspar Lavater (1741–1801). He defined physiognomy in a way that Paracelsus would have appreciated: "the handwriting of nature upon the human countenance." Old aphorisms instructed a new generation.

Readers of *The Pocket Lavater* would learn, for example: "A gently arched forehead, without a single angle, evinces a mild disposition; and often, that the mind is destitute of energy"; "small eyebrows always accompany a phlegmatic temperament"; and "an aquiline nose designates an imperious temper, and ardent passions." Lavater's influential physiognomy was firmly about reading the face, both in life and in portraiture. Yet Lavater also looked at hands. He thought them highly individualized: "Particular hands *can only* belong to particular bodies." Hands were also read as signatures of the self, portraits of sorts.[21]

Physiognomy was a popular enterprise, but it was also formative of orthodox medical and biological sciences, evident in the work and ideas of the most distinguished anatomists and clinicians. Charles Bell, for example, ventured a physiognomic aphorism that was indistinguishable from any in Aristotle's *Physiognomonica* or Lavater's *Aphorisms on Man*: "a person of feeble texture and indolent habits, has the bone smooth, thin, and light." And after his focus on the hands, Bell also attended directly to the face, the classic countenance of physiognomy, in *The Anatomy and Philosophy of Expression, as connected with fine arts* (1847). This project was doubly physiognomic, since it was concerned with facial expression *and* with its representation in portraiture. It is well known that physiognomy broadly, and Charles Bell's work on expression specifically, inspired Charles Darwin, who set out his own *Expression of the Emotions in Man and Animals* (1872) in evolutionary and physiological terms. Just like early physiognomists (and chiromancers), Darwin was reading the visible surface of the body for an interior state, the pressing outward, the "ex-pression" of an interior phenomenon. The vocabulary *was* different, however. For Darwin, this was not the expression of "the soul," or "sentiment," or "character," as in earlier physiognomies, but instead of "emotion," a new word circa 1800, awaiting its great place in the discipline of psychology that Darwin himself was anticipating.[22]

Some anatomists simply *were* physiognomists. Others deployed physiognomic ideas, and on occasion relied on the signature term "character." For Purkinje, the hand, as the special instrument of human labor, might show the character of an individual. Humphry also thought the hand an "index of character." One scientist-palmist reasoned backward from physiognomy. Rather than the shape or signs of the hand revealing a certain internal character, the "habits and characteristics" of the subject produce those shapes and signs in the first place. And so "as certain habits and characteristics produce certain developments of bone and muscle, so from the appearances of those developments in a hand may the habits and characteristics of a subject be unmistakably inferred." Darwin offered passing observations that were not dissimilar. He wondered, for example, whether handwriting might be heritable, partly due to "the form of the hand and partly on the disposition of the mind." We shall see in future chapters how the physiognomic term "character" became part of the vocabulary of the new science of genetics.[23]

Two anatomists turned Lavater's general physiognomy into the more specific discipline and practice of phrenology in the early nineteenth century. The German neuro-anatomist Franz Joseph Gall and his sometime anatomy assistant, Johann Spurzheim, developed a method that correlated skull shapes and personality traits, building on the idea of a "muscular" brain, in which precise areas were larger or smaller or differently angled, revealing particular attributes that could be discerned externally. Heads were "read" for internal character, as hands had been for centuries. And just like Richard Saunders's geometry of the hand, nineteenth-century phrenology was sometimes called "mental geometry." In Liverpool, there was even a Museum and School of Mental Geometry and Physiology. Heads and hands, physiognomy, phrenology, and chiromancy were entirely connected (fig. 3.5).[24]

This was rendered explicit by the author of *The Hand Phrenologically Considered* (1848), who explained that if, as was obvious, the form of the hand can show "age, sex, and race," we must appreciate that it is "not less affected by the particular kind of organisation, the mental disposition, and the temperament of the individual" (fig. 3.6). The phrenologically read hand was not about palm lines, however, rather about shape, thinness and fatness, strength or weakness of muscles, tendons, and joints, and the relative size of fingers, thumb, and palm. Published anonymously, *The Hand*

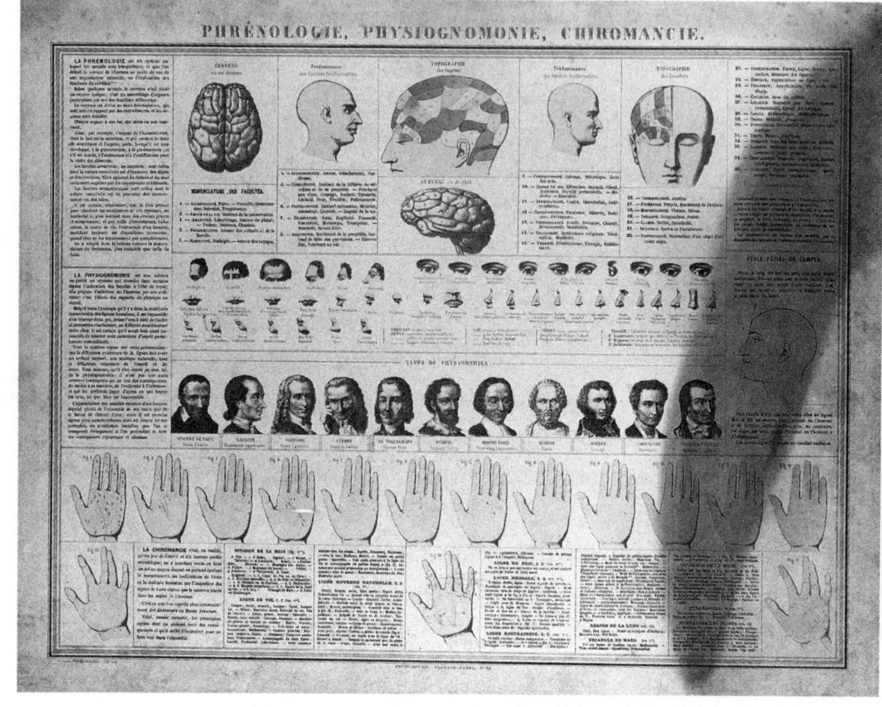

3.5 Phrenology, physiology, palmistry, and the anatomy of the brain. "The basic elements of phrenology, physiognomy and palmistry, with diagrams of heads and hands, and portraits of historical figures." Color lithograph. Paris: Maison Bouasse-Lebel edit imp, [1866]. Courtesy the Wellcome Collection, London, Ref.: 28593i.

Phrenologically Considered might also have been titled "The Hand anatomically considered," since the author worked carefully through all orders of animals from *Annelida* (worms) to *Mammalia*, repeating the argument that "from the structure of an extremity we may obtain a complete insight into the entire organisation of an animal." In the constant crossover of disciplines and knowledges, here phrenology borrowed a zoological argument, after Bell and after the style of natural theology.[25]

It is not entirely clear, but highly likely, that the author of *The Hand Phrenologically Considered* was Richard Beamish, neither an anatomist like Bell, nor a minister of religion like Lavater, but a distinguished engineer and Fellow of the Royal Society. He was

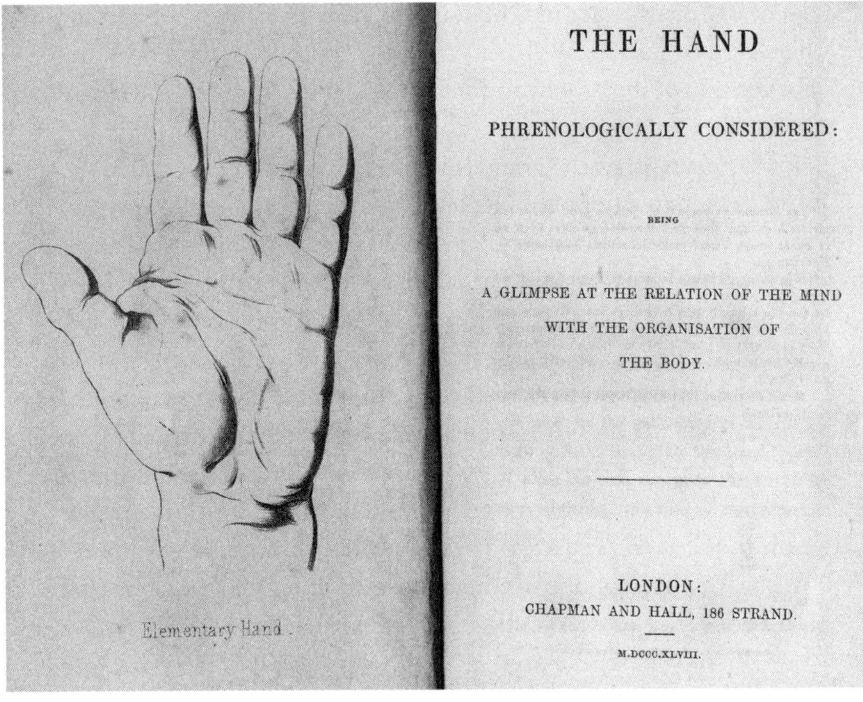

THE HAND

PHRENOLOGICALLY CONSIDERED:

BEING

A GLIMPSE AT THE RELATION OF THE MIND
WITH THE ORGANISATION OF
THE BODY.

LONDON:
CHAPMAN AND HALL, 186 STRAND.

M.DCCC.XLVIII.

3.6 The Elementary Hand. [Anon.], *The Hand Phrenologically Considered* (Chapman & Hall, 1848). Title page and frontispiece. Copy in author's possession.

said to have an intuitive power of reading character from countenances, that he had cultivated this capacity through studying Lavater, and that he was a particular follower of Spurzheim. Indeed, notwithstanding his admiration for Charles Bell, Beamish tried to set various anatomical records straight. It was not Bell, he protested, but the anatomists Gall and Spurzheim, who originally presented the idea of separate functions and subdivisions of the nerve. Beamish was known also to be "a believer in the mesmeric second sight and the spiritualism of mediums." He was not the only Fellow of the Royal Society to so believe, the most outspoken on such matters and the most distinguished just then being the great naturalist Alfred Russel Wallace, the other half of the natural selection breakthrough.[26]

In the summer of 1856, the British Association for the Advancement of Science met in Cheltenham. The great philosopher and historian of science William Whewell was there, speaking about the motion of the moon, and so was the far less illustrious Richard Beamish, speaking mundanely on the "Statistics of Cheltenham." Not to be put off by the grand Master of Trinity College and sometime vice chancellor of the University of Cambridge, Beamish approached Whewell, asking if he could take a tracing of his hand. He followed up with another request. Would Whewell agree to its publication as an illustration of the "philosophic type"? Beamish's point was not to demonstrate some kind of magic or wonder or special mysterious insight, rather simply to show evidence of correlations between inner character and physical shape. He explained that "I have obtained so much evidence of the value of the hand as an index of moral and intellectual development that I am induced to prepare a work upon the subject for the press." This was a physiognomy of the hand, paradoxically *less* enchanted (and enchanting) than Bell's or Humphry's anatomy.[27]

William Whewell consented and not just his hand, but his very constitution, internal and external, was now on the record for all to assess. Beamish diagnosed Whewell's hand as full, strong, and elastic, his temperament as Nervous, Sanguine, Bilious. Its symmetry expressed "large vitality, and high intellectual capabilities." The tracing of this auspicious appendage appeared in *The Psychonomy of the Hand: or, The Hand an Index of Mental Development* (1865), to which Beamish did put his name. There was Whewell's hand, life-sized, accompanied by those of other individuals and types: a gorilla's hand, the hand of an "English navvy," a Hanoverian hand, that of an "African Negro," the hand of "an idiot from Cork," the hand of a Russian, a Dane, a "Hindoo," and the hand of a mummified Egyptian from Thebes.[28]

In pursuing such individual and type specimens, Beamish's physiognomy and phrenology of the hand were inspired by a French classification of human hand types that had been set out in the 1840s, just before the work of the occultist Desbarrolles. In *La Chirognomonie* (1843), upstanding military officer M. Casimir D'Arpentigny—neither medical nor scientific nor hermetic—created an idiosyncratic typology. Evidently vain about his own handsome and well-shaped hands, D'Arpentigny classified seven types, a loose progression from least to most desirable: the Elementary Hand; the Spatulated Hand; the Artistic or Conical Hand; the Useful or Squared Hand; the Mixed Hand; the Philosophic, or Knotted Hand; and the Psychic, or Pointed Hand. This proved to be an enduring classificatory system, certainly for modern palmists of all kinds, but it was also deployed by anatomists, as we shall see. Beamish was following just this method, as it was not lines or creases that D'Arpentigny observed, but hand shape and proportions. In this sense, it was physiognomic more than chiromantic, signaled in the very word *chirognomie*. This all built from an Aristotelean principle of the harmony or correspondence between mind and form, the idea borrowed also by Richard Beamish as a maxim for hand reading: "As is the mind so is the outward form."[29]

As *La Chirognomonie* moved into various English editions and translations, this 1840s French set of hand types retained older understandings of the body, the senses, and the human constitution, temperament, and character. In many ways it was far more eighteenth century than nineteenth century in its references and reasoning, ideas which, while antiquated, were still fully comprehensible. A Galenic doctrine was repeated, for example, hands betraying sanguine, phlegmatic, choleric, and melancholic temperaments, which formed the four primary constituents of the human body and the four elements of nature. The sensitive hand expressed a sanguine temperament, a smiling florid countenance, a character excitable, easily moved, and quixotic. The Elementary Hand revealed a phlegmatic temperament, a body loaded with "cellular tissue and fat," one that is pale and unexcitable. A choleric temperament is built from a body with little fat and sharp features, creating an energetic, passionate, and determined character. This is seen in the so-called Motive Hand.[30]

Richard Beamish's *Psychonomy of the Hand* was at core a loose translation of D'Arpentigny's work, and he described hands in just this way in the 1860s. For him, the *feeling* of the palm corresponded to humoral balances. One palm, for example, was "full, elastic and flexible," with a temperament "Nervous, Sanguine, Lymphatic." Another was a Spatulous Mechanical hand, palm thick, strong, and elastic, with a temperament Sanguine and Nervous. The palms and character of individuals were depicted and diagnosed, that of a fellow engineer, the magnificently named Isambard Kingdom Brunel, for example, builder of the Great Western Railway and the first transatlantic steamship. His full, strong, and firm hand revealed his temperament to be Bilious, Nervous, Lymphatic. Beamish traced hands incessantly, collecting hundreds from friends and acquaintances, and from museum specimens, bones, and casts. Yet the result is visually disappointing, the least designed by far of the thousands of representations of hands I have looked at, from impossibly rich illustrated medieval manuscripts to the delicate printed hand of Mok the gorilla, dead in London Zoo's morgue. Beamish's hands were simply outlines, with a blank interior: no lines, no mounts, no creases, no stars, no signs, no triangles named for magical *séphirotes*. On one measure, these physiognomic hands were literally traced as would a child. On another measure, more generously, the outline *was* the sign to be decoded, the profile of the hand revealed, just as the solid black card of a traditional silhouette caught the profile of the nose, the forehead, the chin, against its white background. Beamish was insistent that the very shape itself correlated to internal character. We might see Whewell's "calm and contemplative, yet industrious and practical" character in "the fuller development of the third phalanges of the fingers." And we only need the shape of Whewell's thumb to know his great will, his originality, and his logical acumen. It is a measure of how immaterial the palm lines could be in this tradition of a physiognomy and phrenology of the hand.[31]

HANDS NATIONAL AND RACIAL

Himself a philosopher of "the classificatory sciences," Whewell held one question in common with all natural scientists, and indeed with the physiognomically inclined Lavater, D'Arpentigny, and Beamish: how a "type" related to an "individual," how divisions, subdivisions, and varieties of those types were to be nominated, and on what criteria. Across all of living nature, just what was an "individual," a "genus," a "species," and a "race"?[32]

Whether one should read a countenance (or a palm) as that of an individual or a type was already a question in Aristotle's original philosophy of physiognomy, and it endured into the eighteenth and nineteenth centuries, the great age of classification of everything. Lavater's tendency was to favor the reading of an individualized character in the face, yet he recognized *national* physiognomies too. D'Arpentigny followed through in his typology of hands. Attaching hand types to particular geographies, climates, and latitudes, not just national character could be discerned, but also, and in an entirely confused way, religious and racial character too. "The form of the hand is indicative of the character of the people," argued Beamish, by which he meant national types in the first instance. The Elementary Hand, for example, with thick fingers, large palms, coarse skin, and short and thick nails, was associated with instinctive temperament, an apathetic person with dull sensations. It belonged to the superstitious. In ancient times it was found in Egypt, in modern times in Brittany, so proclaimed D'Arpentigny. It was also common in the polar latitudes, abounding among the Lapps, the Finns, and the Icelanders, and among "nations of Tartar or Sclavonic origin." It is uncommon in India, he asserted, but can be found "artificially produced" in the Parsis. As the century progressed, versions of this Elementary Hand became more primitive: "It belongs to the lowest grade of human intelligence, and seems only to be gifted with the amount of intellect requisite to provide the merest necessities of life." A kind of civilizational progression was being written into hand types, in which the Elementary Hand was low and rudimentary. This was then folded into a popularized version of Darwin and Wallace's evolutionary theory by Edward Heron-Allen, whom we shall meet in the next chapter: the Elementary Hand "retains throughout adult life the character which it presented in infancy, and it strikingly resembles the hand of those of the monkey tribe most nearly allied to man in their organisation and outward form—a hard, thick palm being joined to short, rudimentary fingers." Also a Fellow of the Royal Society, as it happens, Heron-Allen thought that the very opposite of this crude hand and human type were Moslems and Hindoos: "Among these poetic, cunning, romantic, sensual peoples the Elementary Hand does not exist."[33]

The Motive Hand was described as large with prominent joints, strong, thick, and with a square tip, the palms hollow and firm, a large strong thumb with a muscular root. In both sexes it denotes a "masculine or reasoning mind" and is connected to right and authority, organization and classification. It was the hand of Plato, of Milton, and of Goethe, we are told. A rare hand, in Asia the Motive Hand was most commonly found in India, in Europe most commonly in Germany. It was an aristocratic rather than a democratic hand. The Spatulate Hand, by contrast, has smooth fingers and a large thumb associated with movement, activity, usefulness, frankness of character, resolution, and self-confidence. It was most common in those latitudes in which inclement climate and sterile soil made action, movement, and locomotion necessary, another environmentally determined argument. It was described as belonging to the north, in Scotland more than England, in England more than France, and in France more than Italy or Spain.[34]

The Psychical Hand was highly prized, rare and refined, for all obvious reasons often favored and much drawn by palmists, both occult and scientific. To his delight, Beamish found this hand in a collection of casts held in the museum of India House. He learned that German explorer Adolf Schlagintweit, commissioned by the East India Company, had appreciated the "ethnic importance to the form of the hand," and traveling through Tibet, the Himalayas, and Central India, had taken the casts of hands and feet. Thirty-eight casts had ended up in India House, which Beamish took the time to trace. Already, as casts, they were viewed, valued, and presented as "types." But Beamish placed them into D'Arpentigny's typology of cheirognomy and thought two of them especially rare because they were representative of the Psychical Hand, "the most beautiful of all hands."[35]

In this strange mid-nineteenth-century phrenology of the hand, such climatically produced "national" hand types were linked to different political systems, a kind of physiognomy of latitudinal statecraft. The Psychical Hand might be beautiful, but it was the Spatulate Hand that denoted action, movement, self-reliance, and energy. This highly desirable Spatulate Hand, with its enterprising character, reliably produced liberal democratic polities. Indeed, the Spatulate Hand "is essentially Protestant," the laudable organ of industrialization, empire, and economic expansion. The Duke of Wellington's hand became a kind of type specimen, although he was hardly a liberal, still less a model democrat. It was skillfully illustrated in pencil by the artist Rosamund Brunel Horsley in a manner far superior to engineer Beamish's crude outlines. It is notable that such illustrations depicted the dorsal side of the hand only. Even in such fine drawings, the hand, not the palm, was decoded, its shape, not its lines (fig. 3.7).[36]

In this world typology, there were English hands, American hands, and Jewish hands. The Chinese national character was diagnosed by the square hand type that attached itself to the law, "however harsh and unrelenting." For Beamish, as for D'Arpentigny, "the masses there yield willingly to the exactions of a hierarchy." This was sometimes called the "Useful Type" of hand, showing politeness and industry and a character inclined to the exact observance of ceremonies. A political system followed: the centralized polity of a "square-fingered race." The American hand betrayed another national character altogether, which one English hand reader decided was reserved, selfish, and suspicious, with a disposition thoughtful, gloomy, but equable. "His attitude is without grace, but modest . . . his ideas are narrow, but practical." The good news, though, was that where the prevailing hand type was spatulate, as with "Saxon" Americans, "the political institutions are free."[37]

At one level, this is so much nineteenth-century nonsense, but at another level we mock at our peril, because these ideas were also formative of some anatomists' reification of "race" as a way to think about human variation. Edinburgh's Robert Knox, for example, was another anatomist who worked closely on hands, not least in anatomy works for sculptors and artists, in which images showed both "dorsal" and "palmar" hands. But it was hands of another kind that he incorporated into his infamous *Races of Men*

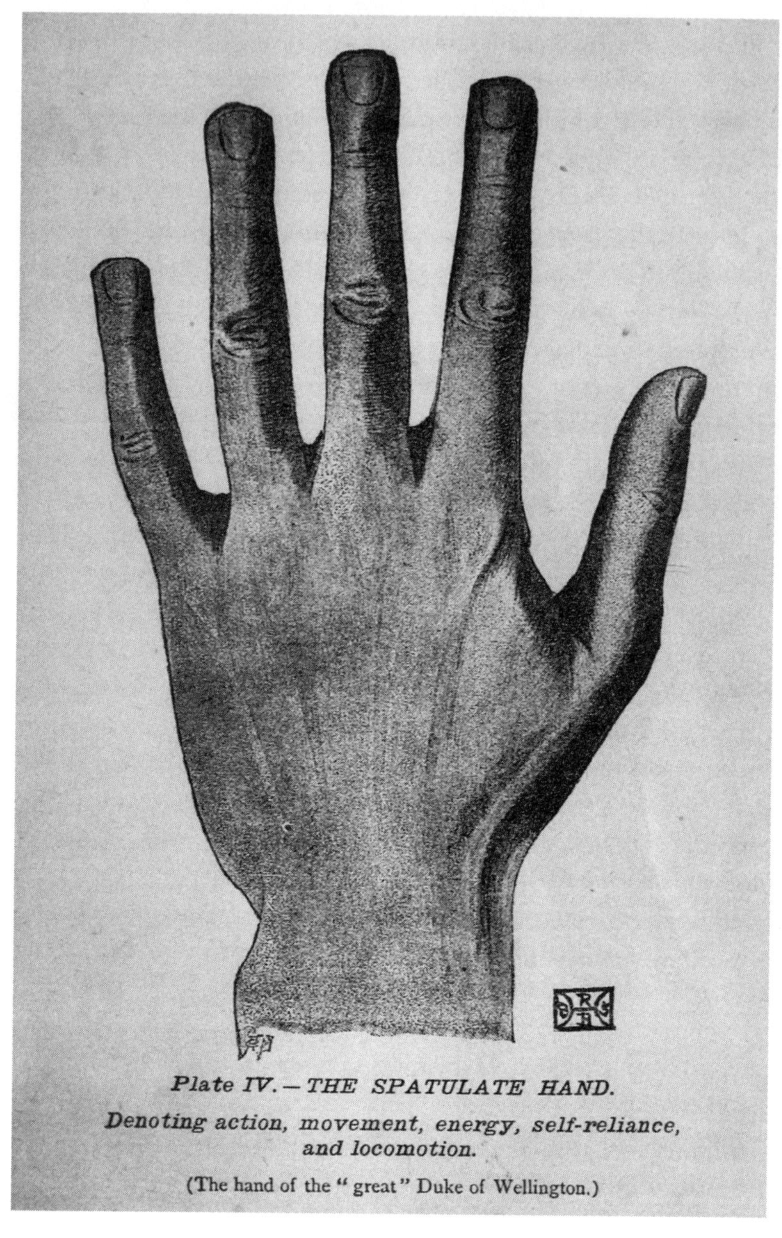

3.7 The Spatulate Hand, "essentially Protestant." Illustrated by Rosamund Brunel Horsley, in Edward Heron-Allen, *Practical Cheirosophy. A Synoptical Study of the Science of the Hand* (G. P. Putnam's, 1893), 67.

(1850). D'Arpentigny's cheirognomy gave him ways to think about a correlation between hands, race, political inclinations, and institutions. It was the "Saxon" race that Knox most sought to explain and promote, the "hard-handed, spatula-fingered Saxon utilitarian." A lover of labor, the Saxon is "large-handed," and this has paid off, so Knox's hand-based argument went. This is why a small group of Englishmen can hold military possession of India, referring to the East India Company as "large-handed, spatula-fingered Saxon traders." The global "Saxon" was the favored race of the brash Edinburgh anatomist, on the cusp of being "the dominant race of the earth." Not only that, but wherever the Saxon planted himself, the politico-legal manifestation of his peculiar racial physiology (we are not quite told how) was planted too. The Saxon was a racial opponent of the military despot, "nature's democrat." Indeed, as Knox wrote in 1850, he saw in America a great physiologically produced Saxon "republican empire, destined some future day to rule the world." D'Arpentigny was his source for all this wisdom, and Knox borrowed the idea of "elementary hands and minds" too, especially applied to the apparently primitive Celt, a race inferior to the Saxon, he asserted. This was betrayed by "hands, broad; fingers squared at the points." He disagreed with the Frenchman's views on other matters: the Jewish hand, for example, and D'Arpentigny's claim that they were the same as the (French) Normans, "with the palm altogether less developed, and the fingers, as it were, square."[38]

From our present, Knox's theory of race is as nonsensical as D'Arpentigny's *chirognomie*, but it doesn't follow that it was therefore comprehended to be nonsense at the time, and that is the historical point that needs to be taken seriously. Indeed, the first English translation of D'Arpentigny's *La Chirognomie* appeared in the *Medical Times*, retitled as *The Philosophy of the Hand*, no less. And there it was reviewed alongside Knox's lectures on the *Races of Men*. Across the anatomy, physiognomy, and phrenology brotherhood, the shape, feel, and structure of hands were correlated to race. Strange, shocking, or even amusing as such claims are, we need to take the typologies of modern physiognomists of the hand at least as seriously as did the medical profession of the time. And it is important to appreciate that this typology of the hand and race was in part the work of anatomists themselves. This is not a history of a "pseudo-science" or "quack" medicine. It *is* the history of science and medicine.[39]

It is not just familiar, but in many ways correct to understand the great advances in medicine over the nineteenth century as part of a progress away from belief toward facts checkable, repeatable, and falsifiable by a scientific method, away from magic and superstition toward a skeptical science. But such a history of progress from faith to evidence is also insufficient. Even among the strictest anatomists of the nineteenth century, systems and organs of the body, including the human hand, were at once microscopic and expansive, physiological and philosophical, anatomical and theological.

The idea that medical practice became increasingly scientific is also insufficient because older medical models endured. Indeed, antique medicine and modern medicine converged on the hand. Galen's ideas were refreshed by fringe dwellers like Beamish and D'Arpentigny, making their way into vernacular health and hygiene, as we shall see in chapter 6. The Hippocratic doctrine of airs, waters, and places was renewed as "race," the constitutional understanding of humans in frigid and torrid zones, latitudes, and geographies turned into differences apparently marked in the hand. Physiognomy endured both as a strict if idiosyncratic method to read character, and as a disseminated ambition to read the human interior by its exterior in other ways. As the organ of touch, of expression, and of the will, the anatomy of the hand related to the inner human as well as to the structure of the skin, the bones, the muscles, and the nerves. It is hardly surprising that an ancient physiognomy resurfaced and that some Fellows of the Royal Society thought the hand a modern wonder, the form of which could and should be decoded to reveal the interior self and the individual will, as well as diseases, conditions, and predispositions.

Leading anatomists and evolutionary biologists of the era continued to wonder about inside and outside, and increasingly about how a distant species' past might be retained on the surface of the human body, evidence of transmutation. The hand was there at the birth of evolution by natural selection, "the descent of man," and primatology. As we shall see, it did not go away. Successive generations of twentieth-century anatomists, physical anthropologists, and population geneticists continued to look at hands for signs of race, signs of criminality, signs of disease, signs of mental illness, the mark of the individual and of the type. Out of this nineteenth-century anatomy and evolutionary biology, the sign on the hand was to become, entirely technically, a phenotype.

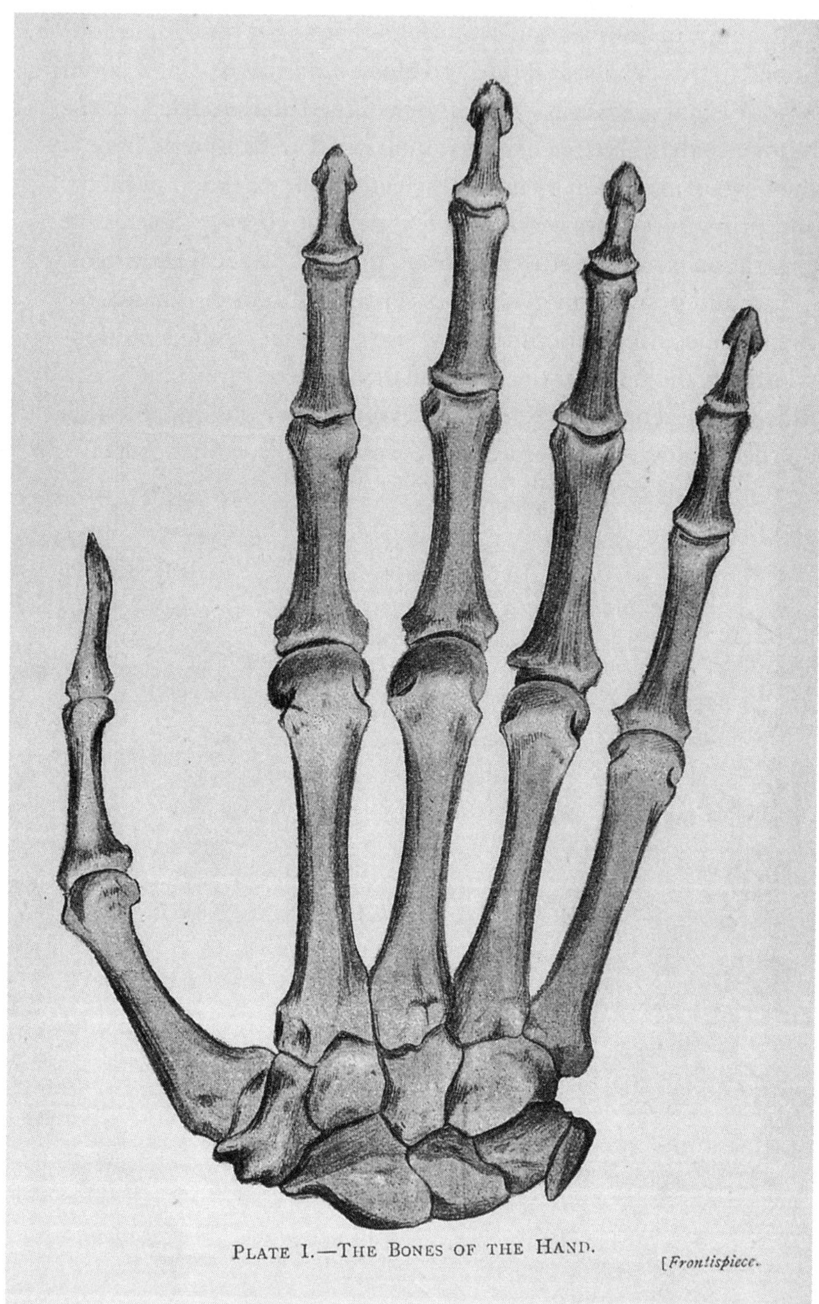

4.1 The bones of the hand, illustrated by Rosamund Brunel Horsley. Frontispiece, Edward Heron-Allen, *The Science of the Hand* (Ward, Lock, 1886). Courtesy the Wellcome Collection, London.

PART II

THE MODERNIZATION *and* MEDICALIZATION *of* PALMISTRY

FOUR

The Cheirosophy of Edward Heron-Allen

"The materialism of the age stands in our way," complained celebrity palmist Edward Heron-Allen (1861–1943). "The materialism which seeks to 'account for' everything, and which commences the process by doubting everything." We might expect a palmist to say just that in 1885, but Heron-Allen was to become Fellow of the Royal Society, Fellow of the Linnean Society, Fellow of the Geological Society, Fellow of the Zoological Society, and Fellow of the Royal Microscopical Society. His statement was directed at the greatest materialist of the age, agnostic Thomas Henry Huxley, whose scientific naturalism was just then inciting mass doubt on "the existence of mind, of soul, of everything." Heron-Allen resented this, and yet he had just authored his latest book on the hand, with an argument that "cheirosophy" should rank as a physical science. In so doing, he filled it with the work of some of the great modern anatomists: Charles Bell, Richard Owen, and indeed materialist Thomas Henry Huxley himself.[1]

If in the seventeenth century the occult sciences were formally astrology, alchemy, and natural magic, in what ways were "occult" and "science" connected in the late nineteenth century? The question raises foundational matters through which nineteenth-century "moderns" sought to distinguish themselves from the superstitions and magic of their own past, and of "traditional" societies in their present. By what means and methods were belief and truth, faith and proof to be adjudicated? Thomas Henry Huxley invented "a-gnostic"—without knowledge—with reference to that which is not necessarily untrue, but which is unknowable. Neither the Christian God nor any other supernatural being can be verified through repeatable experiments. Huxley was the great nineteenth-century ringmaster of disenchantment, questioning not just religious faith, but also his generation's belief in phenomena beyond a natural or material world. What was the place of palmistry in this conversation?

There was a great range of hand readers attempting to modernize and even medicalize palmistry in the late nineteenth and early twentieth centuries. Some we shall meet traded almost entirely upon the esoteric. "Cheiro" and "Keiro" (chapter 5) were squarely figures from the familiar occult revival. Others are far better understood and analyzed within a history of medicine. They emerged out of traditions of popular health and hygiene, especially those that linked bodies and minds, like phrenology. Entrepreneurial Ida Ellis is exemplary in this regard (chapter 6). Still other readers of hands decried occult chiromancy altogether, aligning themselves with orthodox medicine, even clinical research, at least aspirationally: Katharine St. Hill embodied precisely that in her own special ambitions for medical palmistry (chapter 7). We have seen that reading signs of the hand had long been performed routinely by physicians and by anatomists, and so the physiognomic tradition of reading character or health was not *necessarily* occult at all. Indeed, quite the reverse. In short, by 1900, palmistry *could be* occult and esoteric, but it need not be.

The idiosyncratic work of the marvelous Edward Heron-Allen borrowed from each of these traditions, his writing, ideas, and practice as anatomical as they were astrological, as occult as they were scientific, as orientalist as they were occidentalist, as scholarly as they were practical. "Polymath" is an overused term. Few truly earn the accolade, but there is no doubt about its applicability here. The most workaday thing about Edward Heron-Allen is that he was born into a legal family, and worked in his father's firm in Soho, central London. Thereafter, he was a noted violin maker, bibliophile, scholar of Persian literature, author of science fiction and fantasy novels, and a scientist celebrated for decades of microscopic work on tiny foraminifera, single-celled organisms whose shells layer the seafloor. It was for that diligent and extensive work that he was elected Fellow of the Royal Society in 1919. Over the 1880s, however, he was a palmist and scholar of chiromancy, as dedicated as he was celebrated. With the lively, charming, and eccentric Edward Heron-Allen, we can begin to chart the lesser-known path of chiromancy toward the twentieth-century clinic and lab, toward psychology and even genetics, as well as its better-known path toward the pier, leisure, and pleasure.

In a magnificent riot of different knowledges, Edward Heron-Allen brought together each of the traditions we have explored in the first three chapters of *Decoding the Hand*. First, he was a serious collector not just of rare books, but also of ancient manuscripts. He knew the classical, medieval, and early modern history of chiromancy well, its origins and its tradition of occult secrets, reading across multiple classical languages, and then acquiring more. He researched documents in the British Library in London, the Bodleian Library in Oxford, and collected many precious early books and manuscripts for the vast Heron-Allen personal library at Large Acres, his eccentric home in coastal West Sussex. He owned, for example, a precious calfskin and gilded copy of *Opusculum Chiromanticum* (Greifswalde, 1625), Adrian Sicler's *La Chiromancie Royale* (Lyon, 1666), and *Chiromantia Medica*, printed on vellum in The Hague in 1667. Second, he had an orientalist's knowledge of "Eastern" practices and connections, was fluent in Persian, and was a distinguished translator and interpreter of classical Persian literature. And finally, he actively incorporated the most recent anatomical and medical knowledge of the hand into his own practice and his own books. To understand him "simply" as an occultist in the tradition of Lévi or Desbarrolles is to fail to appreciate his self-description as scientific. All this he managed to align with a revival of the occult in its high *fin-de-siècle* moment, a culture and commerce in which he traded lavishly and liberally, in print and in practice, in libraries and in parlors. As we shall see, his generation saw handwriting, as well as hands, as an imprint of the self. If so, I venture here a rare reading of my own: Edward

Heron-Allen was as gregarious and magnificently embellished as the flourishes of his ostentatious signature, which he almost always offered both in Persian and in English (fig. 4.2).[2]

PALMISTRY IN THE 1880S

Edward Heron-Allen recounted that while he always had a sensitivity to hands, he was a hopeless physiognomist, inept and untalented at reading faces, although any physiognomist would surely have appreciated his own. Energized by the occult revived by Lévi and Desbarrolles, he began to offer hand readings to his own upper-middle-class London circle in the early 1880s, both men and women of an artistic set, primed for their inner secrets to be disclosed in private, or in careful company, by the dashing Heron-Allen. In truth, however, his ambition was already to be a writer more than a palmist. It was as researcher and author of books on chiromancy and on the hand that he sought to make a living. "Today, after having been admitted a solicitor for a year and a half," he wrote in his journal, "I have thrown up the law and devoted my time, and my talents such as they are to literature." He walked out of his office in Soho Square with the firm intention to publish commentaries, novels, nonfiction history books, drama, and poetry. Indeed, Heron-Allen packed his box of solicitors' things and proceeded straight to the British Museum "to make the notes for the article on Hand-superstitions." His already-established minor notoriety in hands made for a lively and marketable subject. This "inaugurates my literary life." It was a new start, and he never looked back.[3]

Edward Heron-Allen announced his interest and skills initially in the privately printed *Chiromancy, or the science of palmistry* in 1882. The next year he produced another book on hands, gestures, and "chirography"—reading autographs and handwriting—and in its appendices discussed dactylomancy, or divining by rings, something that Desbarrolles also focused on in his later books on the mysteries of the hand. Heron-Allen's book had a curious title: *Codex Chiromantiae: Being a Compleate Manualle of y Science and Arte of*

4.2 Edward Heron-Allen, c. 1887. West Sussex Record Office, Edward Heron-Allen Papers, EHA 1/6/2/4. Reproduced by permission of the County Archivist, West Sussex Record Office, and with kind permission of Sam Jones.

Expoundynge y past, y presente, y future and y Charactere by y Scrutinie of y hande, y gestures thereof, and y Chirographie. It was privately printed by the Sette of Odd Volumes, a dining club of bibliophiles established in 1878. Heron-Allen was introduced into the circle after meeting the bookseller, Bernard Quaritch, at an auction of books on violins. In summer 1882, Quaritch invited him to talk to the club on violin histories (he thought them "oriental," not European, originally), and in 1883 he was elected. His self-appointed name for the Sette of Odd Volumes was "Brother Necromancer," the wizard, the magician.[4]

Like so many others in London, Heron-Allen was actively reaching back to the *magi* of earlier centuries, and both reliving and retelling their secrets. With vague reference to hermetic symbols and ceremonies, more ironically than seriously enjoyed, the Sette of Odd Volumes was partial to affected titles, chants, and rituals. For example, a foreword to Brother Necromancer's privately printed *Codex Chiromantiae* was an old oath by which Vettius Velens, the Astronomer, swore his brothers to secrecy "by the Starry vault of Heaven, by the Circle of the Zodiac, the Sun, the Moon and the Five Wandering Stars (by which Universal Life is governed)." The London-based members played along: William Murrell, physician, became brother "Leech," James Roberts Brown, brother "Alchymist," and Frederick York Powell, Oxford's Regius Professor of History, brother "Ignoramus."[5]

Yet this was also a time and place when more serious reincarnations of hermeticism, Rosicrucianism, and Freemasonry were thriving. The Sette of Odd Volumes was at the playful end of a more genuinely occult continuum. The Hermetic Order of the Golden Dawn, for example, included Arthur Edward Waite, the great English translator of the work of Éliphas Lévi, and Samuel Liddell MacGregor Mathers, translator of the seventeenth-century *Kabbala Denudata*, now in English as *The Kabbalah Unveiled.* These hermetics were sometimes learned, as with Waite, often multilingual, as with Mathers, but they also dressed up in Egyptian costume, folding orientalism and occultism, "Egyptian," and Jewish and Christian hermeticism into impossible late Victorian shapes. Taking oxygen, energy, and ideas from French *occultisme*, from

American spiritualism, and from nineteenth-century Egyptomania, this Anglophone occult was certainly the cultural counterweight to the scientific materialism of the age.

By comparison, Edward Heron-Allen was straightforward. Later adopting a more scientific approach to the hand, at this point, he was temperamentally and intellectually partial to the original Gnostics (preferably the Persians), more than dry Agnostics like Huxley who were ushering in the secular Age of Unbelievers. Enchantment, not disenchantment, was Edward Heron-Allen's natural inclination. And he lived in an era, in a class, in a gender, and in a nation where that could be eccentrically indulged.

Edward Heron-Allen was part of another set in 1880s London, a circle of novelists, playwrights, actors, and singers. He was constantly visiting Blanche Roosevelt, the onetime opera singer who by 1885 had herself turned to novels and journalism. He was close friends with Oscar Wilde and his wife Constance Foy, in truth besotted by the latter, a love affair: "Our first point de depart was her interest in Cheiromancy," he diarized in December 1885, "she is now my greatest confidante." He read the hands of such special acquaintances, but this was an exchange between friends, not yet commercial. After one dinner in December 1885, Blanche Roosevelt asked him for a reading, and he warned her: "Everything in your life is written in your hands, so if there is anything you don't want told, don't show them to me." She nonetheless invited him to proceed (he noted in his journal that that is how everyone responded), holding out the most extraordinary pair of hands he had ever seen. Some "awful home truths" were revealed. Edward Heron-Allen had been engaged in mesmerism also, one evening taking "Mrs. Oscar" to a "mesmeric séance." This involved what he called "the simpler forms of muscular mesmerism," as well as some interesting experiments in "will direction without contact or sound," each of which he described as completely successful. At parties and dinners, he was called upon to read hands of both men and women, even as, during the day, he was translating D'Arpentigny, *La Science de la Main*: "I must get it finished & cash the cheque for it!" The pen was the income he had exchanged for the law, and thus far reading hands was part of his social life, not his work, his confidences with special friends. That was about to change.[6]

Heron-Allen traded on hands with enough notoriety in London to be engaged by the agent James B. Pond for a North American tour, sailing from Liverpool on October 23, 1886. Bram Stoker—whose Gothic *Dracula* was to epitomize a genre in which Heron-Allen himself was later to write—happened to be a fellow saloon passenger, and a sociable friend in New York. There, Heron-Allen rented rooms at the grand Everett House in Union Square, and almost immediately invited the press to a palm-reading reception, an ostentatious announcement of his arrival. He delivered a four-part series of lectures at Chickering Hall to great acclaim, then toured Boston, Philadelphia, and Chicago, reading the hands of the famous, lecturing on ancient and modern chiromancy, and publicizing his several books. In the United States, Heron-Allen found himself in an incessant round of what he called séances, for which he was much in demand, and at least initially well paid. Yet it is clear from his journals that it was all dragging him down, even though he operated in impressive circles, reading hands at the homes of prima donna Clara Louise Kellogg, reforming clergyman Henry Ward Beecher, and poet, journalist, and editor of the *Century Magazine*, Richard Watson Gilder. It was an American version of his London circle, less the eccentric Odd Volumes. But the novelty of Edward Heron-Allen wore thin, and he was participating in his own devaluation, agreeing to talk on palmistry and perform hand readings for schoolgirls and for people he simply didn't like. To add insult to injury, he started to admit to himself that he wasn't getting paid enough to do so. By 1888, the shine had definitely tarnished, a few journalists called him a fraud, and it all started to come undone, with unpaid bills, too many dubious social connections, and soon enough even his agent dropped him unceremoniously, no longer a financially viable prospect. Edward Heron-Allen returned to Britain in early 1889, determined to leave palmistry behind and to resume other semi-fictional, semi-scientific enterprises. He was hardly at a loss for things to do, tales to tell, languages to learn, novels to write, violins to make, books to collect, and new species to find and name.[7]

Heron-Allen's momentary success was in part about his near-perfect embodiment and performance of a certain classed masculinity. His beauty helped, but it was underwritten by a new style of occult intimacy, deriving from *La nouvelle chiromancie* announced in Paris in the 1860s. As regards the famous triad of class, gender, and race, Desbarrolles's chiromancy was a different configuration altogether to the traditionally feminized Gipsy encounter. That was typically between a Roma woman and a credulous servant girl or between the fortune teller and a middle- or upper-class woman, the latter in some kind of protective custody of a middle- or upper-class man, an exciting but safe encounter depicted in countless pastoral scenes. The new chiromancy was certainly practiced by men *and* women, and yet there was a particular significance ascribed to the masculine palm reader. The jacket of Desbarrolles's *nouvelle chiromancie* caught how the mysteries of the hand and the self might be revealed by a man in a respectable encounter with a bourgeois woman (fig. 4.3). Yet this was in a private and interior context that bordered on something else; too revealing, too

4.3 The new chiromancer: bourgeois and masculine. Adolphe Desbarrolles, *Les mystères de la main révélés et expliqué* (Dentu, 1859). Title page. Copy in author's possession.

intimate with regard to knowledge of character, or the soul, or the self. This was a re-gendered tableau, the very opposite of the crude work of *la gitane*, the apparently deceitful Gipsy. Desbarrolles and Heron-Allen represented a new caste of *magi* and of "wise men," their knowledge received from a learned and text-based masculine tradition, not a folk and feminine one.[8]

In London, Edward Heron-Allen certainly read the hands of both men *and* women, but they were all of his own class, members of his political, artistic, and literary circles. When he toured and lectured in the United States before his fall, he gained both celebrity and followers, men and women falling into a kind of love, born of the intimacy and charge of holding hands, and the possibility that this modern *magi* might, indeed, know the secrets of one's own self.

Edward Heron-Allen sought and read the palms of the famous. When he did so, he typically collected and kept the ink impressions and drawings of their hands, asking his subjects to verify the print with their autograph-signatures. Taking, keeping, and even publishing celebrity hands became a common practice that created, in turn, "celebrity palmists." It was an effective way of accruing credibility: the more scientific the subject, the better. Richard Beamish had done this a generation earlier in his phrenological study of the hands of scientist William Whewell, engineer I. K. Brunel, physician Dr. Haurowitz, the Right Hon. Lord Brougham, and the engraver John Martin. Desbarrolles also made a special point of displaying the cast of the left hand of Victor Hugo. Edward Heron-Allen's albums of American hands include the palm prints and readings of journalist Kate Field, writer Julian Hawthorne, jurist Oliver Wendell Holmes, caricaturist Thomas Nast, and actresses Helena Modjeska and Dame Ellen Terry. In the process, celebrities' hands served as specimens of sorts, in the human typology that palmistry, in one version, was becoming.[9]

Another hand read by Edward Heron-Allen belonged to Richard Burton (1821–1890), the Middle Eastern traveler, explorer, and Persian linguist who had detailed fortune-telling among the Roma of Persia, Syria, Morocco, Hungary, Spain, Brazil, and more. Burton and Heron-Allen met each other at a meeting of the Sette of

Odd Volumes in the summer of 1882, an enduring friendship built not least on the delights of classical Persian literature and the pair's obsessive interest in the quatrains of *Rubáiyát of Omar Khayyám*. Two other famous palms that Heron-Allen read belonged to Oscar and Constance Wilde. In June 1885, they also sought his skill in divining by numbers, casting a horoscope for their newborn son. "It was born at a quarter to eleven last Friday morning," Wilde informed Heron-Allen. His wife was "anxious to know its fate—and has begged me to ask you to search the stars," a poignant request, because the child was to die in the terrible war, 1915 (fig. 4.4).[10]

We can see in this chart the extent to which Heron-Allen's palmistry was entirely astrological at this point. And yet, the astral body, natural magic, and *fluide astral* at the core of a new French

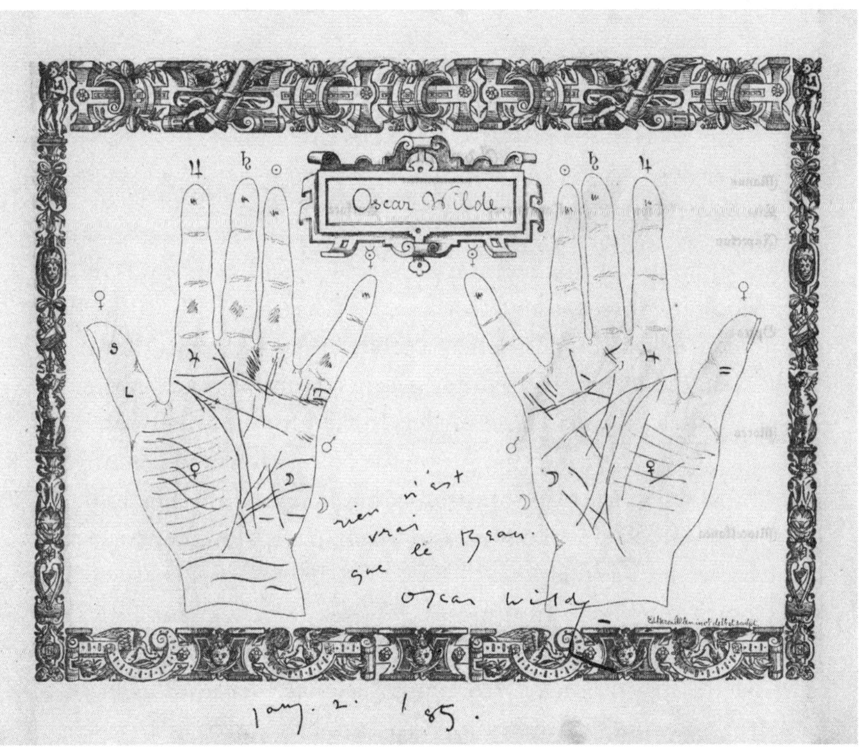

4.4 Edward Heron-Allen, palm reading of Oscar Wilde, 1885. Courtesy the Houghton Library, Harvard University, Chyromantia Manuscripts 1884–1893, MS Eng 1624.

chiromancy, like the Tibetan secret doctrine that underwrote theosophical clairvoyance, became simply too much for Heron-Allen. He began to insist on a palmistry more scientific than esoteric. It is telling that later in the 1880s he translated from the French not Desbarrolles's Kabbalah-inspired study, but the more sober and physiological C. S. D'Arpentigny's *La Science de la Main*. As we have seen, this was physiognomic, not "astral," and through it, Heron-Allen was to distinguish and refine his own style of scientific cheirosophy. He was ultimately irritated by Desbarrolles's effort to wrap chiromancy in "a proto-plasm of gnostic mysticism," and to enrich it "by means of the Kabbala." Over time, this came to be his least preferred manner and method. Indeed, Heron-Allen nominated two species of hand reader: the "astrologic cheiromant" and the "physiologic cheiromant." Desbarrolles was key instance of the former, while he was himself most certainly the latter.[11]

THE PHYSIOLOGIC CHEIROMANT

Edward Heron-Allen was one of many to claim "cheirosophy" and "cheirognomy" as medical and physical sciences, but he was more brazen than most, calling his *Manual of Cheirosophy* "the principia of a science," casting himself as a modern-day Newton. They certainly shared an eye for palmar geometry. In his own geometry of the hand, Edward Heron-Allen looked for an upper angle, a middle angle, and a lower angle on the palm, each of which might be blunt, short, acute, or obtuse. In that long tradition, he looked for quadrangles and triangles, and for shapes within them: a cross or a crescent or a star.[12]

Cheirosophy was an uppercase "Great Science" for Edward Heron-Allen. It was a bold claim, but in many ways also routine for occultists grandiosely to claim "science." From Blavatsky's "secret doctrine" of theosophy, to Steiner's Austrian-born anthroposophy, to Mary Baker Eddy's US-born Christian Science, nineteenth-century occult revivalists and new religions were all apparently "scientific." So were more humble individual practitioners of the esoteric arts. "Professor Melini" (Thomas Morgan) of Edgware Road, who worked as palmist and phrenologist, was typical when he explained: "Mine is essentially a scientific business . . . I am a member of the British Institute of Science." These were all modernizing gestures, but the British Institute of Mental Science, as we shall see, was simply an organization dreamed up by the entrepreneurial Ida Ellis to authorize her self-taught phrenology and palmistry. In short, everyone claimed "science." That hardly made it so, yet the very word could perform its own magic.[13]

Edward Heron-Allen had a more substantial claim than many, incorporating anatomical and medical material fully and knowledgeably into his books on the hand, as the "physiologic cheiromant" he wanted to be. His *Manual of Cheirosophy* opened skeletally, not cosmologically or astrologically. The frontispiece is an illustration of the "Bones of the Hand," a base knowledge of stripped-back nature, not complicated multilingual culture (fig. 4.1). A delicate pencil drawing by Rosamund Brunel Horsley—close friend, medical illustrator, and sister, as it happens, to the surgeon Victor Horsley—this frontispiece was not meant to be art, but anatomy. These bones were as far from *le fluide astral* of the Francophone occult, or the astral body of Anglophone theosophy, as a human hand could possibly be.[14]

Heron-Allen drew heavily on the anatomical tradition. He detailed the hand in the human embryo, for example, drawing on Quain's *Elements of Anatomy*, explaining how the hand is apparent even at one month after conception, although it then resembles a fin. It becomes entirely recognizable by seven weeks. Heron-Allen built up a picture of the human hand from bones, muscles, nerves, arteries, and veins, as would (and did, as we have seen) any anatomist. Drawn to the microscopic, he also illustrated his science of the hand with images of muscle fibers and magnified sections of skin. There are more nerves in the hand than in any other part of the body, he claimed, alongside so many anatomists, and this made the hand a principal seat of the sense of touch, and gives it a "direct and constantly apparent connection with the brain." Heron-Allen detailed well-recognized pathologies signaled by signs and symptoms on the hand: how leprosy was signaled by the stiffening of the first finger; how a thick first finger and short nails were seen in the scrofulous; how "Filbert nails" were a sign of the "tendency towards consumption." And he even dismissed similar but erroneous teachings of "old doctors," very old doctors. Avicenna's claim that short fingers indicate weakness of the liver was simply an antiquated superstition, he clarified.[15]

We have seen how directly anatomists were also decoders of hands, and so there was almost too much information for Heron-Allen to work with. He drew from German physiologists Julius Bernstein and Ernst Heinrich Weber on the sense of touch and the sensation of pain, from Humphry on the muscles of the hand, from Arthur Kollmann on dermal patterns and fingertips, and from Charles Bell on the sensory and motor nerves of the hand. He appreciated that anatomists made arguments about a whole animal, including whole humans, from small pieces thereof. From a fragment of bone, an entire animal might be reconstructed, the famous claim from the Victorian paleontologists so enamored with bones and fossils. Heron-Allen asked why this principle might not also work from the hand: Why "should we not be able to 're-construct' the Man, his habits, and his characteristics, from the examination of the Hand, the prime agent of his character and of his will?" It is a matter of common knowledge, he offered, that as all dispositions are different, so all hands are different. Why would we not be interested in how one related to the other?[16]

Nominating himself firmly as a physiological and not an astrological cheiromant, then, what did Edward Heron-Allen actually think about the astrological tradition? He certainly deployed astrological terms for fingers and for the mounts of the hand, but was not convinced about claims that the positions of stars and planets at birth determine prospects, character, health, or fortune. Over the 1880s, he reserved his judgment: "Much has been said (though little has been definitely known) for and against the influence of the sun, the moon, and planets upon the earth and the people inhabiting it." As we have seen, he agreed to read various sets of hands in an astrological and horoscopic manner that seventeenth-century Richard Saunders would have broadly recognized, but Edward Heron-Allen thought all this not so much right or wrong, as lacking. A sole focus on the distant influence of planets and stars on the individual body seemed to ignore entirely the more proximate influence of what we would now call "genetics": what he called parentage, inherited characteristics, and "the physical effects of the mental and physiological conditions of our progenitors." The "astrologic cheiromant" believes that a life is determined by heavenly natal circumstances, and yet the "physiologic cheiromant" is interested in a study of *tendencies*: "the *tendencies of a man's nature* are the result of his ante-natal and ancestral circumstances." His generation looked backward in time for embodied meaning, rather than outward from the body to celestial structures or even earthly atmospheres and environments. It is one of the great distinctions between early and late modernity. Planetary arrangements might be an ultimate cause, but inherited tendencies were surely a proximate cause, Heron-Allen believed. Ideally both should be considered, and, more than just entertained, should be experimented upon. While his own focus was on the anatomical, physiological, and inherited body, he nonetheless thought it misplaced to try to separate out "cheirognomy" and "cheiromancy," the physiologic and the astrologic approaches that together make up the fascinating science of "cheirosophy."[17]

By 1899, Heron-Allen seemed to have backtracked even more, actively questioning the popular alignment of cheirosophy with astrology. He well knew that occult-oriented palmists relied on, traded in, and even fully believed their astrological system, but they were confused: "The use of the astrological and Zodiacal signature in connection with the science of Cheirosophy was and is simply nothing more or less than a convenient system of stenography or shorthand." Those magical signifiers so full of meaning were diminished to the status of attenuated signs. In doing so, Heron-Allen knew he disappointed those whom he called "astral palmists" with his now-firm view on the matter. It was a disenchantment of sorts, an explicit undoing of magic, and a retreat from his own astrologically informed hand reading of the 1880s.[18]

THE PHYSICAL AND NATURAL SCIENCES OF THE OCCULT

If those leading occult revivals and new religions of the late nineteenth century routinely claimed "science," the reverse was also true. All manner of philosophers and physical and natural scientists hypothesized about, and actively experimented on, the nature of the matter that seemed to enable various unexplained phenomena: mesmerism and hypnotism, clairvoyance and spiritualism, thought reading and palm reading. Lévi's *fluide astral*, Desbarrolles's astral chiromancy, and Blavatsky's clairvoyance were all unmistakably occult. Yet their interest in the air and ether, fluids and rays that seemed to transmit mysterious powers and energies was the *matter* that a surprising number of scientific actors also investigated.

Dozens of theories circulated that linked "physics and psychics," through which the power over human bodies and minds of electricity in the air, or rays in the sun, were explored. When Desbarrolles wrote of the fluid that flows between a planet and a particular mount on the hand bearing its name, this was not unlike the many orthodox investigations of the matter that seemed to flow between a mesmerist and the mesmerized, or between animate and inanimate objects. Were these magnetic forces, physicists wondered? They were enchanted by similar mysteries: how an object, or another mind, or part of the body could hold and communicate information on a substance other than itself, on other planes, or in other dimensions. In this case, how might the hand be a literal medium to another place or time, or, in plainer palmistry, to the soul or character or interior, or to the past or the future of the mind-person-body whose hand it was? "Sympathetic hands" were revived, the object of modern investigation and experiment.[19]

This is why when Edward Heron-Allen turned to consider "astrologic cheiromancy," he linked it immediately to sciences as much as the occult, and to the physical sciences as much as to the medical sciences, to studies of light and heat, to atmospheric and meteorological studies, to claims about sound waves and light waves, and to a longstanding dispute over the material particles of light that Newton had proposed, long before. What is the nature of a possible quasi-solid, quasi-fluid matter "beyond the atmosphere in which the entire solar system floats"? How might that same matter be a force, or convey a force, that operates between humans, between humans and certain objects, and between bodies and minds? What constituted the "aether" that so many physical scientists had sought to identify? Well might he ask such questions, since scores of physical and natural scientists did so too. Indeed, they had been busy with these matters for decades.[20]

By Edward Heron-Allen's time, conversation and experimentation on human bodies were often about nerves and electricity, a question inherited from the eighteenth century, not least from philosophers of mind for whom a traditional division between the physical and the mental might well be undone. Another palmist of the 1880s, Rosa Baughan, thought the modern interest in magnetism and electricity was the latest rethinking of ancient Kabbalah scholarship on astral influences of the planets. But Heron-Allen looked to more proximate intellectual efforts, far easier to understand. He set out how the German philosopher Herder had wondered about a "sensitive fluid" in nerves, which he had likened to electricity, but considered "as a more subtle, more sensitive essence, to which we owe our lives, and which for want of a better term is known to us as our soul." In the next generation, Scottish physician John Abercrombie, like so many, built on the anatomical study of the brain and the spinal cord to engage with sense, perception, and the mind. He also speculated on the *matter* within nerves. The German physiologist Johannes Müller wondered whether a "sympathy" exists between the nervous system and electricity, similar to that found between electricity and magnetism. At that point, nobody knew. Was this electricity, or matter that was simply *like* electricity?[21]

This philosophical and scientific conversation indicates how and why practices like magnetism, mesmerism, hypnotism, galvanism, and all manner of electrical treatments were not just popular therapies, but were also experimental objects of inquiry among physicists, physicians, and natural scientists. Palmistry was proximate to these investigations, as was the long history of the idea of "sympathy," "sympathetic hands," and various means of "thought reading." Not deterred by the implausible or the incredible, scientists of real distinction actively researched the means and matter of such bodily and mental forces. Alfred Russel Wallace did so most famously, the great biologist who set out evolution by natural selection, alongside Darwin. "We know so little of what nerve or life-force really is, how it acts or can act, and in what degree it is capable of transmission from one human being to another," he commented in 1866. This insufficient knowledge did not mean that therefore phenomena such as "the apparently miraculous cure of many diseases, or perception through other channels than the ordinary senses, ever take place." In his *Scientific Aspect of the Supernatural*, Wallace prepared the way for Heron-Allen's late nineteenth-century generation expansively to consider possibilities that might make "occult"—the hidden or invisible or imperceptible—something open, visible, known, and comprehensible. Thomas Henry Huxley the materialist agnostic might even be persuaded.[22]

Some of them formed a Society for Psychical Research. That was in 1882, just when Edward Heron-Allen was embracing palmistry. Unsurprisingly, he was a member. Their original objects of inquiry included the study of "hypnotism, and the forms of so-called mesmeric trance, with its alleged insensibility to pain; clairvoyance, and other allied phenomena"; and a "critical revision of Reichenbach's researches with certain organizations called 'sensitive,' and an inquiry whether such organisations possess any power of perception beyond a highly exalted sensibility of the recognised sensory organs." Palmistry and chiromancy were not priority topics for this Society. The hand, however, appears and reappears throughout the records of its meetings, the special organ of touch, perception, and will, just as anatomists had declared. They described cold hands, clammy hands, and hands of the dead. There were even casts and molds made of the hands of visiting ghosts and spirits. Hands could be taken over by another's will, and made to unconsciously write, transmitting information between minds “without speech or sign,” the inverse of John Bulwer’s *Chirologia*. Hands were also metaphors for hidden and secret knowledge, for fraudulence and its opposite, truthfulness. Sleights of hand, so to say, were investigated, as was the significance of firsthand, secondhand, or thirdhand accounts—what counts as evidence. The significance of hands always being visible and open—the opposite to hidden and occult—signaled veracity.[23]

Versions of Glanvill’s sixteenth-century sympathetic hands were discussed, what even then had been called “a new chiromancy.” In one experiment, a person placed his hand on some part of the body of another, “and while so touched the latter finds objects or performs movements according to the silent will of the other.” This was witnessed and interpreted by philosophy professor Henry Sidgwick, who led a subcommittee appointed to adjudicate the alleged marvelous phenomena connected with the Theosophical Society. Part of this investigation was about the authenticity of handwriting, how one’s “hand”—script—betrayed one’s true self. This focused on letters that “the Mahatma” ostensibly wrote and sent to Madame Blavatsky “phenomenally.” But the diligent committee researcher considered these to be in most cases the handiwork of Madame Blavatsky herself, which is to say, written in her hand, and by her hand, traces of which were obvious even though she “acquired by degrees greater skill in the practice of the disguised hand.” It was a forensic examination of handwriting for deceit, a version of Desbarrolles’s examination of handwriting for character, and indeed a version of increasingly popular legal and criminal investigation of handwriting for legal identity. A kind of signature, handwriting could be individual and unique, like fingerprints. Some experiments were new, some were old. The Society for Psychical Research reopened an experiment from 1810, for example. Furthering their mission to investigate the “the existence of a magnetic sense,” it revealed enduring ideas about a hand’s connection to planetary forces. It all may seem a long way from science, but we should note that the Society’s own prompt to discuss a “magnetic sense” arose not from the occult world, but from an article in *Nature*. And at the 1856 meeting of the British Association for the Advancement of Science, where Richard Beamish traced the hand of William Whewell, it was noted that modern discoveries were constantly redeeming "the fancies of medieval times from the charge of absurdity." If, as had recently been shown, a bit of steel suspended near the earth was influenced by the position of the moon, 200,000 miles away, the old claim that "the stars might exert an influence over the destinies of man" could hardly be deemed "preposterous or extravagant." Rather, argued Oxford's botany professor and president of the Association, this was something to research.[24]

Many of these scientists and philosophers of minds and bodies were trying to *explain* various wonders, rather than explain them away. They sought to disenchant, rather than dismiss. In this sense, the Society paralleled the work of lay palmists circa 1900, who aimed to expose specific fraudulence and tricksters, while never presuming or claiming that *all* "psychic" phenomena (or palmistry) were fraudulent. Indeed, the reverse was the case: this left entirely open for experimental adjudication the possibility that such phenomena were (or were not) demonstrable, and that such demonstrations were repeatable.

AN ORIENTALIST'S CULTURAL HISTORY OF THE HAND

There were thousands of English practitioners of the hand over the 1880s and 1890s, naming the discipline palmistry, chirognomy, chiromancy, or cheirosophy, but none of them matched Edward Heron-Allen's special learning as a distinguished scholar of languages and literatures. He had acquired Turkish initially, taught to him during three lessons a week starting in 1885, by Garabet Hagopian, the Armenian envoy in London, and by Charles Wells, a Turkish lexicographer. Heron-Allen was a friend of Nāẓem-al-Molk, the Persian minister in London, and later studied colloquial Persian with Mirza ʿAlinaqi of the Persian Legation. Edward Denison Ross, professor of Persian at University College London, also refined his vocabulary, intonation, and capacity elegantly to translate from the great ancient canon. Heron-Allen was especially well known in these circles for his translation of the work of Omar Khayyam, from a manuscript in the Bodleian Library. As with everything Victorian and gentlemanly, there was a club of which he was a key member, the Omar Khayyam Club, established in 1882. Edward Heron-Allen even wrote to the shah of Persia to let him know that the Bodleian Library held an ancient Persian manuscript, "believed to be the oldest manuscript in existence." He was about to translate it into modern Persian, and into English (with notes): Would his Imperial Majesty like a copy? Kindly forward the book when it is published, was the curt response from Tehran.[25]

While other palmists in this period were writing workaday textbooks and "grammars" aiming to simplify palmistry as knowable by all, Edward Heron-Allen was producing works of high learning. His studies parade page after page of the most detailed commentary on Latin, Greek, Arabic, and Persian history and literature. Just one example: toward the beginning of his *Science of the Hand*, we are told how white hands are often granted a "mystical loveliness." We can read this in Exodus 4:6, but the King James Bible was just the routine beginning for Edward Heron-Allen. He followed this up with the exact *Persian* translation of the verse on white hands, quoting also from Omar-i-Khayyam, printing that Persian script, and even adding an arcane note on a peculiarity of translation that only his fellow Persian linguists—"Orientalists"—would appreciate.[26]

Edward Heron-Allen's scholarship on the hand should really be read as a superior cultural history of the body; learned, researched, sophisticated. Interested in performance, custom, and gesture, he was at pains to demonstrate how hands were highly symbolic. He documented how gesture worked culturally, the significance of the hand in everyday conduct, in legal conduct, and in religious conduct: the meaning of the pricking of the thumb; how Cicero's assassins cut off his hands, as well as his head, and put them all on display, "a homage to the powers of the hand in oratory"; the cutting of the fingers of forgers and seditious writers, as late as the sixteenth century. He detailed the hand in Christian liturgical choreography, the precision of the fingers in blessing and benediction as performed by different churches and by Eastern and Western denominations, the gesture a holy language. Persian, Arabic, and Sanskrit texts served as repositories for Heron-Allen's cultural history of the hand, as much as learned chiromantic texts. He described how in Persia it is customary to fold the hands, or hide them in the presence of a superior, "thus symbolizing an abrogation of the will." The Turks consider the hand to represent the Deity, he explained: "The fourteen joints being to them, as it were, the beads of a rosary, and the five fingers each representing a pious precept: viz., 'Belief that there is but one Allah, and that Mahammud is his prophet,' 'The necessity of prayer,' 'The righteousness of Almsgiving,' 'Observation of the Ramadan,' and 'The Journey to Mecca.'"[27]

Although commentary on India, Persia, Syria, and Egypt peppered Edward Heron-Allen's work, he barely referenced Roma palmistry or the thesis that other linguists—including his friend Richard Burton—were just then attaching back to those traditions. When he did, it was to dismiss: "the charlatans, thieves, rogues, or vagabonds who have unfortunately trespassed in all ages upon the fair domains of Cheirosophy." This was to become typical of nineteenth-century hand readers of Heron-Allen's class and milieu, who asserted their authority and learning *against* a Gipsy tradition, not in the least deriving from it.[28]

Really, Edward Heron-Allen was the most shameless intellectual show-off. Every page of his own books on the hand became an opportunity to boast of his learning, his languages, his intricate scholarly tracking and corrections. Never university educated, he was more classical than the classicists, more Eastern than the orientalists. A long passage from Galen that opens his *Manual of Cheirosophy*, for example, was a translation drawn from John Kidd's treatise, *On the adaptation of external nature to the physical condition of man* (1833). Yet an early nineteenth-century printed text was too easy. Heron-Allen also made sure that we knew, that he knew, precisely how Galen's passage had been translated by scholars in twelfth-century Spain, fifteenth-century Florence, and eighteenth-century Paris. Typically, his work ostentatiously displayed centuries of telling and retelling, and publishers indulged his ventriloquizing of a hundred wise men, from Aristotle to Newton, from Galen to Bacon, from Tacitus to Rabelais.[29]

The notes and explanations and corrections that overtake each page of his *Science of the Hand*—what in fact becomes the bulk of his text—are at once riveting and somewhere near obsessive, deeply learned but also approaching unhinged, a little like George Eliot's Mr. Casaubon chasing the "key to all mythologies" in her novel *Middlemarch*, only less pompous, and less stuck. Heron-Allen's authority was as real as it was substantial. He retold the Aristotelian origin story for chiromancy that had Egyptian roots, for example. "It is said that Aristotle, when travelling in Egypt, found an Arabic treatise on this science of the hand, graven in letters of gold, upon an altar dedicated to Hermes, and that he sent it to Alexander, as being a study worthy of the attention of the highest savants, where it was translated into Latin by one Hispanus." Yet, unlike most palmist-authors who repeated this Aristotelian and hermetic origin, Edward Heron-Allen had himself studied the earliest Latin manuscript of just that, preserved in the British Museum.[30]

The hermetic story was especially resonant because it referenced Egypt. For Heron-Allen and his London milieu, everything that could possibly be connected to Egypt was. This had been so for some generations, in Britain, in France, in the United States, all sparked by Napoleon's Egyptian enterprises and inclinations at the beginning of the nineteenth century. By the mid-Victorian period in England, giant statues of Rameses II and whole recreations of Egyptian palaces could be enjoyed by those visiting the Crystal Palace exhibition. And a generation later, Egyptology and Egyptomania had an organizational dimension with the Egypt Exploration Fund founded in 1882 to encourage archeological expeditions, coincident as it happens, with British occupation of Egypt that year. Edward Heron-Allen joined in, interested in Egyptian symbology and a recognized expert on the "Nefer" sign, a hieroglyph that was sometimes understood as a representation of a lute, but later as an anatomical representation of the heart or perhaps the trachea. Closer to his physiognomy of the hand were those ancient Egyptian bodies, suddenly available for inspection by the curious of Bloomsbury, Edinburgh, Liverpool, Paris, or Chicago:

so-called "mummies." They were even available for purchase, and Edward Heron-Allen bought at least one.[31]

For some time, mummies' skulls had been interrogated as phrenologically interesting. So too, Egyptian mummies' hands and fingers were occasionally traced or printed. Richard Beamish included not one but two tracings of an "Egyptian Mummy Hand from Thebes" in his *Psychonomy of the Hand*. He found them among the collection of antiquities at the Hartwell House museum. Likely his tracings were items 500 and 501 from an 1858 catalogue, the right hand of a man and the left hand of a woman. These "fragments of mummies" were listed with the precision of any occult palmist. Of the man's mummified hand, severed from his body, "the nails and the ends of the fingers are excessively wide, and there is still to be seen some of the gold-leaf with which it was customary to adorn the extremities of mummies embalmed after this [bitumous] process." Of the woman's left hand, also separated, "the nails and the ends of the fingers have been stained with henna . . . It is still customary for women to stain the nails, and the tips of the fingers and toes, and the palms of the hands, and the soles of their feet, with henna, in Egypt and throughout the East." Even Lionel Penrose, the geneticist examined in chapter 11, referred with some fascination to the still-visible dermal prints of Egyptian mummies in an article in the *Lancet* in 1973.[32]

Edward Heron-Allen traveled to Egypt in the spring of 1903, after his first wife's (Mariana Lehmann) death from consumption the year before. Grief turned into joy when he met his second wife, Edith Pepler, on that Egyptian journey. His Cairo accommodation, the Shepheard's Hotel, was located near the new Museum of Antiquities. On March 28, he went to its *Salle de Vente*, a sales room built to service the phenomenal trade in Egyptian artifacts, including human and animal remains. There, he "positively bought a Mummy!" he exclaimed in his diary, "A full sized real unexplored man-mummy, case and all complete." Shipped back to his home in Sussex, it gained a new life by the seaside, standing upright (the Egyptian was bipedal) in Heron-Allen's large library, in the corner near his desk, set off at the best angle to his grand piano. Much later—in 1930—he donated it to the Hull Museum, where local members of the British Medical Association had it X-rayed, revealing the Egyptian inside. In 1903, Heron-Allen was excited about "the chance of finding treasures in him when he is unwrapped (if ever)," and yet the mummy remains in the Hull Museum, as yet unfurled, its hands still wrapped.[33]

In the end, Edward Heron-Allen was more a custodian and a historian of hand wisdom than he was a practitioner of palmistry. He owned a precious fifteenth-century Arabic manuscript from the Maghreb, northwest Africa, that dealt with divination from the hand and used symbols of the sun and of the moon.[34] He owned an opusculum from 1490, twenty-two pages that made up *Cyromancia Aristotelis cum figuris*, from which he could cross-check Latin manuscript accounts of Aristotle and Hermes in Egypt. He was Keeper of the knowledge. We can see this in his choice to commission a portrait of Johann Hartlieb for the title page of his *Manual of Cheirosophy*, another illustration by Rosamund Brunel Horsley (fig. 4.5). Heron-Allen was especially enamored of the fifteenth-century compiler of *Die Kunst Ciromantia*, partly because the book itself was so special, predating printing in mobile type; it was a "blockbook," and the bibliophile and antiquarian knew just how rare and precious that was. More than that, however, Heron-Allen saw himself as a modern-day Johann Hartlieb, quill in one hand, observing the palm of the other.[35]

The Victorian iconoclast known as Edward Heron-Allen couldn't keep his many passions apart. His book designs, dedications, and prologues routinely displayed ostentatious Persian inscriptions. His very name was presented as a golden Persian flourish on the back of *Practical Cheirosophy* (fig. 4.6). This was his own Paracelsian doctrine of signatures. But he couldn't stop there, also illustrating his thumb and fingers holding his book of the hand. This Persian signature, so proximate to his thumb, was akin to another identity mark that appeared that very year and within his own social circle. Arthur Conan Doyle's invention, the detective Sherlock Holmes, could now recognize a criminal by "the callosities of his forefinger and thumb." And that was

4.5 Rosamund Brunel Horsley's illustration of Johann Hartlieb. Edward Heron-Allen, *A Manual of Cheirosophy* (Ward, Lock, 1885), title page. Courtesy the Wellcome Collection, London.

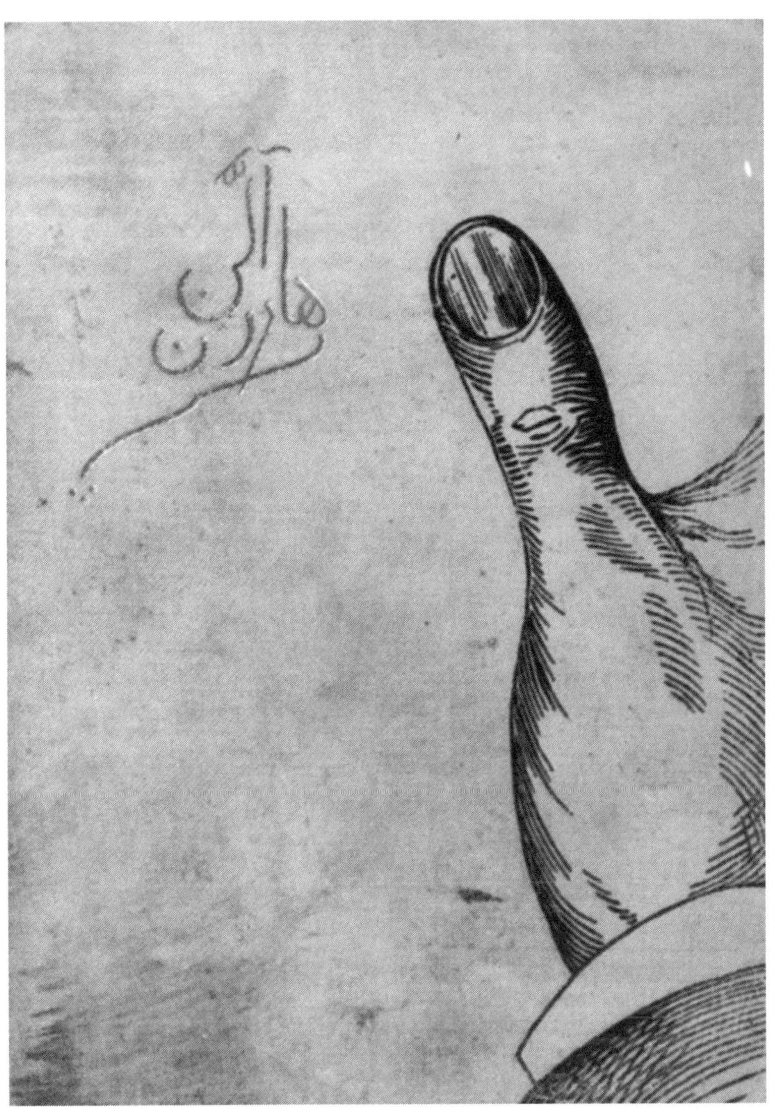

4.6 Edward Heron-Allen's Persian signature. Edward Heron-Allen, *Practical Cheirosophy: A Synoptical Study of the Science of the Hand* (G. P. Putnam, 1897), back cover. Copy in author's possession.

inspired by the extraordinary work of statistician Francis Galton, who just then was proclaiming that fingerprints were individually unique, so much so that they could now substitute for signature identification.[36]

It is characteristic of modernity to try to force a distinction between magic and science, truth and belief, but doing so inadequately catches the borderlands that engaged so many, even in the materialist nineteenth century. This is so not just because many occultists claimed "science," but also, and rather more importantly, because so many scientists were themselves researching the mechanisms and the substances of apparent magic, apparent wonders, and of the occult in its sense of "unknown" and "unexplained." In the same way, it turns out that it's not even possible reliably to link scientists to disenchantment, and palmists to magic enchantment. We have seen that some of the nineteenth century's foremost anatomists explicitly connected the hand and its anatomy to the Divine. And counterintuitively, Edward Heron-Allen the palmist and orientalist effected the reverse: he worked hard to disenchant chiromancy *away* from its connection with wonders, magic, and the occult, to make it scientific and modern. And yet he layered successive meanings of the hand, showing how its nature—we might say its base anatomy—was always imagined through culture, indeed through many cultures. His special skill was to tell the stories through which these various knowledges *could* make sense, side by side: on one page of his books, the hand as it was understood in the Persian originals of *Arabian Nights* and in the quatrains *The Rubáiyát of Omar Khayyám*, and on the next page, the hand as it was understood in *Quain's Anatomy*. That was the Necromancer's real magic.

"The antecedently incredible may nevertheless be true," declared the celebrated biologist (and spiritualist) Alfred Russel Wallace. This was the mode in which Edward Heron-Allen was a scientific cheiromant, and as time moved on, he began to thrive on the incredible as well as the true. He gave up palmistry—cheirosophy—but stayed in touch with the ever-growing world of palmists, astrologic and physiologic, reading and occasionally contributing to the professional *Palmist's Review* edited by Katharine St. Hill (chapter 7). Edward Heron-Allen was irrepressible, turning to any number of other intrigues that lay somewhere between science and fiction. He loved the fantastic, and relished the mysterious borderlands between truth and myth.

Even as he was looking down microscopes at tiny marine shells and naming new species, he was writing a fantasy short story under the pseudonym Christopher Blayre. "The Purple Sapphire" was a cursed gemstone.[37] But he claimed it to be real, and presented to the world not just the story, but the actual gem. It had come to him after being looted in the so-called Indian Rebellion in 1857, but "from the moment I had it, misfortunes attacked me." He took elaborate precautions, even more elaborately recounted: "I had it bound round with a double headed snake that had been a finger ring of Heydon to Astrology, looped up with Zodiacal plaques and neutralized between Heydon's magic Tau and two amethyst scaraboei of Queen Hatasu's period, brought from Der el-Bahari (Thebes)." The gem was cursed, dangerous, and he had it packed away inside seven boxes, never to be opened until thirty-three years after his death. But in 1941, divesting herself of a lifetime of her father's eccentric collections, Edward Heron-Allen's daughter gifted it to the Natural History Museum, London, where it sits to this day along with his dramatic letter of warning. The truly wonderful thing is that it sits now among other treasures that once belonged to this palmist-turned-orientalist-turned-scientist: his vast collection of foraminifera photographs, and his books on micropaleontology, housed in what is still called the Heron-Allen Library.

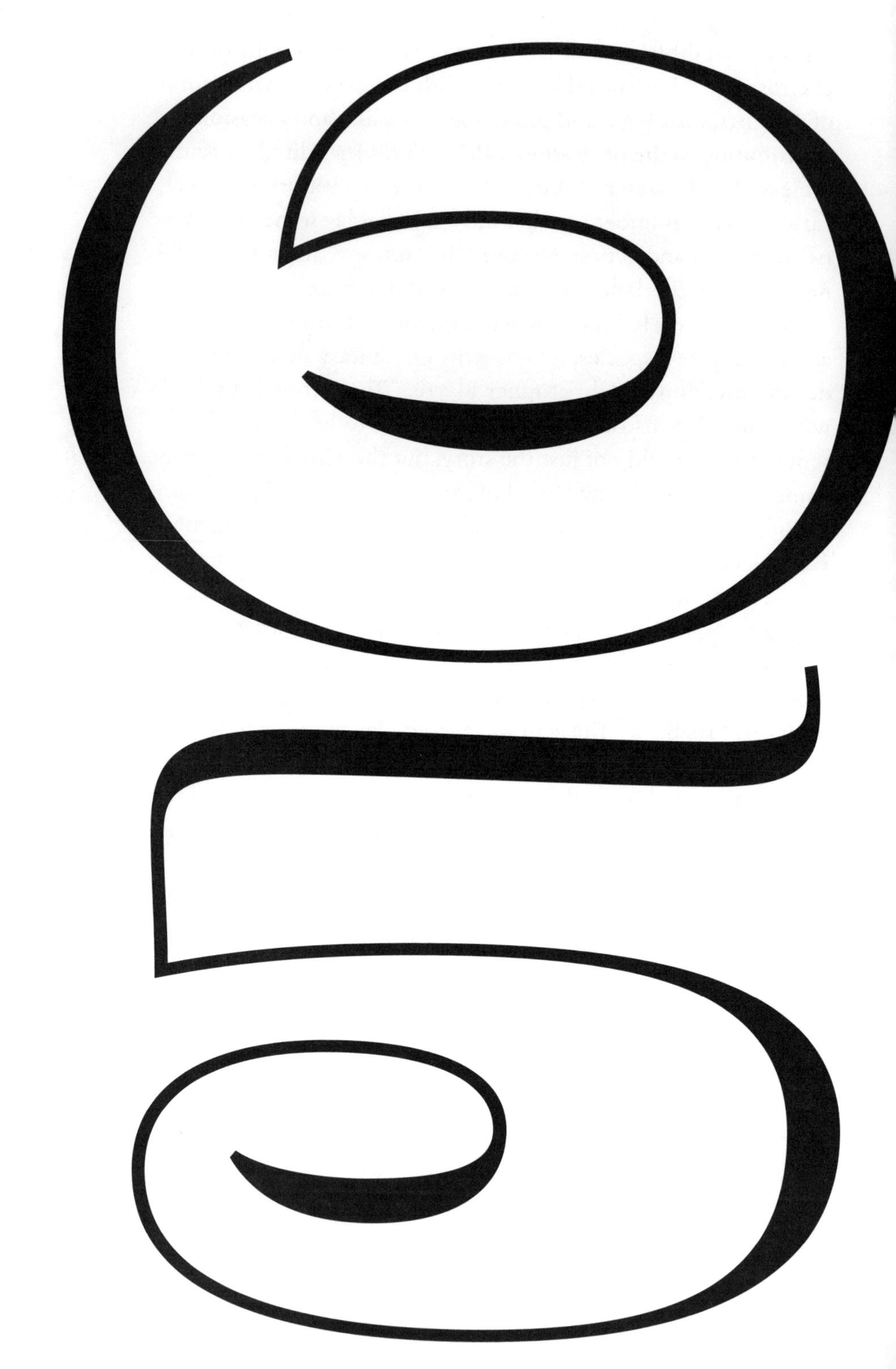

FIVE

Fortune and the Law

CHEIRO AND KEIRO

"Cheiro" was the Irishman William John Warner, who became a well-recognized palm reader in New York, Paris, Rome, St. Petersburg, and London. Born in Dublin in 1866, he was dead in Hollywood in the earliest hour of October 8, 1936. The clock struck thrice at 1:00 a.m., the press reported. Cheiro traveled across the world, the great cosmopolitan occultist, reader of celebrity palms, and a minor celebrity himself. He told the past and futures of prime minister Gladstone and the shah of Persia, of Mark Twain and Sarah Bernhardt, and then capitalized on their confidences by publishing the same. A showman in the era of showmen, he was ridiculously self-impressed, holding his contribution to be both unique and unrivaled. He had personally made palmistry respectable, uplifting it from "an art relegated to gypsies." From a Dublin Catholic seminary to India to Hollywood, Cheiro's life story does indeed signal something unique, but he is most interesting for us because he inhabited—squarely and unconditionally—the kind of "enchanted" world of marvels and wonders that sociologist Weber theorized at the time to be disappearing, but that so many historians insist simply changed shape and form.[1]

One of his London-based competitors assumed the sobriquet "Keiro." Cheiro even used this to his advantage, boasting that he had so many imitators that some went so far as to copy his very name. Charles Yates Stephenson and his wife Martha Stephenson—Madame Keiro—worked in the West End of London for many years, steadily practicing and publishing in palmistry and clairvoyance and crystal gazing. The press magnate Sir Alfred Harmsworth—owner of the original tabloid, the *Daily Mail*—sent his sales into the stratosphere when he brought a case against the "Keiros" in 1904. They were charged with fraud, and Harmsworth's lawyers thought the time had come to test the old Witchcraft Act. Would it be more effective than the early nineteenth-century Vagrancy Act that was conventionally brought against fortune tellers? Like Cheiro, Keiro practiced palmistry explicitly within an occult tradition at the high point of its parlor-based revival.

Cheiro and Keiro were operating within strongly classed worlds of business, showmanship, and self-promotion. Their West End base was urban and bourgeois, metropolitan as well as cosmopolitan. Their readings were intimate, like Edward Heron-Allen's, but more commercial. Cheiro especially recounted individual sessions in a manner that hints at palmistry's power as a therapeutic encounter. We can begin to perceive how, in the next half generation, hand reading would be folded into formalized psychology and psychoanalysis.

Just like seventeenth-century chiromancers, the many orders of palmists around 1900 went to considerable lengths to legitimize their work within a range of respected knowledge traditions, some ancient, some proximate. Far more routinely than Edward Heron-Allen, Cheiro nominated biblical origins, signaling the antiquity of chiromancy and divining from palms. Keiro also opened his books with scriptural endorsements of hand reading from Exodus, Deuteronomy, Job, Proverbs, and Revelations, just like seventeenth-century Praetorius. Both referred to Charles Bell's *The Hand* (1833), safeguarding their palmistry even more strongly within a Christian genealogy. Cheiro recapitulated the divine design of the hand one hundred years after the distinguished natural theologian. Keiro was even more florid: "We are all carrying with us a most beautiful map, in close connection with the soul, or brain (the seat of the souls), and on that map is the writing of God." And yet in truth palmistry arrived in the West End, circa 1900, through multiple traditions of practice and knowledge. Cheiro authorized his work via Sanskrit texts too; both his palmistry and numerology were taught and refined in northern India, or so he claimed. He was both the custodian and the translator, as he would put it, of an art even more ancient than the Judeo-Christian Scriptures.[2]

CHEIRO

William John Warner took the name Cheiro, but he also went by Count Louis Hamon, sometimes Count Leigh de Hamong, derived, he made sure everyone knew, from the Hamon of Normandy, whoever they were. He recounted himself as a born student of the occult, who could always see into "the soul life of things." From his father he learned poetry, from his mother esoteric knowledge, including the names of the lines of the palm. And yet his father tried to deflect this inclination by sending him to a Catholic seminary. The upshot was a young man even more persuaded that palm lines were "a more tangible chart of life" than the Scriptures he was made to memorize. Still, the Scriptures held an enchantment regarding destiny, a poetry of fate that appealed and inspired (fig. 5.1).[3]

Far from understanding his work as a science as did Edward Heron-Allen, Cheiro understood his powers to be intuitive, "given." This gift was an aptitude naturally to see patterns and meanings in inanimate things like names, places, numbers. He was a self-described "seer," whose sensitivity for messages was only available to those, like himself, who have "ears to hear" and "eyes to see." He thought of his work as an expansive comprehension of the senses: "To believe is to perceive—either by the sense or the soul." It was not quite clairvoyance, he explained, but rather he had trained his naturally intuitive mind to concentrate in a particular way, thus able to "feel" coming events. This approximated the

5.1 Cheiro with his own hands concealed. From *Cheiro's Language of the Hand* (Herbert Jenkins, 1900), plate 1. Copy in author's possession.

way that "animals sometimes feel the approach of danger hours in advance." Cheiro sometimes described himself as a translator; not as a medium between the living and the dead, but as a kind of interpreter for those "who came to listen to the message of life which it was my privilege to translate to them."[4]

Cheiro traded explicitly on "Eastern" knowledge, but with a far greater interest in India than Edward Heron-Allen, not least with a claim to having been taught the art of palmistry by Brahmins in "the land of the birth of the study." Reading palms was as old as language itself, he declared, linking this to the then much-admired "Aryans" of Indo-Europe. At the dawn of civilization, "the *first evidence of a word language* belongs to the Aryan race." Their descendants, he explained, the ancient Hindus, discovered planetary knowledge such as the precession of the equinoxes that takes place every 25,600 years. It was these people who founded the study of the hand in his account. Cheiro traveled as a young man to Bombay, where he met Brahmin descendants of scholars who had thousands of years earlier developed *Samudrika*, a science that interpreted the meaning of the expression of the lines of the body. This was refined into the higher and more specific study of what he called "Hastirika," or "the Science of the Lines of the Hand." Cheiro nominated his teachers as descendants of the Joshi caste, famous for their occult knowledge. They knew the workings of the heavens, and created both "Fadic" numbers and the science of astrology, which traced the influence of the "Planetary System" across the lives of all people, whether they were aware of it or not. He brought the East to the West End, recreating it in his "Indian Room" where he received his clients or "interviewees," an enclosed and dense space heavy with Persian rugs (fig. 5.2). Cheiro also brought to the West End many accounts of the occult powers of the Mystics of India—"wonder workers." In a reverse gesture, if a far lesser one, Cheiro read the palms of learned men of India, including that of Swami Vivekananda, who was just then bringing yoga and Vedanta to the west. His print of the "philosopher's palm" is even now read by Indian astrologers and palmists. The Anglo-Indian exchange so characteristic of this period intensified with Cheiro's palm reading of Indian-based Annie Besant, president both of the Theosophical Society and, later, of the Indian National Congress. Indeed, Cheiro's own apparently Indian-born astrological knowledge was returned to India in a suite of Bengali translations of his work, published as late as the 1980s, and still sitting on the shelves of many Indian homes today.[5]

5.2 Cheiro brought the East to the West End. "Cheiro's Indian Room." From *Cheiro's Language of the Hand* (Herbert Jenkins, 1900), final plate. Copy in author's possession.

Cheiro was unusual in the London occult scene—perhaps closest to theosophists—in combining his palmistry with the reading of "Fadic numbers," an ancient practice that was just then coming to be called, in English, "numerology." The orientalist Max Müller, Oxford's acclaimed professor of Sanskrit, owned and consulted a copy of *Cheiro's Language of the Hand*, and sent an impression of his hands to the author. But it was the "occult mystery of numbers" in which Müller was most interested, asking Cheiro how he had developed his own system from ancient Indian numerology. Likewise, when former prime minister William Gladstone invited Cheiro to his home in Wales, the seer was asked to explain not just his palm lines but his numbers. Cheiro obliged, applying his own method for telling time on a palm by his "system of seven." It turned out that like Müller, William Gladstone was also already schooled, owning *Cheiro's Language of the Hand* as part of his personal library, entered into his vast catalogue as a book by

"Louis Hamon." And when Cheiro met with the aging and ailing theosophist Madame Blavatsky (she thought him a reincarnation of eighteenth-century magician and occultist Count Cagliostro), he foretold her death through a combination of palm signs and numbers. Blavatsky herself could easily enough point to where her own Line of Health cut through the Line of Life, but could Cheiro give that some precision, using his knowledge of Fadic numbers? The series of fours and eights held her secrets, he explained, calculating from the date of her birth, aligned with the House of Saturn, otherwise known as "the House of the 8." He calculated a year's reprieve, but also offered that "your indomitable willpower may carry you a little beyond that age, especially as at your date of birth your Sun, the Giver the Life, was then entering the House of Mercury negative." Cheiro was a believer in willpower. Both Madame Blavatsky and her willful successor, Annie Besant, pressed Cheiro to join the great cosmopolitan diaspora that was theosophy. But he resisted association with any group, society, or movement: a "lone wolf," he described himself, to the end.[6]

Cheiro was at pains, like Heron-Allen, to stress that his occult knowledge was not to be confused with vernacular superstition. He set himself apart, implicitly and occasionally explicitly, from "Gipsies." He noted, for example, that taking such signs as the red cross on the finger of Mars or the black spot on Saturn to indicate murder were simply "relics of that black age which once claimed palmistry as its own." Instead, Cheiro presented himself as a custodian of a far higher knowledge, not inherited through a folk culture, but through book learning, the great avalanche of hermetically inspired early modern chiromancies. It is clear that Cheiro, again like Edward Heron-Allen, accessed both manuscripts and early modern books in various British and European libraries. He relayed one incident in Rome, meditating in St. Peter's, when an aging man instantly recognized Cheiro as a fellow seer. The old man entrusted him with a manuscript that was his lifetime's work spent carefully copying ancient Egyptian occult manuscripts that had been apparently lost in Julius Caesar's burning of Alexandria's library in 48 BCE. In India, too, Cheiro was given access to ancient scripts and permitted to take notes from sacred books, including one made of human skin. At another more mundane place and time, a Surrey auction, he and another bidder competed for Culpeper's *Arcana of Astrology*, this copy printed in 1602. Striking a friendship, the Londoner hosted Cheiro many times in his disheveled top-floor flat off Oxford Street, piled high with early astrological works, which Cheiro consulted and borrowed.[7]

Cheiro himself added to this book learning in the massive industry that was occult publishing in the decades around 1900. *Cheiro's Book of the Hand* (1892), *Cheiro's Language of the Hand* (1895), and *Cheiro's Guide to the Hand* (1898) was a quick trilogy. And in the next century he democratized the rarefied tradition, or really commercialized it. *Palmistry for All* (1916) and *You and Your Hand* (1931) both sold well on his world tours, especially in their US editions. The American press deemed him, his books, and his lectures wildly successful, and his celebrity was far more enduring than Edward Heron-Allen's. Chickering Hall in Boston was packed, while members of New York's literary Lotos Club listened carefully, evidently agreeing more soberly that palmistry was a "reliable guide." Americans could see for themselves that the illustrious, the famous, and the influential were persuaded, since Cheiro accompanied his lectures with "Stereopticon Views" of hands of the famous. In a unique twist, he even popularized palmistry in film. The hand impressions of such clients as Lord Kitchener were brought to the new cinema public in the 1920s, accompanying his book *Read Your Past Present and Future: Teaching the Study of the Hand by Pictures as on the Cinema* (1927). It was the new phenomenon of moving pictures that was the real magic.[8]

In his West End Indian Room, in parlors of the rich into which he was invited, and on his intercontinental tours, Cheiro read the palms and numbers of the monied, and of some of the world's most famous and influential figures. In his final period of work, the earliest generation of Hollywood screen actors and directors sought him out. Sometimes their identity was unknown to him, as they put their hands through heavy curtains especially set up for an evening's entertainment. One turned out to be Edward VII. Others sought him out with their identities known—the shah of Persia, for example—and Cheiro appreciated the high geopolitical stakes of that hand reading. Mark Twain approached him skeptically, but liked and praised him. Others were already believers. W. T. Stead, the journalist, editor, and publisher (not just of the liberal *Pall Mall Gazette* but also of the spiritualist *Borderland*), had his hands read by Cheiro, who predicted his death on the *Titanic*. Much of Cheiro's authority derived from carefully collected and displayed endorsements that he kept in an "autograph" book alongside prints of hands. Like Edward Heron-Allen, he asked the famous to add their signature and date to the palm print, and most obliged. This became a prized stamp of verification, and each enhanced the value of brand Cheiro.[9]

In Cheiro's 1912 *Memoir*, he extracted the best from thousands of testimonials, partly to self-verify, partly to name-drop, and partly to demonstrate how his palmistry and numerology could serve a range of needs and purposes. He explained the past and predicted the future. And like the Ellis family, and a thousand phrenologists, Cheiro read "character" in a way that intimated the beginning of a system for interior vision, on the cusp of becoming a formalized psychology. Some endorsed his capacity to tell their previous lives with "wonderful accuracy," while others comprehended him rather more as an interpreter of their selves, approaching a twentieth-century "personality." Mark Twain declared that "Cheiro has exposed my character to me," while Judge William A. Vincent was shown "the mirror of self." Another signaled a session that was therapeutic: "Cheiro's advice helped me through the hardest battle of my whole life." US poet, spiritualist, and theosophist Ella Wheeler Wilcox endorsed him too: "Cheiro *helps* as well as astonishes."[10]

It is curious to consider a showman-entrepreneur like Cheiro within the modern history of the therapeutic encounter. Yet this is how a suite of his clients described their sessions, *and* what Cheiro himself disclosed. In his own telling, for example, the actress Sarah Bernhardt "sobbed" as he revealed various truths of her life. One woman confessed: "My emotions were analyzed and traced back to their beginnings." For another, he read "one's disposition, one's inner self. He tells one of weak points, points to guard against, as well as those to cherish; of fancies, ambitions, and aspirations which we thought hidden from everyone but ourselves." Despite his showmanship, we can see in Cheiro an antecedent to twentieth-century analytic psychology. Or perhaps this was part of his showmanship; since the palmist (the psychologist, the counselor) must attend to their own conduct, they must also perform. "Speak honestly, truthfully, yet carefully," he instructed practitioners.[11]

Cheiro's moving encounter with Oscar Wilde on the banks of the Seine is another instance. Wilde was tired, worn, and poor after his imprisonment from 1895 to 1897 for homosexual acts. He had consulted Cheiro much earlier, in London, before his trial, when a break, a fall, was predicted. When they came across each other unexpectedly in Paris, years later, Cheiro said that he went to Wilde and held out his hand. "In this terrible loneliness he held it for a moment and then burst into tears." It was a different performance of human touch altogether. Through it, Cheiro released a breakdown of torrential weeping and then "we talked—talked till the music ceased . . . He went through the trial again—the mistakes he had made, the life in the prison, the joy of liberty—all." In his own telling, Cheiro saved Oscar Wilde from suicide and gave him courage to live. And as a final therapeutic acknowledgment of the repressed returning and resolving, Wilde told him: "Your presence brought the dead past out of its grave." It is obvious that Cheiro applauded his own role here, self-presenting as intimate savior. But even granting that expedience, the handsome and cosmopolitan Irishman was not just a "seer" who looks into the future, but a therapist who looks into souls.[12]

Cheiro enjoyed the idea of special powers of mind, mysterious but real forces that manifested with intensity in some people. Folklorist Leland, by contrast, had tried to de-exceptionalize—to disenchant—such powers in the context of Roma fortune-telling. Powers of perception were not to be understood metaphysically, but "in strict accordance with the soundest conclusions of modern physiology." Leland thought that the Gipsy fortune tellers' quickness of perception was a practiced exercise, neither fraudulent nor miraculous, and a skill latent in most people that might be brought out with training. "We are all sorcerers and live in a wonderland of marvel and beauty if we did but know it." Cheiro, however, thought it all far more special. Both powers of perception and unusual willpower were rare gifts. They may yet hold a physical, not metaphysical, explanation, but that had evaded researchers. It was a sense, perhaps, a sixth, seventh, or eighth sense, but to which organ of the human body such a sense might be attached, no one yet knew. Matters of mind and body, sense and perception were certainly the business of physical scientists and physiologists, as we have seen, many of whom shaped and researched both "psychology" and "parapsychology" in the early twentieth century. For Cheiro, the interesting new idea of the "sub-conscious brain" was worth considering. He did so with Mark Twain, as it happens, the pair trying to think through physiological explanations for the capacity to see a future. Together, they examined palm lines under a microscope, even the minute furrows in the tips of the fingers that had intrigued Purkinje, and that Galton was just beginning to reclassify, as we shall see (chapter 8). But the microscope was an antiquated machine that simply read the body. What about machines that might read the mind?[13]

In Cheiro's era, there was much experimental paraphernalia on offer that might explain, or at least record, the power of divining and of "reading thought." He favored one particular "Thought Machine" invented by Professor Savary d'Odiardi and shown to the Academy of Sciences in Paris. In 1896, Cheiro paid to be tested by it and—naturally—his stunning powers of mind were confirmed. Quite simply, it could be made to move by a willpower that radiated outward "through the atmosphere." He called it a "Register of Cerebral Force" that responded to "hidden forces of the body—and perhaps also of the spirit." D'Odiardi was an exponent of medical electricity, and ran an "electro-medical hospital" at Notting Hill Gate, where Cheiro assisted with experiments over the summer of 1896. They compiled hundreds of records with an improved version of the thought machine, and Cheiro arranged for one to be brought to his Bond Street rooms, this quintessential object of *fin de siècle* technology sitting oddly with the orientalist *objet d'art* that created his "Indian Room." He arranged studio photographs to be taken and published, in which his hands are notably concealed (again); this was all about brain focus, not touch (fig. 5.3). Cheiro traveled with his mind-reading paraphernalia too, an unlikely companion. Summoned by William Gladstone to Wales, Cheiro set it next to him on the train, and he found the former prime minister profoundly interested in the machine, subjecting servants to its test, sadly powers shown to be impotent. But then Cheiro and his machine registered prime-ministerial willpower. And behold! Gladstone's was the most remarkable "will force" of all the thousands of people tested by Cheiro and medical electrician d'Odiardi. Not only was he able to make the needle move, he could will it to suspend.[14]

What was it that moved with such force between particular minds, through the atmosphere, to the needles of such a machine? There were antiquated explanatory terms and scientific research about which Cheiro wondered. "Odic force" was one, the term that German chemist, geologist, and (we might say) neurologist Carl Ludwig von Reichenbach had invented in 1845, later actively investigated by the Society for Psychical Research. This was a highly sensory theory, developing the German's interest in earthly magnetism into a vitalist explanation about human sensitivity, including "Odic sensations of Touch and Light," "Odic Polarities," and the "Speed of Odic Conduction." It emerged from, but superseded, the older idea of "magnetism," including animal magnetism. For modern Cheiro, this was all possible but a little crude and old-fashioned. Was there not something more subtle going on? Cheiro occasionally referenced "nerves" and the experimental work of not a few of his contemporary physiologists. They had shown that "the brain cannot think without the hand feeling the influence of the thought," because so many nerves connected the two parts of the body. Might the power of will that registered on his machine, and that neuroscientists were beginning to recognize, not also be the force "that in its continual action marks the hand through the peripheral nerves"? This may also be "the unseen force" that writes on the hand both "the deeds of the past" and the "dreams of the future." There may be a natural law, a physical law of palmistry, after all.[15]

5.3 Cheiro and d'Odiardi's thought machine, c. 1900. *Cheiro's Memoirs: The Reminiscences of a Society Palmist* (Rider, 1912), facing 154. Copy in author's possession.

Cheiro's business was primarily the future. Palmistry could be a kind of preventive intervention, since once we know what the future holds, he wrote, we can likely redirect "fate" to some extent, and sometimes in high-stakes ways. A man might have the "the mark" of a murderer, and act upon it. But we must recognize that he was once an innocent child whose palms were nightly clasped in prayer. If only those tiny hands had been read, the warning could have been given, and perhaps heeded. This was Cheiro's message about the future: it was not necessarily predestined.

The sometime Irish Catholic seminarian—like most late Victorian and Edwardian Protestants in England—would recognize the doctrinal implications of such a claim. The possibility of redirecting fate came up against theologies of predestination and of providence, the power of divine foresight and guidance. But God rewards knowledge, Cheiro reassured his clients and readers, with emphasis: "The man, or woman, who studies his hands with the desire to know what are his tendencies *is obeying the highest instinct of creation.*" Humans have a free will—the power to do otherwise—and although Cheiro was hardly a moral philosopher, this matter was at the heart of the telling of fortunes. Palmistry offered not just the possibility of doing otherwise (to fate), but enabled something of the power, ability, and even *responsibility* so to do.[16]

Cheiro secularized and even biologized the problem of the future by occasionally substituting "fate" with "tendencies," thus shifting from something like God's predestination to something like Nature's inherited characters. "Tendencies" was the term *du jour*, one deployed in popular eugenics, as well as within scientific eugenics, and concerned with biologized "characteristics" (or, as geneticists were beginning to put it more strictly, "characters") that people, like peas, pigeons, or horses, inherited and passed on between generations. In this era, these might include alcoholism, feeblemindedness, homosexuality, or criminality. The more technical word and idea of "heredity" also crept into Cheiro's palm reading, anticipating the expert absorption of some palm lines as phenotypes that Lionel Penrose of University College London's Galton Laboratory spent decades researching in the years after Cheiro's death (chapter 11). Cheiro paraded his own theory of heredity, and indeed just how or whether patterns of palm lines were inherited from one generation to another was becoming a standard question, even for esoteric palmists.[17]

Like most readers of the hand, Cheiro declared that the left showed inherited tendencies, while the right showed more individualized "developed or attained characteristics." Sometimes identical palms were evident between generations, to be read as a continuity in life patterns across time. When interviewing prominent statesman Joseph Chamberlain and his son, Austen, for example, he saw a near-identical "line of Individuality" or mark of Destiny that traveled from the wrist to the first finger, the sign of "the Lawgiver or Dictator," possibly not the best mark in a parliamentary democracy. The son would live the political life of the father, he predicted in 1894, although the sensible would put eldest son Austen Chamberlain's stellar subsequent career as chancellor of the exchequer, foreign secretary, and leader of the Conservative Party down to the inheritance of a pre-prepared social and educational status. French sociologists were just then beginning to put a more refined word to this: "habitus." This signaled the "intuitive" mannerisms, tastes, and conduct that individuals acquired from their (class) context, and that powerfully structured future opportunities: a kind of embodied politico-cultural predestination. The next-generation palmist, continentally trained psychoanalyst Charlotte Wolff (chapter 10), would bring French and German phenomenology and psychology to hands, minds, gesture, and ways of being, with far greater intellectual sophistication than Cheiro. And yet they were both in the business (like emerging medical geneticists) of reading bodily signs and correlating them to the past and the future, to life predictions, or at least probabilities.[18]

For Cheiro, outcomes were not altogether fated, and this heightened rather than diminished the significance of those who could read "tendencies." They can be changed, and so "actions in the future altered." He rendered some of this profundity into poetic form. A magnificently set collection from 1895 was titled *If We Only Knew*, opening with his poem of that name. Another was titled "Fate," and a third, simply "If."

> If Fate were naught—and we were wise,
> How calmly would we plan the earth!
> There'll be no sorrow, tears or dearth;
> Nothing but joy would fill our eyes
> If Fate were naught—and we were wise.[19]

It turns out that Cheiro was a published poet before he was a published palmist. And in his major work, *Cheiro's Language of the Hand*, crude commerce was transformed into emotive stanza-epigraphs, the magic of words meeting the magic of the language of the hand. We could and should be *more* wise than we typically are, Cheiro believed. Or better, people might purchase that wisdom from the likes of himself—an investment in foresight reliably acquired from those who could decode signs of the hand, and so read the future.

THE FORTUNE TELLER'S DEMISE

In the West End of London in 1892, Countess Cardelli was warned by police about the illegality of advertising fortune-telling, and so she went to the trouble of changing her signs from "palmistry" to "cheirosophy." But she was disgruntled because a "man called Cheiro" was continuing to advertise freely and liberally as a "palmist." Why was this caution not extended to all, and equally? Then little known, Cheiro already worked from his room at 106 Bond Street. As he became famous and then traveled Britain and the world divining futures, more and more of his competitors and compatriots of the occult in the West End, like Countess Cardelli, were being warned, charged, fined, and on occasion, imprisoned. Fortune-telling was the specific offense. This was not new, as we have seen, but the sudden rise of a highly visible occult trade, especially in the West End, brought clairvoyants into the criminal domain, of interest to politicians, police, lawyers, and the owners of a new press phenomenon, what was soon to be called "tabloid journalism," sold in compact, cheap dailies and aimed at an urban working class.[20]

The reasons for late nineteenth-century criminalizing of palmists are counterintuitive. This was not, as we might expect, a targeted regulation of Roma fortune tellers, but was rather built on a defense of them. The problem was a new class, and to some extent gender, of fortune teller, the rise of "fashionable charlatans of the West-end," often enough men. Many asked: Why were they not also being held to account for the fraud for which poor Gipsy women were imprisoned when they sold a fortune to a credulous servant girl? Versions of this question were put to the House of Commons from the early 1890s onward, and with some regularity. One member of Parliament asked why palmists in rooms in comfortable Regent Street, Bond Street, and Oxford Street could "openly advertise and carry on the trade . . . while poor persons in respect of similar offences are severely punished by imprisonment?" And "why a black person, believed to be a West Indian, was prosecuted for occultism, or the like, while other persons are permitted to openly advertise their dealings with futurity?" Another parliamentarian doubled down in 1893: Had the secretary of state seen advertisements for palmistry in the *Daily Telegraph*? Why wasn't this advertising of fortune-telling prosecuted as were people in the street, and "will he take means to put a stop to this form of fortune telling, as to treat all classes alike?"[21] In 1900, again, it was noted that all manner of palmists, fortune tellers, and "other necromancers" advertised and practiced in the West End, but "seeing that poor gipsies who practice the same calling are prosecuted and punished for obtaining money by false pretences [does] the Public Prosecutor intend to take a similar action against these fashionable soothsayers?" This response to commercial occultists was as sudden and sharp as the rise of the trade itself. The problem was that fortune tellers were suddenly urban, literate, and educated, as likely to be men as women, and were highly visible—too visible—on West End streets, even "in and around Mayfair." The silent agreement about acceptable class and gender encounters over fortune-telling was being crossed, and visibly so.[22]

It was class distinctions in the application of the Vagrancy Act (1824) to which these members of Parliament referred. As we have seen, this law retained a clause to prosecute palmists and other occultists from the Tudor Vagrancy and Egyptian Acts. Section IV nominated "every Person pretending or professing to tell Fortunes, or using any subtle Craft, Means, or Device, by Palmistry or otherwise, to deceive and impose on any of His Majesty's Subjects." Fortune-telling specifically was the deception in question. This power was certainly used over the nineteenth century to fine or warn Roma women telling fortunes across England and Wales, part of the legal toolbox with which Travellers could be moved on. But in the decades after 1890, it began to be used in London to warn, charge, and prosecute a different breed of chiromancer altogether, who were making good money, mainly in the West End. This was a suddenly growing and visible sector who "advertise openly the profession of the supernatural."[23]

In 1912, there were estimated to be about 700 occultists practicing commercially across London. Both men and women, they covered the spectrum of expertise and identities, but with differently gendered patterns of self-presentation. Grace Colyns was a clairvoyante and a "magnetic healer." She diminished her special powers and the reasons why people came to see her: "I am a trained nurse, but too ill to carry on my profession. In this business I do not deal with poor people of the servant class but only with the well-to-do who pay me 2/6 or so to entertain them." Madame Chandra of 91 Regent Street, on the other hand, insisted her expertise be taken seriously: "I have practiced my wonderful gift since I was 7 years of age," but "I am not a palmist, I am a clairvoyante." And another, Catherine Rosemond Hagerty, who practiced as Madame Lawrence, drew her own authority from the famous Cheiro: "I practice palmistry. I do not believe in clairvoyance or crystal gazing. The palmistry I practice is taught in Professor Cheiro's book. Is he to be proceeded against?" He never was.[24]

Many men assumed a professorial authority. Professor Pickens of 57 Conduit Street was a palmist who, when warned by the police, tried to reassure them: "I don't tell fortunes, I am a psychic. I have made a deep study of it." Professor Kerlor of 1 Piccadilly Place, palmist and clairvoyant, also complained. "I have made a serious study of my profession and sold a great number of books dealing with it. I am endeavouring to build up here an occult library but will not advertise again." Many evaded the offensive fortune-telling fraudulence by explaining to police that they read character only, not the future. In Bournemouth, Professor Alexander Davies requested that his sentence be quashed, and to be informed whether he could still practice scientific palmistry without prosecution. He explained that he neither told fortunes nor used "subtle means to deceive," and that none of the evidence proved that he did. Rather, he was learned in discerning the meaning of marks "in a scientific manner." Indeed, he was a professor of phrenology, and had practiced as such for twenty years, having studied under the masters in France, Germany, and Italy. In other words, he was neither an occultist nor a Gipsy: "I never did pretend to tell fortunes—I could never descend to do anything of the sort." The *Bournemouth Observer and Chronicle* subtitled their article on the occult arts as "Fraud or Entertainment?" But clearly Professor Davies thought neither, and held the very suggestion in disdain.[25]

While there were fines and imprisonments, the metropolitan police over several decades were more reluctant than observant in their duties. Indeed, they were instructed by the Home Office to curb any diligence. There was enough political pressure to worry about the numbers of clairvoyants who might have to be charged. Repeatedly the Home Office advised the Metropolitan Police that "proceedings should only be taken in cases where there is reason to suppose that the young or ignorant are being defrauded by persons professing to use superhuman agencies." Nonetheless, palmists of many kinds were warned, prosecuted, fined, and occasionally imprisoned.[26]

Why were these idiosyncratic soothsayers and professors so problematic? Who were they crossing? Complex late Victorian cultures of gender, class, and ethnicity were at work, and in both public and private spaces. This new chiromancy confused too many social boundaries—or just enough to create its own cultural and commercial success—of class, race, and gender, of intimacy and the interior, on Regent Street, Bond Street, and Oxford Street.

To some extent, this was about the public space of London's West End, where a comfortable urban geography was challenged by a newly conspicuous occult commerce. Although fortune-telling itself took place privately, in rooms, the streets were filled with "sandwichmen," each advertising a particular practitioner. At least one magistrate was uncomfortable with their proliferating number, a matter that "had recently been spoken of at the Clubs he visited." He brought the matter to the police in a private capacity, he was sure to say, not as a magistrate. This advertising is what the police tried to shut down, or at least moderate. It was usually boys and men who gathered up a few pennies by wearing the sandwich boards, and they were formally warned by police that advertising fortune-telling crossed the law of the Vagrancy Act. Herbert Hinckley of Brick Street Piccadilly carried a board for Abdul Malak, and when warned by police said he had been doing it for a year: "I suppose you will see my Guv'nor." Frank Taylor of Brick Lane, Spitalfields, worked for Keiro, and was a little more forthright. "I shall hold my Guv'nor responsible for what I am doing. He pays me for what I am doing."[27]

Just who and what was being protected by this new policing? The essence of the clause in the act was that the young servant girl needed protecting from her own credulity and from the Gipsy fortune teller. At one level, the credulous poor were still deemed to be vulnerable in the West End crackdown. Professor Zodiac—Robert Scott Blair—was charged and imprisoned in 1912, the judge noting him guilty of "an act of vulgar imposture which stamped him as a rogue and vagabond . . . that he had cheated poor people to the tune of about £600 in 15 months." But this was not just about the poor being defrauded. Middle-class women and men were also buying their fortunes in the rooms of the West End, and fortune-telling in private but commercial rooms crossed class and gender in a different way than encounters in the countryside or on the pier. There was a suggestion that men posing as clairvoyants were especially adept not just in their fraudulence, but also in their capacity to blackmail. The confidences drawn out and the information derived from palmistry could be intimate and usable, and it was perhaps the new masculine clairvoyant in close encounter with women of any age and class who was most suspect. Middle-class wives, including of politicians, also needed protecting from their own credulity, so it was thought, and from the intimate spaces and close encounters in which too-private secrets might be disclosed.[28]

While the synthetic orient of Cheiro's Indian Room approximated the interiors of many a fashionable West End parlor, the occult revival also brought both mythic and real cosmopolitan encounters of an intimate order to the center of the Occident. The press occasionally noted that the traders were "chiefly foreigners, both men and women." And yet explaining the increasing application of the Vagrancy Act as white, masculine Britons asserting their own race, class, and gender is insufficient. Most of those charged *were* white Britons from Putney or Brixton, not from Budapest, Cairo, or Calcutta. All the players in the criminal and legal systems who charged Madame Ziska, for example—just like her clients—knew that her turban, crystal balls, and magenta curtains were as fake as Mr. Rochester's in *Jane Eyre*, or at least as affected. Madame Ziska of the magical address, 21a Electric Avenue, Brixton, was charged in December 1912. She practiced alongside Madame Le Didra and Madame Astrol, and they were all summoned for pretending to tell fortunes of the police witnesses Annie Betts and Florence Barnes, who were hired as private detectives. Le Didra's immediate response was to shun the police warning: "I must do as I feel and put up with the consequences." Madame Ziska—who advertised as "seventh daughter of a seventh daughter"—was less defiant and tried to evade the fortune-telling charge. "I am a phrenologist, and scientific character reader," she said, agreeing to remove her advertising boards, which included the words "Palmistry" and "Clairvoyance." She proceeded to advertise as "Madame Ziska, delineator of character." The physiognomic tradition saved her.[29]

A competitor, Hadji Mahommed, was also warned by police because he advertised as "the old original Egyptian clairvoyant, crystal gazer, sand diviner, and palmist. Under Royal patronage." Yet it so happens that "Hadji Mahommed" was no sobriquet. He *was* from Egypt, where his father, grandfather, and grandfather's grandfather had long been telling fortunes and character from palms. He was arrested in late 1912, wore his red turban to his hearing, and salaamed to the magistrate. Hadji Mahommed lived in the upper rooms of 209 Oxford Street, with his wife and seven children. How was he to keep them and pay the rent, he asked the magistrate, if he could not continue his work, which he had been doing for forty years? He advertised as a seer, a palmist, and an occult scientist. But it was the crystal ball and crystal gazing, as much as any reading of hands, that were the offensive object and practice in his case. To no avail, Mahommed explained that crystal gazing "had been going on for thousands of years." He had traveled to many countries making his living from that teaching, had been in England for twenty-three years, but now would like to return to Cairo at the first opportunity. He was both fined and imprisoned for three months under the Vagrancy Act.[30]

The police engaged private detectives—often but not always women—to pose as clients, have their fortunes told by suspect occultists, gather evidence, and report back to Scotland Yard with every detail. Wonderfully, the forensic reports lie in musty folders in the National Archives in Kew. These detectives were then brought in as witnesses at various trials. At Hadji Mahommed's hearing, detective Annie Betts went to his rooms, placed one guinea into a bowl of sand, after which "he pressed her hands, palms upwards, on a crystal which was in the sand, and said something in a foreign language." Then he counted the impressions her fingers had made in the sand, and pronounced on her future of two husbands, a death in the following year, and "a long life to 85, coming and going from England." Another detective-witness to fortune-telling fraudulence was a former palm reader herself. She declared to the court that she was not, in truth, a believer, and never had been, and so the expertise of this self-confessed fraud was especially valuable in proving—or at least intimating—the fraudulence of others.[31]

There was confusion, though, about the law. Was it the exchange of money, or the fortune-telling itself, that was illicit? Madame Athene (Marie D'Beaulieu) asked the police: "Is palmistry done free an offence?" Some complained. Hashnu Hara of 211 Oxford Street objected when police told her to stop advertising: "I have children to keep. I think it is very hard." Madame

Taun of 169 Oxford Street was defiant and brave: "This is my living and I will willingly do a month's imprisonment." On the other hand, some saw the policing as a clean sweep of an unregulated and unscrupulous trade. Madame Stars of 43 Oxford Street fully supported the crackdown: "There are many impostors."[32]

The persecution of fortune tellers in London came to be part of palmists' self-story, their internal narrative of misunderstood gifts. Cheiro noted that Henry VIII passed an act against "Palmists, Astrologers, Witches and Workers of the Devil, condemning all such 'rogues and vagabonds' to lose their possessions, to stand one year in the public pillory, and then expelled from the country." Much later, in a 1969 reprint of Cheiro's *You and Your Hand*, palmist-editor Louise Owen made a point of updating the persecution of her kind: "The new Act of 1951 repeals the Ancient Witchcraft Act of 1735 and makes substitutions for certain provisions of sect S.4 of the Vagrancy Act of 1824, but the position of all who practice palmistry and the associated arts remains uncertain." It was a self-serving narrative, but it was also true. The Witchcraft Act (1735) *was* replaced by the Fraudulent Mediums Act in 1951, which made monetary gain by spiritualists and psychics an illegal deception.[33]

Still, as with Salem, the early twentieth-century West End witch hunt stirred accusations among the persecuted, when occultists started to provide incriminating intelligence about each other to the police. One was disgruntled that while good occultists like herself complied, and removed her advertising of fortune-telling, others were allowed to continue with alacrity. She pointed the finger, and they were named: Abdul Malak was one, and another went under the name of K̲eiro. Underscoring the *K* was a particular effort, indicating an additional point of deception. "By this means he has led people to believe that he is 'CHEIRO' who made so many famous persons such as Gladstone, Balfour, Chamberlain etc, believe in his work." Professor Abdul Malak S.D.A.E.L. (Cairo), "Egypt's greatest sand diviner and occultist," was accused and then cautioned. And so was Keiro of 131 Regent Street, who advertised as "the genuine and original" (that is, vis-à-vis Cheiro). Keiro, however, had been through it all before. And it was, indeed, the still-current Witchcraft Act (1735), not the modern Vagrancy Act (1824), that brought him before the police, the law, and the new world of a sensationalist tabloid press in 1904.[34]

KEIRO

"Keiro" insisted that his was a Scottish name, but it is clear that he shamelessly profited from Cheiro's intercontinental fame. He had practiced in Philadelphia, Boston, Sydney, Hastings, and in London since 1899, sometimes working as a "medical electrician," and occasionally teaching hypnotism, though he no longer practiced in those capacities, he told the court in 1904. Nor was he a phrenologist or a clairvoyant. It was the hand that was the "only true guide to a man's character: the face was not a reliable index," and so Keiro advertised as "the leading and oldest established Palmist and Psychometrist in the World (over 24 years experience)." Feigning modesty, he told the jury in 1904 that he never took the title "professor," unlike so many others. Martha Stephenson—Madame Keiro—had worked as a trained nurse for twenty-five years and held four medical diplomas. A clairvoyante from infancy, the gift ran in her family, and she had studied palmistry for the last fifteen years, practicing from 1893. She said she never used a crystal, but she did, problematically, advertise both clairvoyance and crystal gazing. This couple was especially enterprising. They earned up to forty pounds each week from the managers of Swan & Edgar, the department store on Piccadilly Circus, telling fortunes to their customers, and blatantly advertised in the *Westminster Gazette*. Yet in legal terms they traded "gratuitously," acting on advice from lawyers that receiving payment from the company, not the client, would keep both themselves and the department store secure. And so, it was able to be advertised that "Keiro, the World Renowned Palmist and Madame Keiro, the World-Renowned Clairvoyante, will give Free Readings to Customers."[35]

It nonetheless transpired that this husband and wife were indicted "for unlawfully undertaking to tell fortunes." The case was brought by the London press magnate, Sir Alfred Harmsworth, owner of the *Daily Mail*. Harmsworth's motivation, by his own account, recapitulated that of earlier members of Parliament, that "rich and poor might be treated alike." Harmsworth was certainly one of the rich, but he made his money selling papers to the poor, or at least to London's working class. A major Fleet Street press magnate, his *Daily Mail* first ran off the press in 1896, and it is clear with regard to Keiro that his real motivation was to create a paper-selling scandal. The *Daily Mail* published a series of articles on the case even before the prosecution started, which the Keiros' lawyer, the prominent Roger Dawson Yelverton, considered prejudicial. This was a jury trial. Yelverton pursued Harmsworth, implicating the publisher himself with the game, and the profit, of fortune-telling. He had not only "submitted his own hand to be read, but in one of his papers had advertised that any one sending six stamps and marking the lines of his or her hand on a diagram would receive a true delineation of character by a person called 'Zoresta.'" This was Ida Prangley, an author of works on palmistry, and writer for *Heartsease*, another of Harmsworth's papers. Cross-examined, Harmsworth denied this, but he did admit "that he believed that the lines of the palm did reveal character, but not that they disclosed the future." The distinction between physiognomy and chiromancy endured, in this context often expediently.[36]

Harmsworth's lawyers decided that the Witchcraft Act of 1735 may be more useful than the Vagrancy Act in this matter, because police and prosecutors were finding it increasingly difficult to charge the "new class" of palmists as vagrants or vagabonds. They were plainly living in, and working from, West End addresses. *The Times* explained it thus:

> Under the Witchcraft Act of 1735 it was an offence, punishable on indictment, for any person to "undertake to tell fortunes." Then came the Vagrancy Act, under which a person "pretending or professing to tell fortunes" could be convicted by a police magistrate and punished as a rogue and vagabond. This was a more convenient procedure for dealing with the class of people such as gipsies and persons wandering from house to house, who were usually prosecuted and the older Act fell into disuse.[37]

The interesting thing about this Witchcraft Act, newly applied in 1904, is that neither the defense nor the prosecution were actually arguing about whether the Keiros were modern witches, but whether they were deceiving the credulous for money. In 1735, all earlier witchcraft acts were repealed, and the very meaning of "witchcraft" changed in law. It could no longer be presumed or asserted that witches or their magic or their Satan-induced spells were *real*. Instead, anyone who professed "magic" was knowingly fraudulent, seeking to gain from their deception. Far from extending a fact of witchcraft from the early to the late modern period, then, the 1735 Act made real witchcraft a modern impossibility, legally speaking. It marked the *end* of early modern witchcraft trials and executions, not their extension into the modern period, and so far as the law was concerned it rendered witchcraft and professed magic only ever a matter of fraud and deception.[38]

In attempting to establish just whether or how their palmistry was fraudulent, the prosecution and the defense cross-examined a suite of palmist-witnesses. Key were the private detectives whom the police and Harmsworth's solicitors had employed to gather evidence on the Keiros. Miss Dorothy Tempest was instructed by Harmsworth's lawyers to visit the Keiros, report on their process, and describe the environs.

Keiro had his name on a brass plate outside his Regent Street room, she began. Once inside, there were two screens, and behind one of them she was asked to sign a waiver. He definitely told the future, the deception that was the offense. He looked at her hand and told her she was nervous and highly strung; that between then and March she would be in danger from the sea; that in the ordinary way she would live to be forty-four, but as she had a "sister line of life," she would live to be seventy-five or seventy-seven; that she would be married in two years and would have one son. He held a crystal ball in his hand while he spoke. He told her if she wanted to know more to consult his wife, who was a clairvoyant. She paid him a guinea and came away. Miss Dorothy Tempest herself was interesting. She had been a detective for five years, after a career on the stage. But she had been out of work recently "and was starving." The pay of 10s 6d per day that she received as a detective kept her fed and sheltered, but Yelverton implied that it compromised her evidence. In the past she had also read people's hands, but not for money, she was quick to clarify. Indeed, she "did not believe in palmistry."[39]

It was all about belief and fraudulence. The prosecution told the jury that they must be convinced that the defendants "knew the pretenses were false before they could convict." But they clarified that "the mere professing to be able to foretell the future was in itself sufficient evidence of fraud to justify the jury in convicting." And as to the belief of those who sought their fortunes, he was clear that "it did not matter, according to law, whether the consultants of the palmists knew the whole business was a humbug or not." Yelverton, acting for Keiro, argued that no one had been called as a witness who complained of actually being defrauded. But more interestingly, he pursued the question of belief itself. "Before his clients could be convicted," he argued, "it must be proved that they knew the science they practiced was false and misleading." Was palmistry *believed* to be true or not? Indeed, *was* it true or not? It was true according to the Bible, Yelverton challenged. Palm reading was in the Scriptures, something the jury could find detailed not only in God's book, but in both Cheiro and Keiro's books as well: people had *believed* for the last 5,000 years. Not only that, but men like Herbert Spencer and Sir Oliver Lodge had in recent times written of the art of reading character from the hands. If *these* authorities all believed, so could and did the Keiros. And if they *believed*, how could it be fraudulent?[40]

And so, Believers were then brought in one by one, and cross-examined. Elizabeth Winter, authorizing herself with "six medical qualifications," said that she had consulted Keiro, who had read her hands "very correctly." She wanted to know more about her health than she could learn from doctors (herself, it turns out, a nurse). Henry Savid of Brondesbury was Keiro's brother-in-law, and a chemist, who assured the jury that palmistry was a demonstrable science "just like geology." Professor John William Taylor was also called to attest to palmistry's scientific legitimacy, although he identified as a phrenologist. Taylor had been a witness six times in palmistry cases, was the first president of the Occultists' Defense League, and told the jury that Mr. Stead, editor and spiritualist, was once a vice president. Much could be told of a man's character and habits by an examination of his hands, and events likely to happen to him were also indicated to a certain extent, Taylor explained, although this was not the same as clairvoyance. And then, doubling down, Yelverton invited the jury to consider other "wonders" that at first lacked explanation, but which have since become ordinary: the phonograph, the telephone, and wireless telegraphy. These technologies were "quite as marvellous as anything students of the occult claimed to perform." Those who believed in such phenomena—such "marvels" that turned out to be true—should be free to express their faith without fear of criminal prosecution.[41]

Proving belief was surely an enticing prospect for the prominent lawyer from Middle Temple. Indeed, the layers of truth-seeking, legal storytelling, deceit, imposture, and expedient and sincere "evidence" in this moment of palmistry and policing are almost too delicious. It all came to a new level when respected medical palmist Katharine St. Hill was asked to give evidence that palmistry was a science. She absolutely knew this was the case, and as we shall see, spent her life not just practicing but organizing around this surety. But knowing that palmistry was a science did not in itself mean that fraudulence was impossible. She herself was on a mission to hunt out charlatans from the profession, diligently revealing imposture and imposters. And so, when asked about Keiro and pressed for evidence of his fraudulence, she stated that this would be revealed in the hand itself. That is where the clearest evidence lay. "If Keiro was a liar I should see it in his hand." She did. While she knew nothing of clairvoyance or of crystal gazing, anyone could see that his palms disclosed no sign at all that he held powers to read them.[42]

Through these twists and turns on veracity, deceit, evidence, and belief, the jury found Keiro and Madame Keiro guilty. The judge saw it as a test case: "It had been said that the sole object of this prosecution was to ascertain what the law was, and to see whether persons like the defendants were in the same position as poor people of a much lower class who were being prosecuted all over the country. The opinion of a jury had now been taken." But he was lenient, on the understanding that the Keiros should not continue either advertising or practicing.[43]

Keiro happily, even defiantly, immediately turned to writing and publishing: *Palmistry, Clairvoyance and Psychometry* was in the bookshops by 1905. The wonderful subtitle allowed *everyone* to tell their own fortunes: "A Clear and Common-Sense Explanation of the Science by Means of which Everyone May Read his Own Character and Foretell his own Future and Fate." In it, he mused about sensing the future: "There is no doubt the lower animals have this sense to a marked degree . . . The North American Indians' foreknowledge of events . . . are similarly unaccounted for." He put this all as a "sense" beyond the well-known five. It was, perhaps, a remnant. "It is very probable that civilization and the use of reason in man has supplanted and almost annihilated this power, but it is not quite gone." It may be that "the nerve centres are influenced by future events, and through those nerve centres, mark the hand." In 1905, this was a mixture of degeneration anxieties and neurological innovation. It drew on both a popular theosophy of senses, and a new "parapsychology" just then being investigated.[44]

Keiro was defiant, continued to practice, and was cautioned again, with Malak, in 1912. He was still advertising but dropped the word "palmistry." The police wondered how they were going to gather evidence against him since he was now alive to their methods, and their best detective was known to him. That particular deception no longer worked. They would have to get fresh "testers," and the most perceptive and thorough came from a pool of former palm readers; ironically, although perhaps appropriately, charlatan experts. They succeeded, although in 1917 it was not Keiro but Madame Keiro who was sentenced to two months' imprisonment for telling the fortune of a detective.[45]

Cheiro and Keiro both performed their knowledge of the hand as modern "occult" seers. They imagined their expertise not as

Gipsy-derived at all, but as an extension of the learned hermetic tradition largely shorn of its kabbalistic origins, and on Cheiro's part explicitly as a transfer of *Hast Samudrika Shastra* to the West. They operated in a different kind of West End urban geography to Edward Heron-Allen, whose work with the hand was intellectual as much as commercial. When Heron-Allen read the hands of Oscar Wilde, Bram Stoker, and Arthur Conan Doyle, he was one of them, and they were as likely to retreat to a club together afterward, and then return home, too inebriated to write their various novels and plays, or just inebriated enough. When Cheiro read the hands of celebrities, he did so as a commissioned seer, a minor if unlikely celebrity himself. Keiro and Madame Keiro read differently classed hands again, those of the customers of new department stores, and lower-middle-class and perhaps upper-working-class women and men who came to their rooms. Over the 1880s, when Heron-Allen practiced, a new chiromancy was just emerging in an Anglophone world. By the 1890s, it was commercial and lucrative.

It is clear that Londoners of all kinds sought magic and wonders in the decades around 1900. What they received was as likely to be an insight into character, as much as a prediction of fortunes. Certainly, people comprehended Cheiro and Keiro in the language of marvels and of the ancient magi: "a descendant of the old Egyptian sorcerers," "a veritable wizard, a necromancer, a magician." In America, Cheiro was seen to bring old world magic into modernity; in New York, he was even acclaimed as "a male witch who would have been burned at the stake on the days of Cotton Mather." And for the American singer and writer Blanche Roosevelt, he was himself a modern wonder: "marvelous, most marvelous!"[46]

Yet English law had decided that magic could not be actual or real. The legal debate about veracity, magic, and deception unfolded within the terms of an eighteenth-century Act that had undone witchcraft, that disbelieved in magic itself: all magic was deception. But was that deception for gain or for entertainment? People were also prosecuted within the terms of a vagrancy law that was in part a leftover from Tudor times, a response to the earliest "Egyptians," and their palmistry that had long been considered fraudulent.

In this context, it is little wonder that another breed of hand reader tried to attach palmistry strictly to science and medicine, not magic, and tried to regulate and modernize their work. In the West End, but also on Blackpool Pier where the entrepreneurial Ellis family practiced medical palmistry and ran their British Institute for Mental Science, palmistry was shifting from a divining practice back toward its physiognomic possibilities: the telling of character and health.[47]

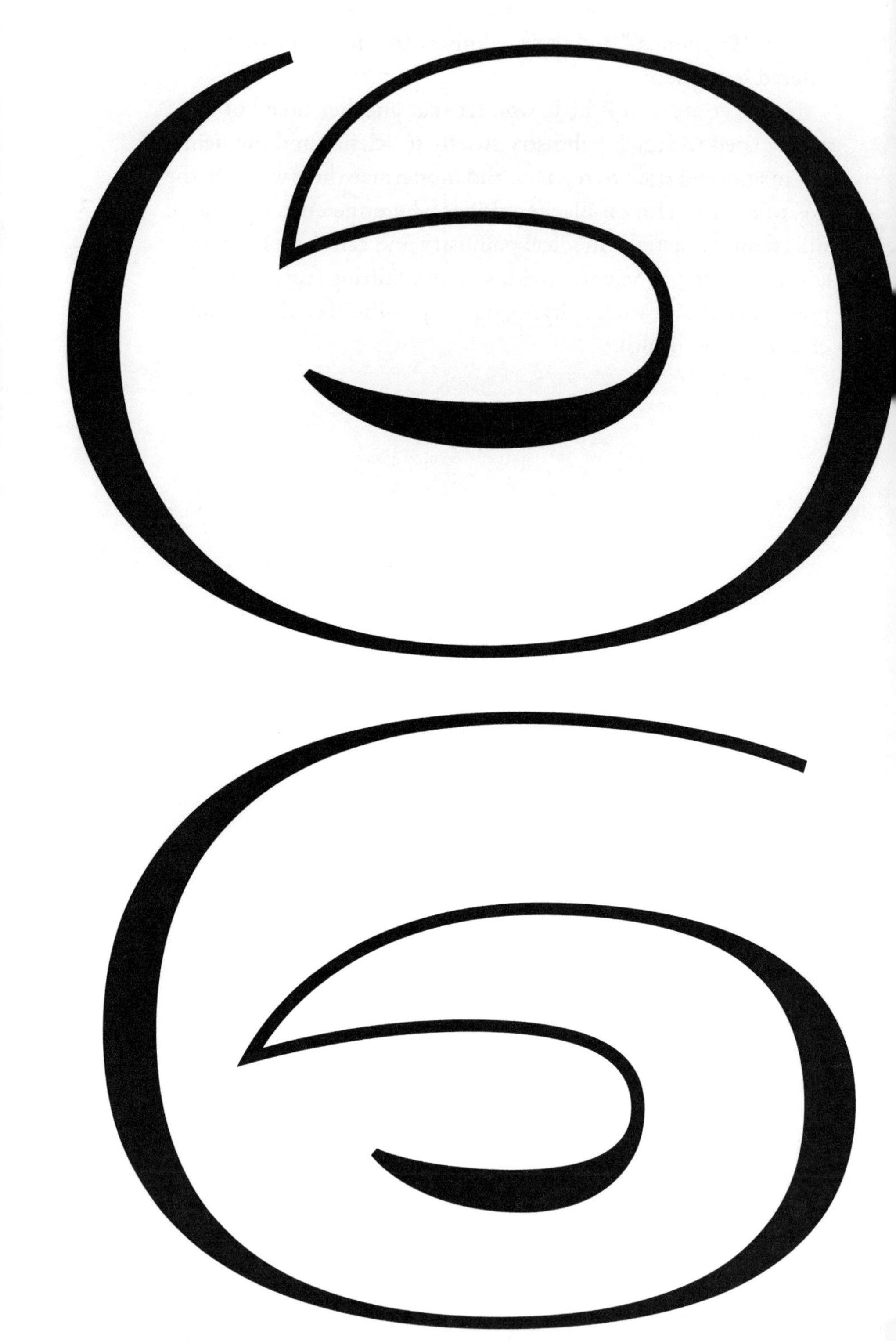

SIX

Commercial Hands

THE ELLIS FAMILY

On the coastal sands of Britain's urbanizing seaside, Romani women told and sold a thousand fortunes. Some were famous, like Blackpool's "Gipsy Sarah" (Sarah Boswell), and her daughters and granddaughters, who profited from a quickly expanding trade with a new working-class generation who had a little spare time and money for leisure and pleasure (fig. 6.1). This landscape for late nineteenth-century palmistry was altogether different to the West End of London. Over the turn of the century, tents on the sand turned into booths, adjacent to a new urban architecture of piers and promenades. There, Gorger palmists—women and men not of Romani blood—also told fortunes. Some borrowed consciously and unconsciously from Gipsy accoutrements, or an imagining of them. Others, however, derived their palmistry knowledge less from Roma traditions, or even the occult revival, and more from popular health practices, in particular phrenology. A vast world of irregular practitioners of all kinds lay underneath and around the smaller and more elite world of university- or apprentice-educated physicians and surgeons: specialists in electric therapies, baths and spas, magnetism, hypnotism, and homeopathy. They offered health and hygiene advice, diagnostics, therapeutics, and

6.1 Postcard of "Gipsy Sarah's Eldest Clever Granddaughter," Blackpool, n.d. Alamy. Collected also in Scott Macfrie Gypsy Collection, University of Liverpool Special Collections, SMGC 2/3, n.d.

prognostics. In leisure towns like Blackpool, palmistry was at least as much an element of this new health commerce as it was a chiromancy of the future. It was part of a suite of ways in which ordinary people sought information and advice about their bodies, their minds, their selves, while enjoying a carnivalesque seaside novelty. When mixed, these modes of reading the hand offered a thriving commercial opportunity for the enterprising.

Ida Ellis and Albert Ellis knew just that. A magnificently brazen couple, their lives and careers show us how phrenology and palmistry were connected within this trade in vernacular health and hygiene. Ida Ellis was an enterprising phrenologist working across northern England before she started to specialize in palm reading in Blackpool, where, with her husband Albert, business boomed. They advertised "advice on health," but the biggest billing was always themselves (fig. 6.2). Their palmistry was different to Cheiro and Keiro's occult practice. It was also different to the "medical palmistry" that Katharine St. Hill was then pursuing in

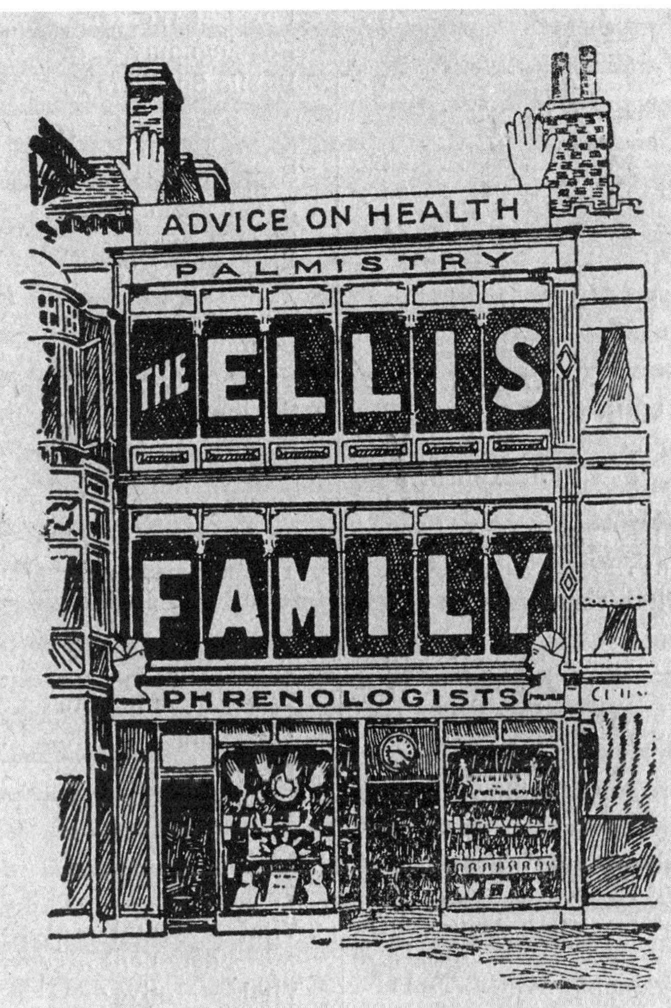

6.2 Phrenology and palmistry, heads and hands. Back cover of Albert Ellis, *Temperament and Character* (Ellis Family, 1909). Courtesy Cambridge University Library.

London (chapter 7), with pretensions to a scientific methodology that eschewed fortune-telling entirely. The Ellis family business was, rather, an expedient catchall that successfully created and satisfied a market across the span of possibilities from the physiological to the psychological to the therapeutic to the astrological.

The Ellis family sold insights to eager seaside customers, in an enterprise whose division of labor capitalized on gender. As we have seen, palmistry was, unusually, an expertise and an occupation in which both women and men could work. But there were inevitably different shades of meaning and possibilities that accompanied male and female practice. Initially a husband-wife, then expanding into a sister-brother team, the Ellis family offers the opportunity to consider how authority and expertise in palmistry was gendered, and how a woman could and did fit into this unregulated domain, not only as a "health specialist"—her sometime self-description—but also as a businesswoman. Ida and Albert Ellis invented all kinds of ways to make money from reading hands, faces, character, and futures. Increasingly, however, this was not about bodies at all, but about information. They wrote, self-published, and sold pamphlets, charts, and books of all kinds; they traded in all manner of health paraphernalia; and, most curiously, they standardized a method of health consultation that was not in person at all, but delivered through the post: early remote health.[1]

Here, then, we see the strange history of the hand unfolding in another cultural landscape altogether, within a different class milieu. Not quite the chaos and fraud of the street depicted in Praetorius's woodblock print of 1661 (fig. 1.1), this was a more structured popular commerce in reading the body for money, not rarefied or special, but ordinary and accessible. The location of the Ellis family operations on Blackpool's shore suggests a carnivalesque or fairground context for palmistry. It was. But it was not just that, either. Indeed, their work is better understood as part of health history, of the thriving industry of irregular practice that was then booming. In the end, the Ellis family's business was more about medical commerce than about fortunes and the future, and more about the modern application of physiognomy than the modern history of chiromancy.

THE ELLIS FAMILY

Albert and Ida Ellis spent their entrepreneurial lives across the length and breadth of England. Albert was born in 1867 in Canterbury, Kent. The son of a whitesmith, he lost his right leg at fourteen years of age after an injury and used a wooden leg throughout his adult life. Ida Eliza was born in Suffolk in 1865. They married in 1889 (she recounted that he was phrenologically chosen by her), the year their son Frank was born. That year they moved to the north of England, to Batley, Yorkshire. In the census of 1891, Albert recorded his occupation as an insurance agent, while Ida already described herself as a phrenologist.[2]

It was certainly Ida who initially pursued a highly commercialized consulting and publishing business, based on phrenology. She advertised in Batley's newspapers as "Professor" Ida Ellis, a student of phrenology for ten years, and a member of the British Phrenological Association. In 1891, she formed the Universal Phrenological Society (no less) "for the investigation of mental science," the first of several such bold moves conjuring institutes and societies out of nothing but her own initiative and ambition. She had been editing a monthly magazine, *Know Thyself*, and now that she had invented her own society, she retrospectively cast this as its official organ: members of the Universal Phrenological Society would receive the monthly, on payment of 10/- per annum. With that subscription also, and with prior verification of credentials, members were able to use the initials MUPS after their name—Member of the Universal Phrenological Society—if their reputation remained good. Certificates might also be granted if they supplied some evidence of expertise and experience, an essay, or a character sketch based on photographs.[3]

The earliest issues of her journal were devoted to "Phrenology, Physiology, Pathology, Pathognomy, Physiognomy, Pleasure and Profit." No palmistry yet. This was an enterprise in health and hygiene. Over 1892, Albert, now intermittently also titled "professor," joined the Ellis phrenological team. They regularly advertised their services in the *Batley News*, as well as in their own magazine, promising to "delineate character from heads and faces, photographs or handwriting daily from 10 a.m. to 10 p.m.," at 115 Taylor Street, Batley.[4]

Professor Ida Ellis was certainly more medical than occult at this point. She called herself a "Consulting Hygienist," in addition to being a "Professional Phrenologist." In 1892 a fellow phrenologist, Professor Dall, delineated her character and announced, expediently: "You would make a good doctor." She both published and consulted on methods of contraception, and advertised this openly. The first issue of *Know Thyself* promoted her own pamphlet, *Sexual Science: the essentials of conception and how to prevent it.* "Startling" and "worth its weight in gold . . . it will save its cost a thousand times over." Here, however, she was conspicuously *Mrs.*, not *Professor* Ellis: "Married Ladies may consult Mrs. Ellis on the limitation of families by legitimate means," daily from 2 p.m. to 8 p.m., or by letter. Her core business was phrenological, however, diagnosing or "delineating" character. This could be discerned not just from observing or feeling heads, but also from objects. Ida's talent for describing "the character, talents, and failings of any person from their handwriting or photo" began to feature regularly in regional newspapers across Britain.[5]

The Ellis partnership moved to 124 Roundhay Road, Leeds, in winter 1892. There, they were announced as "well-known phrenologists, mesmerists, and hygienists," whose specialty was "the cure of fits or epilepsy by means of mesmerism or hypnotism." One newspaper claimed that together they had given advice to over 10,000 patients in a four-year period. It was a short stay in Leeds, however, and in late 1893 they moved again to Blackpool, after a weekend visit to the town suggested promising possibilities. They were right.[6]

Ida and Albert Ellis were astute and bold speculators and promoters, mainly of themselves. Everything they did was purposeful, considered, entrepreneurially fashioned, and almost always successful. While earlier, Ida Ellis had presented her practice as singular, in Blackpool they began presenting as a consulting couple. The more they did so, the more conventionally gendered their pairing became: a division of labor designed to secure as many clients, subscribers, and purchasers as possible for their exploding business innovations. She was now more often "Madame Ellis" in newspapers across England's north, while he was self-promoted to scientific "Professor." Unmistakably by the 1890s, "Madame" signaled a vaguely orientalized occult, both in middle-class and working-class cultures. And indeed, over time, he became the specialist in phrenology, while she took up palmistry, especially after their canny move from Yorkshire to the booming seaside resort in Lancashire.

Blackpool presented another market altogether, its sands, beaches, and piers attracting thousands, millions. Its first pier was built in 1863, and by the late nineteenth century its famous Pleasure Beach on the sands of the South Shore was a named attraction, and it was there, mainly, that Romani encampments formed the cluster and community of fortune tellers in tents. It *was* a place of leisure, of carnival, and of "delight" with or without "delusion." Roma women certainly read palms for the future, but that was not all. There was now a crossover in their commerce that added a more physiognomic and phrenological character reading to the more traditional fortune-telling. "Doreena," for example, promoted herself as: "The world's greatest character reader, clairvoyante, palmist and adviser." Romanichal women were physiognomists too.[7]

It is important to recall that palm reading was not only a working-class pastime. As we have seen, Edward Heron-Allen's clients, and Cheiro's too, were all monied. Closer to the Ellises' geographical world, the Liverpool-based Rathbone family of politicians and high-order businessmen were also having their palms read. Four pairs of hands were presented one evening in September 1872 to Miss E. Smyth, Palmist. The first was "an impractical hand showing strong acquisitiveness, love of approbation and want of self-reliance, not however without decision always tempered by sensitiveness." A second hand "shows an immense practical element . . . a very impulsive but at the same time decisive and steadfast character: liberal but not generous." While another showed "a very, very strong imagination, a creative but indolent and vacillating genius, with immense concentrations and obstinate temperament . . . it is generous but one sided and without great liberality." It is not clear which was merchant and Liberal politician William Rathbone's, but it is clear that the wealthy Rathbone family and friends were having their hands read, and in a manner similar to the Ellis clients; that is, phrenologically for character, more than chiromantically for the future.[8]

Once in Blackpool, Ida and Albert Ellis doubled down on their credentializing, inventing whatever they wanted, and selling whatever would inspire interest, confidence, and commerce. They were brilliant at self-invention, including of their own authority. They created out of nothing, for example, the British Institute for Mental Science. Hedging their bets (or dividing their labor) on the occult and on health, Albert Ellis remained "Professor," announced as principal of the organization, while Ida was "Madame" and vice principal. Albert's brother then joined what was now a family business: Frank William East Ellis, secretary, who boasted "Six Diplomas of Merit," probably all from the previous start-up, the Universal Phrenological Society. The Institute was open 10 a.m.–10 p.m. every day and offered customers advice on health, delineations of character, and lessons on mental science to help prepare them for the diploma of just that Universal Phrenological Society. What brazen and delicious moves.[9]

The Ellises couldn't be contained. They became part of Blackpool, its landscape, its commerce, its politics, its culture. In some ways, the Ellis family competed with the Roma families—the Boswells and the Lees—for palmistry custom. Yet Albert Ellis ended up a defender of the thriving trade on the sands for Gipsy fortunes. When the local council overreached itself, trying to regulate and even clear the seaside of traditional tents and stalls, Albert objected, and formally so. He was investing in real estate but also becoming a "leading spirit in the fight which was made against [Blackpool] Council's endeavour to clear the sands of all stallholders and entertainers." Albert Ellis was elected a town councilor in 1902, by which time he had become, in the words of the *Blackpool Gazette & Herald*, "one of the best known individuals in Blackpool." In the same year, he helped form the Blackpool Sandsmen's Union (sometimes called the Sandites' Union), of which he became general secretary. The union was formed to "watch the interests of persons trading on the sands" and to "watch the progress of any future Bills in Parliament introduced by the [Blackpool] Corporation for further control of the sands." Any ratepayer who was a "sandite" could be a member. By July 1904, Ellis had become secretary of the Blackpool Trades' Council, a local parliament of trade union branches. Though a Liberal in parliamentary politics, Ellis used his connections in the local labor movement to form, in 1907, the Joint Public Bodies, an electoral coalition comprising the local branch of the socialist Independent Labour Party, the Company-house Proprietors' Association, and other small organizations.[10]

In the meantime, the wider Ellis family expanded rapidly into publishing and a business of health merchandise. They were learning that as much or more money was to be made from books, charts, and paraphernalia as from personal consultations. In the 1901 census, Ida and her brother-in-law Frank both listed their occupation as "character reader." In 1911, Ida, Albert, and Frank all described themselves more expansively as "palmist and phrenologist." They indicated on the 1921 census that they worked as "phrenologists" from 82 Central Beach. By that time, considerable wealth had been acquired, not just from their health, palmistry, and phrenology business, but also through real estate investment, highly profitable during the Blackpool seaside boom.[11]

"HEALTH SPECIALISTS"

From a beginning in phrenology, the commerce of health and hygiene was lucrative, and from Ida Ellis's work in Batley to their prominence in Blackpool, phrenology was the family's constant. Phrenology operated more in a medical domain than it ever did as part of an occult. As we have seen, it fell within a suite of therapeutics and diagnostics that moved in and out of conventional medical favor over the nineteenth century, along with mesmerism and hypnotism. A descendant of physiognomy, first systematized in early nineteenth-century Edinburgh, and popularized from the 1830s, this was a longstanding discipline that was by the 1890s familiar across the globe. It was a strange science of "temperaments" through which attributes, qualities, and character could be named, or "delineated."[12]

Phrenology dominated the Ellises' Yorkshire years and formed most of the content of *Know Thyself*. In its first issue, Professor Ida Ellis explained the "temperaments," "faculties," and "organs" of phrenology, which were each classified through the magic number seven: seven temperaments, seven groups of faculties, and seven terms to designate the size of the phrenological organs. The temperaments were Organic, Vital, Motive, Mental, Active, Excitable, and Balanced, a recapitulation of D'Arpentigny's classification of hands. The groups of "faculties" included Domestic Propensities (amativeness, conjugality), Selfish Propensities (combativeness, secretiveness), Selfish Sentiments (caution or self-esteem), Perceptive Faculties (language, individuality, size, weight), Reasoning Faculties (intuition, agreeableness), Refining Sentiments (sublimity or ideality), and Moral Sentiments (spirituality, hope, conscientiousness, veneration, benevolence). At the height of their business in Blackpool, it was Albert who advised and published most commonly on phrenology, authoring such pamphlets and charts as *Phrenological Textbook*, *How to Read Heads*, *Posters of Heads*, as well as *Temperament and Character*. He acknowledged that much of his information was derived from the work of L. N. Fowler and George Combe, respectively the foundational American and English phrenologists. And yet his version was more simplified than Ida Ellis's earlier phrenology, and in a way more embodied and medicalized. Albert Ellis explained not sevenfold classes of temperament but threefold. The first class was the Mental class, to do with the brain and nervous system. Second, the Motive class related to bones and muscles. And third, the Vital class of temperament related to the heart, liver, lungs, and stomach; these "generate life force and give what is termed vitality." This was a way of understanding organs through bodily systems. The human nervous system was illustrated, for example, alongside "the bony system" and "the muscular system," in a way that would have been recognized by any conventionally trained physician. More idiosyncratically, Albert Ellis illustrated a less well-recognized "Vital System" (fig. 6.3). The different "class" of temperament was signaled by different physiognomies. People with large heads and "pyriform" features were to be understood as having a Mental Temperament, and their character was "sensitive," "thoughtful," and "inclines towards study and meditation." People with the Motive Temperament can be identified through their large body features: "They are tall, there is lots of bone. They hold motion, executiveness, and adaptability for hard physical work." A Vital Temperament can be recognized through "big stout bodies and round features." The three are held by all people, but in different proportions, and a fortunate individual will occasionally hold perfectly balanced temperaments, "without excess or deficiency." A balance among the mental, the motive, and the vital might also be cultivated, to some extent, Ellis advised. Indeed, success, happiness, and fulfillment lie in the capacity to harmonize temperamental conditions. It was all a remnant, as much as a revival, of the balanced humoral "complexions" so foundational to the work of Richard Saunders and his fellow seventeenth-century physicians.[13]

For Albert Ellis, "temperament" still signaled an original Latin meaning of a "mixture or arrangement of qualities." Everything animate and inanimate has temperament, he explained. In humans, it makes the mind work in particular ways. Temperament was part of the physical build of humans, the foundation out of which character was both given and could be, to some extent, cultivated. There were agreed rules among experts for understanding temperament. First: "Nature operates always and everywhere by means of organs and never otherwise." It was "organs" that were discerned by phrenology. Second: the brain is the "chief organ of the mind, or the chief instrument in producing all the various powers of man." The rest of the body, including the hands, may be considered an instrument of the mind, subservient to it.[14]

Temperament was also linked to disease susceptibility. Those with predominantly Mental temperaments, for example, were susceptible to "inflammation and congestion of the brain, consumption, spinal diseases, dyspepsia, and various forms of insanity"; those with a Motive temperament to rheumatism, liver

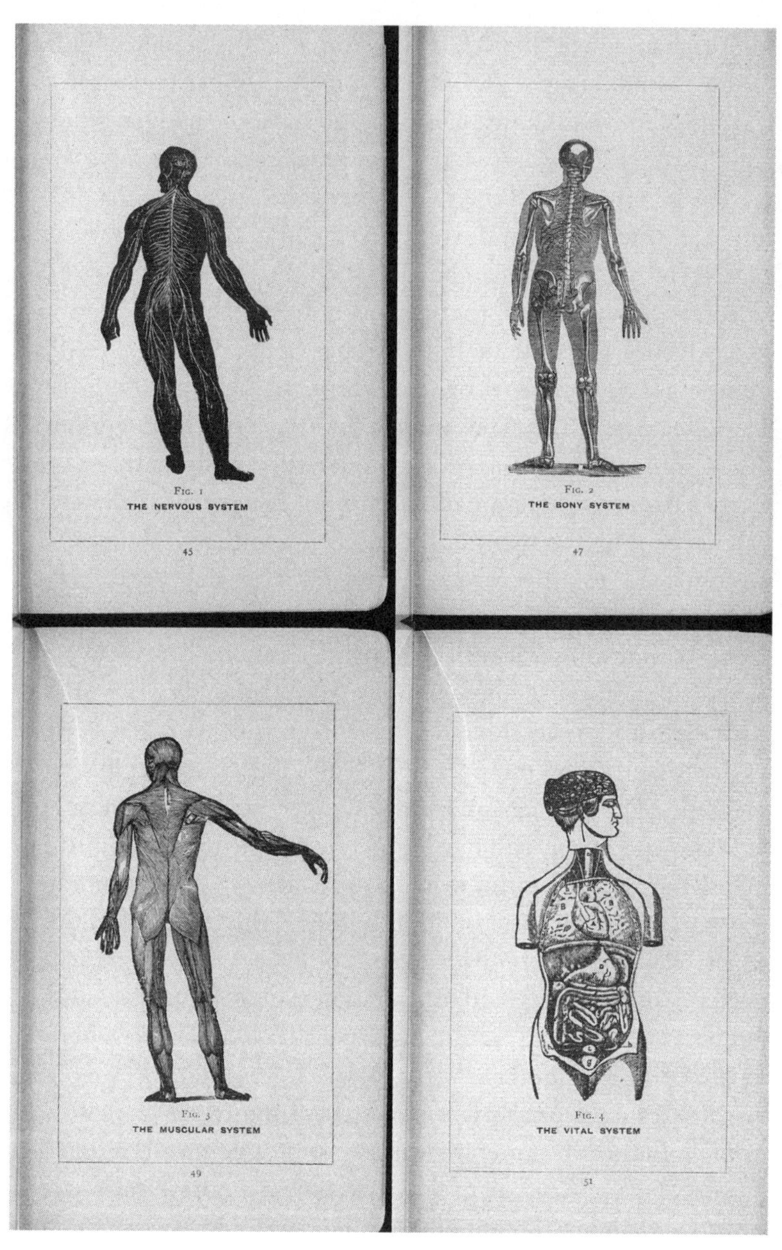

6.3 The bodily systems. From Albert Ellis, *Temperament and Character* (Ellis Family, 1909), 45, 47, 49, 51. Courtesy Cambridge University Library.

complaint, and joint troubles; and those with a Vital temperament to gout, tumors, sciatica, or dropsy. There were therefore better and worse climates in which individuals with specific temperaments should live; better and worse foods they should eat; better and worse types of baths they should take. Here, a diagnostics of phrenology became therapeutic, a regimen for improving health, what would later be praised and practiced as holistic therapeutics.[15]

An embodied practice, phrenology was also a discipline of words. It functioned through a language, a vocabulary, that could be surprisingly eloquent in the nomination of human qualities. It put words to human interiors that drew partly from a Romantic-era constitution of emotional selves, and partly from nineteenth-century valuations of moral character in relation to others, all anticipating early twentieth-century psychological archetypes and psychoanalytic interiors. Phrenology peddled an idiosyncratic dictionary, including Acquisitiveness; Amativeness; Benevolence (or tendency to be kind and sympathetic); Combativeness (or tendency to oppose); Conjugality (or tendency to jealousy); Destructiveness; Imitation; Intuition (or correct first impressions); Self-Esteem; Sublimity; and Veneration (or respect of acknowledged superiority). That these qualities were comprehended as phrenological "organs" is bizarre, but this was nonetheless a refined and even poetic A-to-Z of human being. The phrenological classification of character and self could be surprisingly eloquent, anticipating but also perhaps exceeding twentieth-century classifications of medicalized personality and personality disorders.[16]

From this base in phrenology, the Ellis couple became ecumenical "health specialists," drawing and selling wisdom from anywhere. They published and distributed *Herbal Remedies for Diseases*, for example. Alphabetically arranged, they listed nearly 200 conditions, and their remedies, mostly comprised from the private papers of "Mr Hunt, an able Medical Herbalist who practiced in Nottingham for upwards of a quarter of a century." From Abscesses, Acidity, Acne, and Ague to Water in the Head, Weal eyes, Wheezing, White Swelling, and Worms, the Ellises offered a book of medicines. In another peculiar pamphlet, they advised on one's "Ruling Planet," in a rare astrological gesture, as well as on "Quality of Temperament," in which multiple registers and knowledge traditions beyond phrenology were named and equated: how astrologers, physiographers, pathologists, and Hindu philosophers saw the same bodily condition, but named it differently. "The quality of temperament is classified as Cardinal by Astrologers," they explained, but "Mental by Physiographers, Nervous by Pathologists, Rajasic by Hindu Philosophers." Another temperament was classified as "Fixed by Astrologers, Vital by Physiologists, Lymphatic and Sanguine by Pathologists, Tamasic by Hindu Philosophers."[17]

The riotous montage that made up knowledge in the Ellis universe was designed into one of their Advice Charts (fig. 6.4). A very different compilation to Praetorius's print from 1661, at the top there is no God or Divine declaring "Truth" and "Wisdom," rather the Ellis family themselves. The antique and classical referents endured, however. An Egyptian signifier is prominent, as is the classical column wound by a snake, the longstanding symbol of medicine, sometimes the single snake of the rod of Aesculapius, sometimes the two snakes of the caduceus, the traditional symbol of Hermes and Hermes Trismegistus. Phrenology and palmistry feature, but other methods appear as well, from physiognomy to occult science. Totally excluded is Roma knowledge, the Gipsy families who were their foremost competitors.

Over and above health and hygiene, the Ellis family were entrepreneurs, constantly doubling down on people's interest in, and concern for, their own bodies, their own minds, their own selves. This was true health capitalism in which the value of every Ellis item lay in the opportunity to sell another Ellis item. Their *Guide to Health*, for example, recommended that three other pamphlets be purchased, for 6d each: *How to Improve Body, Brain, and Mind*; *Temperament and Character*; and *Herbal Remedies for Diseases*. They offered further services, for a substantial fee, of course: "The Ellis family attend Bazaars, Garden Parties, At Homes and Entertainments." And to a more educational sector: "They would lecture at Societies." They traded extensively in paraphernalia. If a palmistry reading was sought by post, the client first had to send 6d for a bottle of "Transferine," a liquid that made good impressions of

6.4 Crowded with hermetic, Egyptian, and medical references, notably absent from this self-promoting poster are the Ellis family's foremost competitors, the Roma families of Blackpool. The Ellis Family Advice Chart, I, no. III, c. 1919. Courtesy Cambridge University Library.

hands. They were shameless, really. In one consultation, the client was told he had powers of clairvoyance and crystal gazing, and that if he wished, "globes could be purchased from Mr Ellis" at the end of their appointment.[18]

Another strategy for multiplying income was to leave the slow, sequential, in-person consultation behind—a poor return for time—in favor of quick, simultaneous postal consultations. By post, people received a "personal record" to be completed and returned, a mixture of a medical record and an identification document: hair and eye color; date and time of birth; date of vaccinations and baptism; date of first tooth, first word, first step, as well as weight and height, size of collar, and size of gloves. From this, from photographs, and from a written testimony, their *Guide to Health* would be filled in (by any Ellis) and returned by post. A marked-up template, barely personalized, set out health advice from a limited range of possibilities: advice on what to eat and on how much to sleep; and advice on which type of baths were most suitable, be that a cold sponge, tepid sponge, Turkish bath, hot seawater, cold wet towel, or warm wet towel. And "herbal remedies" were offered for the diseases to which the client was told they were most liable: Saw Palmetto Berries, Snake Root, Borage, Archangel, Butterbur Root, Fluid Extract Skullcap, or Fluid Extract Valerian. They also advised on "hereditary resemblances," noting that "there are certain physiological conditions which enable us to judge whether a person resembles the male or female side of his ancestors." Resemblances were physical and mental, but, they dissembled, "the subject is too extensive for us even to commence an explanation here as to how we can do this."[19] *That* advice could be bought and sold in a subsequent transaction.

PALMISTRY

Ellis family palmistry was an offshoot of Ellis family phrenology. The all-important phrenological Temperament *could* be identified by a physiognomy of hand shape as well as head shape. The Mental Temperament, for example, was signaled by thin, slender hands, the tips of the fingers square, "the whole hand flexible and restless." The Motive Temperament "has large firm hands with broad palms, wide finger tips of a spreading or spatulate shape, and large veins on the back of the hands." The Vital Temperament is signaled by hands that are "broad, plump and soft . . . the tips of the fingers usually turn back slightly." It was all a mix of D'Arpentigny, Beamish, and knowledge amalgamated from a hundred palmistry books. To illustrate, Albert Ellis reproduced photographs of the "Hands of [a] Gentleman" of the Mental-Motive-Vital Temperament combination. This gentleman was born at 5:30 a.m. on July 17, 1867, in Canterbury, Kent, the specific natality a remnant of horoscopic numerology; although astrologically inclined phrenology and palmistry was generally muted in the Ellis world of bodies and signs. The next figure was of the "Hands of [a] Lady" of the Mental-Vital-Motive Temperament, born 10 p.m., August 30, 1865, in Alpheton, Suffolk. In fact, these were self-portraits: the first of Albert's hand, the second that of Ida (fig. 6.5).[20]

Over time, it was Ida who came to be the hand specialist in the division of labor that was Ellis enterprises. Her Lancashire business persona was definitely more palmist than phrenologist. She published *A Catechism of Palmistry* (1898), *Hand Physiognomy* (1899), and the *Key to Palmistry* (1924), and was described in several newspapers as "one of the best known palmists in the North." Like Sarah Boswell, and all her kin then telling fortunes on the sands, Ida read the lines and signs of separation or divorce; loss of husband or wife; numbers of children; good or bad fortune; and whether or not a legacy would be left, as well as length of life, and even the probability of being subject to slander. She decoded marks and signs on the hand that might indicate a broken love affair or only one true love affair, and the signs of matrimony or celibacy. Albert Ellis also put his name to the marriage advice market in *Matrimony: A book for those about to Marry*. It explained "the law of polarity": "Who you should marry from a *physical* point of view; Who you should marry from a *mental* POV; who you should marry from a *harmonious* POV." And it indicated the signs on the hands that indicated "possible love affairs which influence your life." This included a "sterility marker."[21]

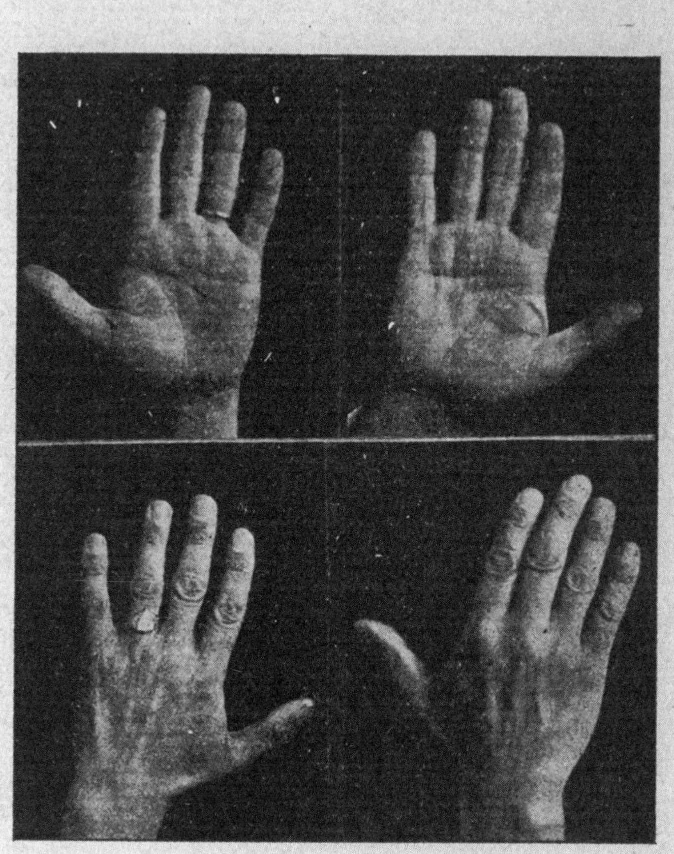

FIG. 7

HANDS OF GENTLEMAN

MENTAL—MOTIVE—VITAL TEMPERAMENT

Born 5-30 a.m. 17th July, 1867
Place: Canterbury, Kent

6.5 Palmistry self-portrait, showing both sides of the hand, its shape, and its lines. Albert Ellis, *Temperament and Character* (Ellis Family, 1909), 38. Courtesy Cambridge University Library.

Unlike Gipsy Sarah, her clever daughters and granddaughters, and all the other Romani women on the sands, however, Ida Ellis increasingly offered this advice by post. Even when she did undertake personal consultations, her report was not offered verbally, but by ticking off the relevant sections of a palmistry template. In 1904, it was recounted that she looked at a client's hands, and after payment of half a crown, "commenced to tick off numbers in the book, and wrote in it." Indeed, at this reading on the Promenade, fourteen clients were examined, between 3:55 and 4:40 p.m., according to a police inquiry.[22]

Even when her palmistry approached medical diagnosis, it was routinized. Lines on the hand may show a health breakdown at a certain age, her forms showed; the ages when your health is best; and perhaps that you will enjoy better health as you get older. There were also lines read for "disease," where she would fill in the blank field on her template, and yet it was all very basic. One might be told that "No chronic disease is shown," but even that would be more satisfying than another option she might circle: "You have no signs on your hands to show if you will suffer from any special complaint." It was outrageously bold, but also vague and crude, compared to the medical palmistry then being practiced by Katharine St. Hill, for example. Photographs tended to present Ida Ellis paying careful attention to her client, her senses as well as wisdom apparently focused. Yet her preferred consultation was not ocular, tactile, or visual, rather disembodied and so time efficient. Her business preference would be no body at all. In this sense, hers was a disenchanted, routinized, standardized, and modernized palmistry. It was far removed from the senses and even from the body that were the business of early modern physician-chiromancers. It was a world away from the touch, sight, and insight that characterized Cheiro's work, no matter how much we must also see him as a modern showman. And it was certainly crude compared to the hand reading of the psychoanalytically trained medical doctor Charlotte Wolff in the interwar years, as we shall see (chapter 10). And yet, it was all phenomenally successful.[23]

The Ellis family wrote fewer books than pamphlets, but the *Catechism of Palmistry* was one, sole-authored by Ida. A "catechism," this was a commercialization and secularization of a genre based in religious instruction. By March 1920, the *Catechism of Palmistry* was being used as the British Institute for Mental Science textbook for candidates being examined for diplomas. In it, Ida Ellis distinguished chirognomy from palmistry and chiromancy. The former was "the science of reading the character and talents of an individual in the form of the hand, its consistency, and from the relative sizes of its different parts, as well as from the configuration of the finger joints, tips and nails." Palmistry moves on from this general form of the hand to consider the lines, "and, in addition to character, deals with events and bodily condition, past, present, and future." Drawing directly from the French D'Arpentigny, she set out the many different hand types that anticipated personality types and archetypes. By the early twentieth century, this had settled into a civilizational progression, a kind of chirological stadial theory onto which were mapped the five senses, themselves organized as evolutionary signs.[24]

The Elementary type/shape was still the lowest, "likened to the first race of human beings on this earth," linked with coarse and crude consciousness, and occupationally aligned with those likely to be "the tool or the servant of others." The "Spatulate Hand" was like that of the "second race of human beings on our earth," with a more developed sense of touch and perception. Spatulate-handed individuals "are capable of self-government," and although loyal to a country, will likely emigrate if freedoms are not forthcoming. In fact, most English-speaking peoples have this hand, borne out by their "colonising tendencies," she noted. The Square Hand shows more self-control and self-restraint, and like the "third race" was connected with the sense of sight. The Conical Hand represents a higher and more ethereal state, associated with the fourth sense of taste; emotional, even erratic, and definitely sensuous. As for the "Philosophical Hand," it shows capacity for metaphysics, love of truth, and deep reasoning. Here she ventured into theosophical territory to capture its significance: "This type may be said to be harmonious with the fifth race of human beings on earth, who have developed to perfection the sense of smell. This is the race now holding the leadership of the nations of the earth, called the Aryan Race." That would seem to be the apparent pinnacle: but no. Finally, there was her own hand type, "the Psychical Hand," recognizable by "its extreme delicacy of contour, being long, slender, nervous-looking, palm narrow and thin, fingers as long if not longer than the palm, and the tips of the fingers sloping to a point." It was special, "a promise of what mankind will attain to rather than what is at present manifested in the average individual." Those with the Psychic Hand displayed an advancement and refinement of all five senses and were evolving a sixth sense: "It will take the form of a higher power than that of logic and reason, and by some is called Pure Reason, but is more recognisable by the majority as Intuition. This type, therefore, represents the Divine world." And, in the manner of too many Aryans, she perceived her own hand type to be superior to other humans.[25]

VAGRANCY AND REAL ESTATE

Ellis enterprises might have unfolded in any northern town, but that it did so in Blackpool places their story adjacent to, and intersecting with, Roma histories. Like any number of Gipsy women, but also like her West End counterparts, Ida Ellis on occasion came up against the police, the courts, and the Vagrancy Act that had links back to "Egyptians." In the first instance, however, the Ellises were policed for their publication of birth control tracts. In September 1891, Ida, Albert, and others were summoned to Batley Borough Police Court for publishing indecent literature. After searching their premises, a police sergeant found around 2,000 books that contained "some very obscene matter." Ida Ellis was—entirely explicitly—the author of a pamphlet that instructed on how to "limit families by legitimate means." The magistrates were not sympathetic, however, and ordered that the books be destroyed.[26]

In October 1905, himself now a councilor in Blackpool, Albert Ellis was involved in an all-male meeting of his supporters in the Brunswick ward of Blackpool, which he represented. The meeting had been called two days before polling day to "expose and denounce" Ellis, and to make voters aware of a pamphlet published in 1893 entitled *Modern views on the population question in relation to the limitation of families by the prevention of contraception, being extracts from lectures delivered by Professor Albert Ellis to gentlemen only and Madame Ida Ellis*. The chairman proceeded to read the entire pamphlet to a lively audience, but it was reported that "the general opinion seemed to be that the meeting had done Councillor Ellis a very good turn," as it would convince electors to vote for him, "out of the English fondness for supporting the bottom dog." The *Blackpool Gazette & Herald* was right, and Ellis was re-elected.[27]

The Ellis duo was less triumphant the summer before, however, when Ida appeared at Blackpool Police Court, charged not for her birth control publications, but for her fortune-telling. Under the Vagrancy Act, she was charged for "using certain subtle means, viz., by palmistry, to deceive and impose on John Leaver," who was a police constable, on Central Beach, Blackpool. Her case was widely reported in the national press, especially as she was convicted and endured a short jail sentence; she was released in November 1904, also the year of Keiro's case. In the north of England, an Occultists' Defense League had been established to counter this new challenge, founded by Joseph Dodson, solicitor, in Halifax in 1898, founder also of the Occult Book Company in 1893. It defended "professors of occult or predictive Sciences against wrongful prosecution, encouraging and promoting the study of those sciences, also prosecuting charlatans and fraudulent exponents of the same." It was not, in other words, offering counsel for Gipsies. For an annual fee, members would be defended by counsel trained for the work, and also guided and advised in the most useful manner. As in the London instances, lawyers generally argued that the practice of palmistry was both legal and scientific, and therefore not fraudulent. And as in the Keiro case, Ida Ellis's lawyer resorted to the Scriptures to try to make her a credulous figure. Two policemen had had their palms read by her, and the defending lawyer challenged one of them to confirm the practice as a "sham." "Do you call it a sham to heal by a laying on of the hand?" he asked the policeman/client, who replied, "Of course." Have you read the Bible? Does it not speak of healing by laying on the hand? The poor policeman nodded, and when asked whether *that* was a fraud, could not answer.[28]

Keiro had also been in contact with the secretary of the Occultists' Defense League, who had advised that he protect himself by requiring that all consultants sign a waiver:

> Keiro hereby gives notice to all who wish to consult him that he has no intention or desire to deceive or impose upon any one; that any consultant is at liberty to believe or not his statements as to character, past life, or otherwise, or prediction of foretelling of the future (if any); and any one who consults him must do so upon the understanding that he has no intention to deceive or impose upon any one, or to obtain money by false pretences.[29]

His clients were asked to sign an additional declaration that they understood these terms, and that they agreed to pay the fee in recognition of them. The Ellis family were similarly careful in their palmistry publications, placing disclaimers at the bottom of every page of their personalized reports: "The above statements refer to possibilities and not to certainties." Their palmistry charts indicated "the past, present, and future," seemingly incriminating them in the fraudulence of fortune-telling. And yet they also clearly stated that "palmistry does not teach that things must *absolutely* occur, but that they *possibly* may unless steps are taken to hinder their occurrence . . . Although we all have certain good and evil tendencies, the making of our future is in a great measure in our own hands." People had free will. The responsibility was theirs.[30]

The many court cases across the country in which palmists and clairvoyants were charged under the Vagrancy Act, or (as we have seen) the Witchcraft Act, constituted a "crusade against palmists," according to the *Manchester Evening News*, meaning those like Ellis, who were part of the great occult-inspired boom, even as she tried to normalize her practice as "health." It was not, at core, a crusade against Roma palmists. In Blackpool, it was known that for many years the police had ignored or tolerated the busy trade on the sands and the South Shore. Until they didn't. When Romani women were brought before the courts, it was a different kind of trial, more about their occupation of space than their fortune-telling fraudulence. It was a device for moving Travellers on, in Blackpool's case, families whose encampment had been more or less secure for many generations. They paid rents (often very significant) for their tents and booths, and often had done so for years. There was no deceit, in any case, they often claimed, since it was plain that signs—advertisements—were always in front of the tents, "and the public knew that fortune telling was practiced. There was no cloaking."[31]

For the "occultists," by contrast, which included the likes of the Ellis family, the legal question of fraudulence was rather more direct. As in Keiro's case, was palmistry (and occasionally phrenology) known to be untrue by the defendant? And, dissimilarly to Romani women's cases, a suite of witnesses would routinely proclaim themselves to be members of the British Institute of Mental Science, the institution Ida Ellis had single-handedly created: Birmingham-based Professor Joyce (Henry Munro), Professor Virgo (Francis Henry Parr), and Mary Davies all authorized themselves as Fellows in court cases. Davies's husband, Professor Albert Edward Davies, was even a British Institute for Mental Science examiner. Alas, such efforts failed. In March 1913, Madame Ridley, a London-based Fellow of the British Institute for Mental Science (and gold medalist for mental clairvoyance), was imprisoned for ten weeks by the Marlborough Street magistrates. In 1916, Madame Palmer, secretary of the Southport branch of BIMS, was fined. And indeed, even being the founder of the BIMS proved insufficient for Ida Ellis, who was arrested and imprisoned in 1904.[32]

Yet there was a great difference between her prospects and entanglement with the Vagrancy Act and those of the descendants of the "Egyptians" from whose circumstance the statute derived in the first place. Ida Ellis, more than well housed and far from even approaching vagrancy, had accrued considerable wealth from buying Blackpool real estate. By 1901, the Ellis family had purchased and lived at 33 South Beach, Blackpool, their publishing and consulting base. By 1904, she and Albert were living at 81/82 Central Beach, Blackpool, where they remained until at least 1911. Ten years later, Albert, Ida, and Frank, as well as Albert's sister (Annie

Isabell Ellis), lived at Leighton House, Lytham Road, Blackpool, retaining ownership of the valuable Central Beach address. In short, in addition to all their other business endeavors, the Ellis family were active and successful real estate investors. Albert Ellis might have defended Roma families and their right to trade on the sands, but he was also collecting rents from many of them. It seems that he owned both the South Shore Fair Ground and the Colosseum Grounds, as he placed advertisements in the local press for the leasing of stalls. In 1910, Albert leased out spaces under verandas in Bank Hey Street and Adelaide Street, and in 1912, during a court case in which he was the defendant, he was reported to be the owner of the Railway Inn. By March 1919, Albert Ellis had purchased No. 5, Back Oddfellow Street, for £451. He was perhaps Blackpool's most successful real estate speculator.[33]

While the Ellis family accrued land and land wealth, in the end, Roma women intermittently arrested for fortune-telling were far more subject to the original spatial policing of the Vagrancy Act, and the history of Egyptian Acts lurking behind it. Eva Franklin, for example, was charged in 1909 for the fraudulence of fortune-telling. Objecting, she told the courts that she was one of the original inhabitants of the sandhills, and had been resident there for forty years (long preceding the Ellis interlopers). She was charged and punished, like Ida Ellis. But far from getting rich from real estate investments and property ownership, the Romani families of the South Shore encampment were evicted in a wholesale way, in 1910.[34]

For all their different relations to "vagrancy," to real estate, and to the commerce of palmistry, neither Ida Ellis nor Eva Franklin was deterred by arrest and imprisonment. Ellis continued to pursue palmistry and publishing in Blackpool and indeed across the country, including by invitation: in 1910 alone, at a "Grand Display" by tradesmen in Croydon, at a trades, foods, and industrial exhibition in Aylesbury, and at another exhibition in Lewisham in November that year. Her prosecution under a vagrancy law became more and more absurd. Ida and Albert were listed on the Blackpool electoral register for 1930, but they recorded their "abode" as 45 Lancaster Gate, Hyde Park, hardly a vagrant's address. By the time of Albert's death on October 3, 1934, they were living at Trewinnard Manor, St. Erth, Cornwall, and his effects were worth a grand £71,855 6s. 3d. Albert's brother, Frank, died in Blackpool in late 1939, while Ida died on February 7, 1940, at Trewinnard Manor, leaving effects worth £36,901 15s. 8d. Much of this fortune arose from the couple's real estate investments in Blackpool, but the initial capital derived from the reading of heads and hands. The phrenology, palmistry, and physiognomy picked up by the Ellis family and practiced across northern England show us what the hand could mean, and where it could reach, across differently classed and gendered contexts. This was all part of a working-class leisure, a seaside distraction, appealing to the same customers who frequented the Gipsy booths and tents. But with great success, the Ellises blended this with a vernacular industry and demand for health and hygiene.[35]

The Gypsy Lore Society that Princeton folklorist Charles Leland had founded in the 1880s was still going strong after Ida Ellis died, populated now by a postwar generation of "Gypsy scholars." One was Liverpool-based Dora Yates, who in 1957 recorded that "the Gypsies of the Northern Counties" held a privileged monopoly over fortune-telling to the visitors who still flocked to Blackpool. Morgiana Lee had been recognized for years as "Queen of the Gypsies," and now her daughter Cinderella (Relli) "is still reputed to be a miraculous seer into the future." For all their phenomenal success and wealth, many years later, the Ellis family were forgotten, while the Lee family of Blackpool thrived.[36]

SEVEN

The Medical Palmistry of Katharine St. Hill

In the late nineteenth century, palm readers often self-described their practice as a mixed science, art, and language. There was a trend—even among the most occult—to apply *logia* and *gnosis* suffixes to their specialist knowledge, authorizing themselves as learned; "chirognomy," "chirognomancy," "chirology," and more. Some *were* learned, though few with Edward Heron-Allen's depth. Others were simply accomplished at displaying the signs of wisdom. Around 1900, a sector of literate, urban palmists, both men and women, conspicuously placed themselves apart, both from fortune tellers and from those who adopted affectations in print, in décor, and in advertising. Invested in rational and professional display, palmistry modernizers policed the boundaries of their field, aiming to regulate their discipline and rework hand reading as systematically learnable, teachable, even examinable.

Katharine Harriette St. Hill (1855–1940) was one, creating the London-based Chirological Society of Great Britain in 1889. She was a tireless regulator, a boundary policer of professional palmistry, and an advocate, lobbyist even, for medical palmistry. "In the older writers," Katharine St. Hill distinguished herself, "the maxims of Palmistry are mingled with the canons of necromancy, astrology, spiritualism and superstition." What she taught and examined through the Chirological Society was so firmly scientific that she felt justification was barely required. What distinguished her own scientific palmistry from chiromancy or hermetically inspired hand reading was a medical thesis about the origin of the palm lines. Early on, she explained, astrologers understood that lines of the hand were maps of the course of the stars or planets that affected the owner at birth. But modern palmists held to a hypothesis that "the lines of the Hand are formed by the action of the brain on the nerves attached to the skin." Through Katharine St. Hill, palmistry became disenchanted at a personal as well as professional level, its magic and mystery swept aside and replaced with a set of testable observations, or so she wished and willed. Her will was considerable.[1]

Katharine St. Hill, *née* Richardson, was the daughter of a medical man. Born in 1855 in Nelson, New Zealand, her father was Ralph Richardson, MD, educated in Edinburgh. A man of some means, he inherited Greenfield Hall and estates in Holywell, Flintshire, Wales. Owing to his wife's health—Katharine's mother, Mary Louisa *née* Seymour—they moved to New Zealand for five years, where he was elected member of the colony's Legislative Assembly, even as that Assembly, and Britain, was at war with Māori. They had eight children, four sons and four daughters, all of whom returned to England around 1860. Katharine and two sisters were still living with their father by the time the 1881 census took place. She married Ashton St. Hill in 1884, a marriage that only lasted until 1886, although she retained his name. Katharine St. Hill died in Berkshire in 1940, then living with her sister, the entrepreneurial Octavia Richardson, who early in the century had co-owned the successful Westminster School of Business Training for Women with Cecil Gradwell, born Cecilia Mary Ann Gradwell. This particular palmist was surrounded by familial medicine on the one hand—including two of her brothers—and bold business initiatives on the other. She decided to put palmistry on a modern path, and in her hands, the new chirology was reorganized by a "new woman"—single, professional, independent. Counterintuitively, if Cheiro and Keiro embraced the often-feminized *fin de siècle* occult, here a woman from a medical family dedicated her life to the modernization of palmistry, in all technical ways seeing to its professionalization and secularization, away from magic, wonder, and secrets.[2]

That disenchantment was itself part of Katharine St. Hill's life journey. Three suggestive portraits display the continuum of possibilities for female palmists in the decades over the turn of the century. The first is of Katharine St. Hill early in her career, in 1894, wearing signs of the occult: a cape or a cloak, elaborate armlets and bracelets, a turban-like headdress with a crescent-shaped barrette (fig. 7.1). These were all signs that connected her to the world of Cheiro and Keiro. She was then active on a circuit of bazaars and charity palm readings, which traded on associations that could be read as "oriental," as "occult," or as "Gipsy." As late as 1909, Katharine St. Hill was busy organizing a faux "Gipsy encampment" for an animal charity fair in Regent's Park.[3] The second portrait is a mature Katharine St. Hill in 1927 (fig. 7.2). The frontispiece to her *Book of the Hand: A Complete Grammar of Palmistry for the Study of Hands on a Scientific Basis* shows the woman who identified as "author" on her census returns. All signs of the occult have gone, and the confident medical palmistry reformer has emerged. That same year, 1927, photographer Fred Shaw of the Gypsy Lore Society caught a piercing portrait of Cinderella or Relli Heron, a "Gypsy fortune teller." Then in Cambridge, she was a member of the Lee family, and may have been traveling from Blackpool (fig. 7.3). Over time, Katharine St. Hill, the great palmistry modernizer, differentiated herself both from the occult and from Roma fortune-telling. She wanted palmistry to be, at the very least, and like herself, a sister to medicine. Within her lifetime, it was.

PROFESSIONALIZING CHIROLOGY

Katharine St. Hill sought to professionalize palmistry in a series of organizational moves that contemporary sociologist Max Weber might easily have deployed as a perfect case study for his emerging theories of rationality in economy and society. Put another way, St. Hill might have borrowed his work as a template for her textbook

7.1 An occult Katharine St. Hill, *Hearth and Home*, December 27, 1894.

7.2 Katharine St. Hill, scientific palmist, eschewed both her earlier occult identity and Gipsy fortune-telling. Katharine St. Hill, frontispiece to *The Book of the Hand: A Complete Grammar of Palmistry for the Study of Hands on a Scientific Basis* (Rider, 1927). Copy in author's possession.

7.3 Another reader of hands, taken in Cambridge the same year as St. Hill's studio portrait, 1927. "Gypsy fortune teller"—Relli Heron [al. Young] daughter of Oliver Lee—Cinderella—June 26, 1927. University of Liverpool Special Collections & Archives, Fred Shaw Photograph Collection, SMGC Shaw P.235. Courtesy the University of Liverpool Special Collections & Archives.

efforts. First, she formed an association in 1889, the Chirological Society of Great Britain, a highly feminized, but not solely female association. She taught scientific hand reading, examined students, and admitted the diligent and talented as Fellows, who could now promote themselves as FCS, Fellow of the Chirological Society. She attended protest meetings organized by the Occultists' Defense League and placed advertisements in newspapers requesting donations to support the defendants. Yet over time, she shifted from an occult to a medical domain, declaring to the world that she was undoing magic, forgoing fortune-telling, banishing "intuition," and standardizing all chirological practice. The society was established for three reasons, she declared: "First for the purpose of raising the study of the hand to the level of scientific research; Secondly for promoting the study of Palmistry in all its branches; Thirdly as a safeguard to the public against charlatans and impostors."[4]

The society met twice a month at their offices, 12 St. Stephen's Mansions, Westminster, between 8 and 10 p.m. St. Hill was the president, supported by an all-woman council. The first meeting of each month was for members only, and visitors were admitted to the second on presentation of a Fellows' card, confirming their good character. Already there was a policing of who was in and who was out. There were regulations for the conduct of members, too. One reporter designated it, without irony, as "the society for the certificated professors of palmistry." Gathering 10 shillings' annual subscription from each member, the Chirological Society set to work. This was organizational, reputational, educational, but also practical business: members of the society could access the latest books in London's new center for chirological knowledge or buy at wholesale price, from the president, the best plaster and wax for casting purposes.[5]

Next, Katharine St. Hill established a journal that recorded the minutes of her society's meetings, gathered articles and insights from its Fellows, and advertised fulsomely from the lively publishing industry that surrounded palmistry. Often enough, these books were far too esoteric for her own scientific taste, but selling advertising copy clearly brought in necessary revenue, covering costs for the society and the journal itself. *The Palmist and Chirological Review* ran from 1892 to 1898, published by the Roxburghe Press. And with the decline of that press, *The Palmist's Review* was co-edited with Ethel Marsh-Stiles from 1899 to 1901.

St. Hill also published a textbook, *The Grammar of Palmistry* (1888), from which she could teach and examine what she hoped would become standardized knowledge. By 1893, some 12,000 copies had been sold. *The Grammar of Palmistry* displayed how the practice was clear enough to be set out systematically, and then taught. She did just that, consolidating this aspirational professionalism by lecturing, including in hospitals, and then examining palmistry students, setting criteria that would differentiate scientific palmistry from "rank imposture." All of St. Hill's books were technical, not sensational, used and treated as texts for teaching and learning. One copy of her *Book of the Hand*, for example, belonged to Lady Atkinson of 74 Oakwood Court, Kensington, a diligent student whose annotations on the teacher's diagrams run carefully throughout.[6]

The title of her textbook, *The Grammar of Palmistry*, is important. A "grammar" establishes the rules of its language, not just to describe but also to systematize. Just prior to St. Hill's grammar of palmistry, for example, the orientalist Edward Henry Palmer produced *The Arabic Manual: Comprising a condensed grammar* (1881). And just after it, the biometrician Karl Pearson published his *Grammar of Science* (1892), and British linguist Henry Sweet his *New English Grammar: Logical and Historical* (1892). St. Hill's grammar invited the consideration of palmistry as a language and herself as a kind of linguist. A good palmist needed to know and think through the meaning of words, she signaled early, in a section titled "Language." The responsibility of giving character diagnoses demanded it. If you say your subject is "conceited," what exactly do you mean? Nature does not mix her terms, or "muddle [her] meaning," even if we do. Rather, Nature provides, in the hand, a "code" that is precise. Her most fundamental rule: "*Each sign has its own meaning, and that alone.*" Palmistry was a language to be explained as it had been in so many early modern texts. If John Bulwer had invented sign language all those centuries before, for

Katharine St. Hill, palmistry still "reduces itself to the likeness of a language." Indeed, her beautifully designed and pressed cover for *Hands of Celebrities* visually referenced Bulwer's hand alphabet drawn in his *Chirologia* (fig. 7.4). In writing a *grammar*, St. Hill was approaching the rules that governed a language of signs and signifiers, of palm lines and their meaning. All palmists, as we have seen, were semioticians of sorts, and so, in a manner of speaking, were all modern doctors.[7]

At the same time as palmist Katharine St. Hill, linguist Henry Sweet, and orientalist Edward Palmer were publishing their very different grammars, Swiss linguist Ferdinand de Saussure (1857–1913) was at work developing his own influential principles for language as a system of signs. His innovation was to set out the bilateral signifier and signified: the sign's form on the one hand (a word, or Katharine St. Hill might say, a line on the palm) and the meaning of that form on the other. Saussure generally thought a sign (a word) was randomly connected to its meaning, part of a long humanist insistence on the arbitrariness of words, a counter to the scholastic idea that language was given by God. Palmists, however, often thought the sign on the palm was physiologically, or mentally, or chemically, or spiritually, or astrologically *connected to* or even caused by something in the mind, or by a bodily event in the past, or by some pathology elsewhere in or on the body, or by the heavens. In other words, the signifier and the signified were not arbitrarily connected. A cross on the mount of Luna means insanity in the family, for example, but not randomly or arbitrarily. This was more like a physician's reading of a clinical sign with a cause, even if that cause was unknown or hidden or concealed (occult in its original use, *occulere*, "covered over").[8]

St. Hill taught not just methods of palmistry, but also methods to *learn* palmistry. She instructed thus: read one hand sign from her grammar at a time, learn it by heart, then take a large manuscript and write out what you have learned on one side. Leave that alone for a week while you observe every hand that comes your way—at least a dozen—drawing them, or taking an impression. Then return to your large workbook, paste in your drawings, writing your own "delineation," that phrenological word again. Then

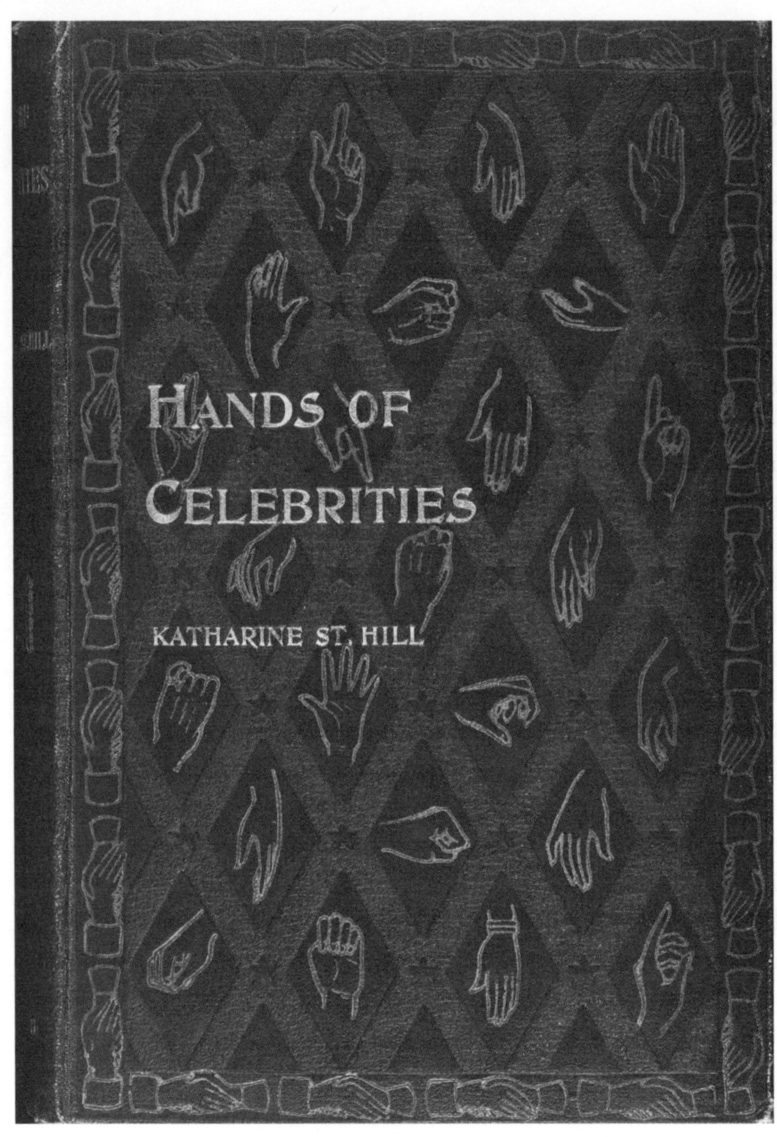

7.4 John Bulwer's *Chirologia* referenced in the design of Katharine St. Hill, *Hands of Celebrities: or, studies in palmistry* (Roxburghe Press, 1896). Image courtesy of the John Rylands Research Institute and Library, University of Manchester.

take your next lesson, slowly building knowledge. Studying the hand itself is a form of self-improvement if done properly, honing "observation, analysis, expression, method, and judgment." Palmistry was to be learned sign by sign, rather in the manner of Cocles's 169 woodcut prints in his *Physiognomiae et chiromantiae compendium* of 1653.[9]

It was an uphill battle to modernize palmistry, methodologically and substantively. Yet this was not necessarily because a scientific world resisted. We have seen the great anatomical interest in the hand that Edward Heron-Allen had extended, and indeed US anatomists were just then inventing "dermatoglyphics," as we shall see. But it was different for a nonmedical woman to assert scientific palmistry. The problem for Katharine St. Hill was the Gipsy association and the investment so many palmists had in occult and ancient wisdoms that might have been "natural philosophy" in earlier centuries, but in the modern world more often than not signaled "magic" *in opposition to* "science."

Katharine St. Hill gradually distanced herself from the occult over a long lifetime of reading hands. In the 1880s, she studied spiritualism, "but it did not satisfy me. I could not see the 'reason why,'" she later explained, or tried to. And her early efforts to modernize palmistry were all made from an institutional and community base among London's esoteric professionals, sharing a business address. She was part of an occult infrastructure that included the hermetic-oriented Roxburghe Press, which revived Richard Saunders in *Moles or Birthmarks, and their Significance to Man and Woman* (1894). The publisher who had produced her *Grammar of Palmistry* was also busy with an esoteric list including *The Angelic Pilgrim: An epical history of the Chaldee Empire* (1882), *Confessions of an English hachish-eater* (1884), *Phallicism celestial and terrestrial, heathen and Christian, etc.* (1884), *The Real History of the Rosicrucians* (1887), and *The Light of Egypt, or, the Science of the Soul and the Stars* (1889), all published by George Redway.[10]

The co-editor of *The Palmist's Review* was Ethel Marsh-Stiles, who herself published a suite of wildly unscientific chiromancies and astrologies from their shared office at 12 St. Stephen's Mansions, Westminster, including *Chaldean Astrology Up to Date: How to*

Cast the Horoscope and Read the Future in the Stars. Ethel Marsh-Stiles, it turns out, was also "O. Hashnu Hara," who wrote on Psychoscopy, Practical Psychometry, Practical Yoga and Persian Magic, Mental Alchemy, and Concentration. She did so as Secretary for England of the Order of the White Rose, and the College of Psychical Sciences and Unfoldment, Syracuse, New York. Under her own name, she also published any number of other journals, including the *Occult Literary News and Review* and *The Predictionist*. Another, *Wings of Truth*, began with a quote from occult-loving, opium-induced Percy Bysshe Shelley, hardly the scientific association sought by Katharine St. Hill:

> Higher still and higher
> From the earth thou springest
> Like a cloud of fire;
> The blue deep thou wingest,
> And singing still dost soar, and soaring ever singest.[11]

"These words so well express the object of the new magazine," declared O. Hashnu Hara, her journal presenting practical "divinity, psycho-therapeutics (mental healing), psychometry, clairvoyance and kindred powers of the spirit." Its first article was "Early Lessons in Clairvoyance," and one of its last "Practical Lessons in Theosophy." This was an epistemological and pedagogical world from which Katharine St. Hill was desperately retreating, and although O. Hashnu Hara might claim that theosophy "now holds a most envious position as the mediator between the various great religions of the world on the one hand and scientific advance on the other," St. Hill held to a different measure of science altogether for the future of palmistry. Over time, she was bedeviled by colleagues clinging to the revival of hermetic traditions, to occult chiromancy, and to "Eastern" knowledges. In her bid for modernity, she increasingly set aside practices like "thought reading" or "telepathy" that enchanted many of her body-and-mind-reading contemporaries, not least Cheiro, Keiro, and O. Hashnu Hara.[12]

This was a personal journey of disenchantment. For a start, palmistry needed to dispense with the idea that it required "intuition" as some kind of special and mysterious gift. She scoffed at the very idea that anything but uppercase Reason did or should underwrite modern palmistry. "Intuition" was a vastly overvalued attribute (a graceless comment, perhaps, as the Roxburghe Press had just published *Scientific and Intuitional Palmistry*, by J. J. Spark). As Dr. Johnson told us all long ago, she licensed herself, intuition was "knowledge without reason, without argument, and without observation." And then she demonstrated her own reasoning powers: "It follows that if you assert that intuition is necessary to the practice of Palmistry, you must abandon the claim of Chirology to be a science." Moreover, such practitioners should not call themselves palmists at all: "It is a misnomer." She came to note, and distinguish between, the "ordinary amateur Palmist" and the professionals. Organizationally, she wanted palmistry to duplicate systems of regulation like those that safeguarded medical practice: "We are very similar in our rules to the medical profession, take our degrees in the same way, and everything be perfectly open and above board."[13]

Amateurs might consider palmistry to be about the future, but this is the first thing a professional dispenses with, she declared. One journalist rightly described the society as heralding "the extinction of the gipsy fortune-teller," and St. Hill's ambition was just that, parading the authority of her class and ethnicity and education, as far distant from Cinderella Heron/Lee as she could. These were all de-orientalizing maneuvers. Yet St. Hill was defining herself not only against Gipsy practitioners, but also against her more proximate competitors in the affluent West End, not least, Cheiro. "Those *dilettante* young gentlemen" brought "a grave and abstruse science into disrepute." Indeed, she charged them with inadequate preparation of the high office, practicing without the diploma and two examinations required for fellowship of the Chirological Society. Cheiro was everything Katharine St. Hill was not. Real palmists, she declared, "simply hate humbug; we never advertise nor allow a nom de plume." And while he was an entrepreneurial sole-player, slipping between London, Paris, and New York, analyzing the palms of prime ministers, shahs, and actors, St. Hill was an organizational leader, efficiently bringing together sensible like-minded colleagues into a society, diligently teaching and examining London's palmistry aspirants, editing her journal, chairing fortnightly meetings, and publishing minutes that could not be more procedurally correct. At Keiro's trial, she was asked: "Who constituted you an examiner of palmists?" To which she replied, defiantly: "I appointed myself. I was the first to start this in England." This was all as proactive and self-initiated as anything Ida Ellis had achieved, but it unfolded in a different class and cultural geography, with far more substantial professionalism.[14]

Deduction, reason, research, and observation were precisely what Katharine St. Hill sought to attach to palmistry; the inversion of the epistemological association of Reason with masculinity and Intuition with femininity. It is thus counterintuitive that male palmists like Cheiro and Keiro traded on the occult and the "Eastern," while a female palmist like Katharine St. Hill was invested in a scientific method of sorts, rejecting orientalist and folk traditions. This gender reversal made her adamant. She had something to prove, both as a palmist *and* as a rational, modern woman: not an easy combination. Precisely because she was professionalizing a feminized world of practitioners and members, she had to overstate her claims for Rationality and against Intuition. She taught her students to take themselves and their knowledge seriously as part of their professional conduct: "Speak authoritatively, do not allow yourself to be contradicted, or to be joked with; then you will find that your manner has its effect, and that you and your Science will be treated with due respect and consideration." She tended to address her students with a masculine pronoun, although they were likely to be women: of forty candidates to her "fellowship degree" in 1898, six were men.[15]

The gendering of palmistry and of authority was taken up explicitly in 1896 by Charles Rideal, enthusiastic supporter of St. Hill and silent co-editor of *The Palmist and Chirological Review*. A relatively obscure publisher and author of idiosyncratic books on Dickens and on *True Detective Stories*, he wrote a long preface to her *Hands of Celebrities* (1896), defending palmistry from its detractors in what we would now call gendered terms. It is men, seldom women, he made a point of saying, who were crude critics of palmistry and who "express themselves in this unpleasantly aggressive manner." This was just as the arrest of palmists and various trials were unfolding across London and beyond. "The Science becomes banned because of a want of perception on the part of those who cannot distinguish between Science and Art-fulness, who cannot appreciate the workings of Nature in her fuller details." In a diminished Paracelsian doctrine of signatures, Rideal likened reading the signs of the body to reading the signs of landscape or of flora, a kind of geology or botany: "the runnings of geological strata, the fronds of the fern, or the lines of the leaf." Rideal also set chirology apart from "spurious" Physiognomy, Phrenology, Astrologysm, and Temperament. Indeed, he was presenting a manifesto: "I demand that Chirology shall be placed on the basis and footing of a representative science." At the same time, however, both Rideal and St. Hill were perfectly happy for their photographs and signatures to be reproduced in *The Palmist* in July 1896, accompanying an article on the link between graphology and physiognomy. The graphologist "delineated" St. Hill's head and hand: "The forehead is decidedly wide and thus indicative of an analytical cast of mind; between the eyebrows the single perpendicular line is to be discerned indicating attention to minutiae." And her "thorough, painstaking reputation" was borne out by "the Liliputian character of the hand-writing." Just what one would want in a society president and the editor of a society's journal, everything shows "the habitual love of system and classification." In 1896, St. Hill's self-consciously modern intervention was an elevated, middle-class, respectable, and metropolitan version of Ida Ellis's self-published *Know Thyself*.[16]

THE PALMISTRY OF KATHARINE ST. HILL

For self-consciously professionalizing palmists, the authority of antiquity was not always helpful. Effective communication of modernity was hardly well served by treatises from late medieval Catholic monks or rabbis on kabbalistic texts, by ancient Vedic manuscripts, or by verses from Job. Like Edward Heron-Allen, she lent her weight behind D'Arpentigny and his physiognomic *La Chirognomonie* (1843), more than kabbalistic Desbarrolles. And yet St. Hill also appreciated that palmistry was as "old as the world," as she put it. "No one knows the age of one, or the beginning of another." Plato and Aristotle studied signs on the hands, the latter acquiring his knowledge from Egypt and Egyptians. The history of the hand was once again told as a history of the book, of the written Word. Of two ancient Egyptian papyri—the oldest manuscripts the world knows—the second, she claimed, was a "treatise on hand-reading." And while the first *printed* book was (quite rightly, she says) the Bible and the second *Morte d'Arthur*, the third was a book on Chirology. But then the Church turned against the knowledge, and students of the ancient Sciences were driven out, and it was practiced solely by "vagabonds and gipsies." This was a familiar narrative. There the science lay, waiting for its return and rightful recognition. The story was different—better in her view—in the Orient, in "the unchanging East." She meant that the study of hands had been tested there long ago, and was both valued and enduring. "It is taught in the schools like Language and Grammar." And so, in the East, chirology or palmistry held its rightful position, its exponents treated "with consideration and respect . . . and there it proceeds with calm and dignity on its way."[17]

Following D'Arpentigny, good palmistry according to St. Hill involved the capacity to read far more than palmar lines. It was a practice about the whole hand and for all senses: Ida Ellis's remote palmistry would have been an impossibility. Never touch the hands at the beginning of a reading, St. Hill instructed; rather, watch their natural fall on the body, including the fingers. Then study the back of the hand, but without touching it, looking at the nails and knots. Then study the front, again without contact, observing the fall of the thumb and fingers. Do this all without speaking, and *then* "touch the palm once gently." Thereafter one might handle them freely—"now the embargo against touching the hands is over"—but don't grab and pinch. Rather, "you need only use the tips of your fingers to press the palm, the thumb should never be used." This was about the palmist's hand as well as the client's, about tactile qualities. But it was also about speech.

Teaching the craft of consultation, the student must learn how and when to be silent, how and when to speak. "The Student should learn not to speak when an idea strikes him: he must be able to put it by, in the back of his mind . . . and to bring it out again in its proper turn." The seer was turning into the therapist.[18]

For Katharine St. Hill, as for palmists preceding and succeeding her, the fingers were still named astrologically: Jupiter, Saturn, Apollo, and Mercury. The relative length and shape of parts of those fingers (phalanges) indicated anything from love of rule and pride, to want of thought and contemplation, to prudence, to love of agriculture. The first phalange indicates reason, the second will, the third love. If the finger of Mercury (the little finger) is just pointed enough, it indicates "eloquence, tact, diplomacy." But if it is too pointed, it signals "ruse, finesse, trickery." The mounts on the palm were to be analyzed too. An excessive Mount of Venus signaled inconstancy, effrontery, coquetry, vanity, idleness, and sensuality. Absence signaled coldness, selfishness, and want of tenderness. She also retained and retold an older reading of "temperaments" attached to astrologically designated fingers. "The temperaments that are most prone to insanity, are those of Saturn when in excess, and of Luna when deprived of a good Mount of its own . . . Venus comes next: she is often hysterical, and liable to the gentler forms of mind disease." On the other hand, "a strong thumb, a good Mars, and a high Mercury, these never go insane." Then she observes the lines themselves: pale or strong, narrow or wide, broken, clustered, crossed with small lines, forked. The lines are to be read not just for their length but also for their depth and color, for their branches, for their direction. Time, or the age at which care should be taken, is told by the hand along the Line of Life and the Line of Head (both counted downward) and the Line of Fate (counted upward). Palm lines told time.[19]

MEDICAL PALMISTRY

The Ellis family had certainly traded on health, folding phrenology and palmistry together. St. Hill, however, set down for more serious readers than those of the cheap Ellis tracts a set of principles and observations in which the hand and physical and mental health were correlated. She saw "medical palmistry" as the future in part because she had long been proximate to medical practice. Her father, born in 1812, was educated in Edinburgh in the 1830s, as it so happens just when Edinburgh phrenology was at its height. Two physician-brothers practiced medicine over the same decades in which Katharine St. Hill herself professed expertise on bodies and minds: Dr. Gilbert Richardson practiced in Putney, and Dr. John Billingsley Richardson, heir to the Holywell estate in northeast Wales, practiced in Torquay, Devon. We know most about the latter, born in 1850 and educated at Edinburgh University like his father, graduating in 1874. He then studied at St. George's Hospital, London, and entered the College of Physicians. In 1875 he was appointed house surgeon to the Torbay Hospital in Torquay, later serving as mayor of the seaside town, during which time he opened, ostentatiously, the Princess Pier. In 1888, the year his sister published her first *Grammar*, he obtained a Doctor of Medicine of Edinburgh University, his thesis on the etiology and treatment of tubercular consumption.[20]

Not only was Katharine St. Hill's family medical, they were also conspicuously healthy. And they gloated about it. It was Dr. John Billingsley Richardson who wrote their father's obituary in 1898, most of which was about the magnificent athletic prowess of the sons. One rowed for Trinity Hall, Cambridge, winning several times at Henley. The next brother, John (composer of the elaborate and ridiculously immodest obituary), was also "a fine athlete," if he did say so himself. He won prizes that ranged from the mile distance "beating the English champion on two occasions at Exmouth," to the Torquay steeplechase, and he was recipient of "no less than nineteen presentation bats for his cricketfield performance." At least he did mention his sister Katharine, president of the Palmistry Society, and, name-dropping, noted that her books included the hand of "our neighbour" the Right Hon. W. E. Gladstone: Hawarden Castle was a few miles' distance from Holywell. Indeed, just like Cheiro, Katharine St. Hill took an impression of the prime-ministerial hand, read his health and character, and published it.[21]

It is little wonder that St. Hill described palmistry as an aspirational sister to medicine, since her family embodied just that. A good palmist, she declared, had the capacity for quick observation as well as "the reason, the analysis, and the tact that go to make up the good doctor." If stock-in-trade palmists worked in "mere affairs of heart," her medical palmistry required research, and developed as a science almost as valuable to humanity as medicine. Indeed, she entreated students carefully to consider their great responsibility just like "a moral medical man in the diagnosis of your case."[22]

Katharine St. Hill laid out the hand signs for the quintessential diseases of her time and place: diphtheria, consumption, influenza, rheumatism, gout, typhus, and typhoid. They all had their typical marks: typhoid signaled by a blue dent near the line of fate; influenza by a blue dent on the head line; typhus a big black dot on the head line, just under the finger of Saturn. Organs had their correlation on the hand. The lungs coincided with the Mount of Jupiter, just under the first finger, for example. And so, consumption is signaled by the heart line rising over that mount and becoming "feathered," or "islanded," or "blurred." The throat was "assigned" to the life line, under the junction of the fingers of Jupiter and Saturn, and there we might read signs of diphtheria, for example. The spine, back, and limbs also belong to the finger of Saturn. The sign of eye disease was located on the heart line, "exactly under the centre of the Mount of Apollo." She thought that the most important line in the hand with respect to health, after the Line of Life, was the Line of Heart. The length did not matter, but it should be narrow and deep, unbroken, and uncrossed.[23]

Then there was the Line of Hepatica. Sometimes this was called the Line of Health, but Hepatica is more accurate, she thought, because this line only told the state of health of the liver and other internal organs, nothing more general or constitutional. The absence of any Line of Hepatica is a good thing, indeed "the best sign of a perfect constitution." On occasion, men and women had different lines, and they changed over a lifetime. For example, the Line of Hepatica was deepened in mothers, just before and after birth, and sometimes a temporary star appeared. Jaundice is shown, by contrast, "by a long double island of the line." Sometimes, differential diagnosis was difficult for her, as for physicians. She once mistook the combination of a blue dent in the head line and a broken loop in the Line of Hepatica to be jaundice, but she stood corrected: it was yellow fever.[24]

St. Hill's work did not just teach what correlated from palm to health, but the reverse. For example, she explained that in several cases of "valvular disease" (of the heart) she found retrospectively the heart line broken in two under Apollo. In other words, she reasoned backward from the condition to see what was happening in the hand, and whether there was a pattern across cases. The purpose was diagnostic, and although Katharine St. Hill largely disavowed fortune-telling, the future might occasionally be told as a cautious prognosis. "We can diagnose a case just like a doctor, and tell a client things which may happen under certain conditions." Her palmistry was still a humoral medicine of a sort, assessing the balance of heat and cold, moisture and dryness, all of which could be felt or seen: hot dry skin signaled liability to fever; hot damp skin a tendency to consumption; cold damp skin liability to liver complaint. Signs of consumption, for example, were skin damp to the touch, with nails observed as curved, thin, and red, with large white half moons. Gout also changed the color of the hand, presenting as yellow, shiny, and soft "with a feeling of satin when touched." Occasionally she could see a dent in a hand, confirmed by her own touch: "There was a perfect little blue scoop, into which I could put the tip of my little finger." This signaled bad previous bouts of influenza. And St. Hill intermittently "percussed" the hand and wrist, a classic medical technique.[25]

Such diagnostic measures and methods, while seemingly antique, were still common for a whole range of regular practitioners in the early twentieth century, from physicians to nurses to midwives. And the Hippocratic triad "airs, waters, places" that directly affected human constitutions was still very much alive in her own period, driving the late Victorian and Edwardian concern for weather, wind, altitude, humidity, all of which could be life draining or life giving. If a client had soft and damp hands and a sallow complexion, she explained, they are liable to infection and should be warned not to take unnecessary risks, "or to live in unhealthy climates." Like all palmists, Katharine St. Hill was necessarily an observer of skin and its color. People whose lines were inclined to redness were "liable to fever and catch all kinds of infection easily." She thought that fair people did not get "agues and malarial fever" as easily as do darker people. Fair people stand heat best—but only if they are not fat—"while rather dark people stand cold." This was an unusual statement for the period, when the physiology of "race," perhaps especially including "whiteness" and its unsuitability for the tropics, was a specific medical topic of research. Most of her subjects *were* white, however, so she was usually speaking of complexion in "white" people. But not always, and occasionally she would proclaim on "dark peoples' races." The line of fate, St. Hill explained, was often understood by French chirologists to be apparent in the hands of whites only. However, from her own experience it certainly appears in "the hands both of Maoris and of Hindoos." In the decades around 1900, seeking physical and psychical difference between groups of humans was the scientific enterprise *du jour*, as biologists and anatomists bedded down physical anthropology. And so, in some ways, the more palmists entered that particular discussion, the more they displayed their scientific credentials.[26]

Katharine St. Hill's accounts of how to recognize a healthy or unhealthy hand could have come from any medical manual on basic clinical observation, or from countless domestic hygiene books that taught women to draw lessons from skin: its moisture, dryness, color, and texture. A healthy hand shows a "firm, rather hard palm, clear skin, inclined to be dry rather than damp, large bright nails." The unhealthy hand might show crooked or twisted fingers, "a flabby palm, a soft damp skin, brittle or too hard nails." Well before one looked at palm creases, cruder signs should be read: something as ordinary as bitten nails, for example. French teachers "say that children who bite their nails are idle and of a *caractère difficile*." But she explained that both the character and the biting were a manifestation of internal ill health. And then came her specialist palmist's knowledge of the lines: a healthy hand would show few lines in their normal place, clear, and unbroken. The fewer the lines, the less troubled the subject, "either physically or mentally." The unhealthy hand might show many small lines across the palm or thickly concentrated in one area. They might be broken or discolored. Fading lines were especially significant since the depth of palm lines signaled "vitality." Lines thus start to disappear "as vitality and life itself fades out," a sure sign of the end approaching.[27]

Katharine St. Hill comprehended her own work as modest medical research, a process of correlation. She would walk along the wards of many hospitals, for example, and look for the mark of a broken leg in each hand of the occupant. Once a dozen or so with a similar mark are found, then there is something to go on, research-wise. She kept a set of wax-cast hands from children in hospital with bronchitis. From them, she demonstrated that "there is the same line in each hand indicating the illness." In light of this work, she claimed that surgeons at Manchester Hospital asked for her knowledge of the signs of diphtheria at one point. She looked at a number of children's palms, her diagnosis was verified, and "they very kindly and honourably acknowledged our reading as correct." At a children's asylum, fourteen subjects showed, to her surprise, perfect head lines, but highly abnormal heart lines. She asked the medical pupil to test them all for heart conditions and found that they did in fact have heart disease: "So I came to the conclusion, based on Cheirology alone, that in these cases the idiots had their brains in a perfect condition, but that the disease of the heart was so acute that the blood never nourished the brain at all, and that it was consequently inert, and hence idiotic." Therein, she thought, lay a possible prevention, curing the heart disease and giving good blood.[28]

St. Hill clearly researched old medical materials too. When considering signs for leprosy, she cited a seventeenth-century text by the French physician to Louis XIV:

> Leprosy has its source in the liver, and one of the first signs by which it is known appears on the index finger; the muscles

> which control the motions of this finger wither and dry up in the space between it and the thumb. The flesh wastes away so that only skin and bone remain. Now this could not happen if there were not some analogy and secret connexion between the liver and this part.[29]

Both the medicine of seventeenth-century France *and* the medicine of 1920s England bore some resemblance to the kind of medical palmistry she practiced and taught. They were each sensory; they looked for a connection between one part of the body and another; they sought signs of an illness that may manifest on a separate part of the body altogether to the location of that illness, or may be general, or constitutional. Here, leprosy literally manifested first on the index finger. This was presented rather more as a symptom of the disease than a sign. And St. Hill added that she had been told that a prominent barrister's first intimation of leprosy was when he dropped hot sealing wax on his first finger, but never felt it burn. She gathered information and evidence, case by case, as medical doctors did.[30]

A decent enough researcher, St. Hill's commitment to evidence was considerable. If she could not locate an illness in a sign of the hand, she said so. This was the case with leprosy. Nor did she know how cancer manifested. It did not seem to have a hand sign. She surmised that that was because cancer itself had no locality: "It attacks all parts of the body." But this also meant that it could affect all and any parts of the hands. And so, masses of lines broken and crossed *might* signal cancer, and this was her hypothesis as she looked at fourteen drawings before her of subjects who had succumbed to it. Instead of fading lines, apparent in many fatal illnesses, cancer intensified the lines. "The vitality seems to be intense right up to the end," possibly because cancer attacked those with strong, not weak constitutions.[31]

St. Hill understood palm lines to change over time, unlike Galton's fingerprints, whose value rested on lifelong singularity, unchanging patterns. Life lines especially might fade with illness, and if they strengthen over time, this signals that the illness itself has had a positive effect. She thought that newborns had their mother's lines at birth, but that these disappear usually at three weeks, gradually to be replaced by their own lines, individuated. Babies are born with many fine lines, "almost always a replica of the lines of the hands of one of the parents." But then they change, as the baby "takes notice" or becomes its own being. One newborn she saw at a fortnight old had its parents' lines replicated, but never developed its own. By seven months, all the lines had disappeared out of both hands, and the baby died. Its own palm lines failed to appear, which she took as a medical sign of failure to thrive.[32]

St. Hill also thought that palm lines altered over adult lives, according to imperfections of the brain due to accident, or illness, or idiocy. If a brain's power was damaged and then restored, palm lines would fade and then grow again. "Madness will alter the lines of hand," for example. But death would not, she claimed. "You can read a dead man's history in his hand," and she had indeed studied the hands of corpses, opportunities afforded by her medical father and brothers. Yet some illnesses would mark a hand permanently, signs not that the person *will get* a disease, but that they have had it: "Diphtheria, typhoid and spinal disease also leave a mark for life." Or, in another example, after the removal of an internal tumor, the Line of Hepatica would be discolored, or darkened. Indeed, surgery itself would result in a "star of shock," although that typically faded out over time. A bad fever would leave a blue dent, sometimes a black spot, often on the head line. If the recovery is good, but also if "the memory is not vivid," one would expect it to fade or disappear, but in a severe fever that left a permanent trace on the constitution generally, it would remain visible, or would revive.[33]

This belief that palm lines changed aligned with St. Hill's philosophy on fate as told by the hand. Quite simply: "Life is not a fixed fatality." She taught that breaks in the life line were not necessarily fatal signs, but they did signal an age at which warnings should be given about special care for health. Still, the line of life broken in both hands was serious. And yet, no matter what the mark or sign, the length and quality of life could be influenced by healthy living and by good medical intervention. Indeed, it is to "the skill and knowledge of the Medical Profession that the Subject owes his longer life beyond the period allotted in the first instance by Nature." If people died earlier than a palm sign suggested, it could easily be because their care of their own bodies was insufficient. They might have recklessly exposed themselves to "some unnatural infection," for example, probably meaning sexually transmitted diseases, and should wear the consequences. She even thought this should be categorized as suicide, since a subject had actively allowed his own constitution to be lowered. It sounds like a self-serving *ex post facto* rationalization for a life that doesn't fit a line. But given St. Hill's equivalent concern for medicine and palmistry, is this not also a health warning, even an attempt at crude health education and responsibilized self-care? For Katharine St. Hill, a good hand reader's duty lay in warning clients of their risks. We know we can actively affect our health future, she was saying, in large part because of medical advancement. It was everyone's responsibility to avail themselves of modern medicine, especially when forearmed with information from the hand.[34]

Palmists' vocabulary of "fate" was increasingly folded into a more scientific language of heredity. St. Hill instructed that the left hand in particular held the signs of "hereditary tendency." This was why hands must be read comparatively and jointly, and in both sexes, countering earlier conventions in which the right hand was read in men, and the left in women. If the left hand showed inherited susceptibility, the right was "more personal," by which she meant more individualized. Two hands together should be consulted for asthma, for example; the left hand might show a family history, while the signs in the right hand show whether the case will improve as time goes on. It was a handedness that ascribed a past and a future to the left and the right. The life line in the left hand, to take another example, is "a record of the Subject's ancestry," or, she explained, "the span of life as intended by Nature in the first instance, the general result of the constitution and temperament." But in the right hand, she tells us, "we have the actual facts."[35]

In *Medical Palmistry*, St. Hill aimed to lay out this knowledge not only for her own students, but also for "other investigators in human physiology." She had herself lectured on palmistry at various infirmaries and asked for "courteous consideration" from the medical profession at the very least. Perhaps more could engage directly? She indicated that doctors already attended her classes, but they often only used their knowledge "*sub rosa*," lest the "charlatanism" damage their career. She wished that a clever and cultured medical man could have written *Medical Palmistry*, instead of herself with only "half knowledge."[36]

PALMISTRY IN THE ASYLUMS

Unlike Edward Heron-Allen, the Ellis family, Cheiro, or Keiro, Katharine St. Hill had access to all kinds of hospitals. Especially when she began to think about mental health and the hand, she visited at least four large asylums. She observed the hands of people at the Dartford Asylum and at the Victoria Hospital for Children in Chelsea, for example, offering examples of the hands of those suffering insanity and idiocy as well as cases of paralysis and diphtheria. "I literally 'walk the London hospitals' and I have studied the inmates of the Blind Asylum, the Idiot Asylum, and the Hanwell Institution," the asylum for the insane in west London. In the asylums for the blind, she confirmed for herself that those born blind, in fact, had no classic sign on their hand for "eye trouble." She went to several asylums to confirm this, she tells us. It was a kind of evidence-led observation that gave rise to a hypothesis: "I can only conjecture, in explanation of this curious fact, that as they have never known the uses of these senses, they do not miss them and therefore the brain does not mark the hands with the memory."[37]

St. Hill took away one or two truths, which she conceded that asylum doctors would well comprehend, "but which the hand alone showed to me": that there are two distinct kinds of brain disease, the first in which the brain itself is diseased, the second when brain disease is an effect of a diseased other part of the body. She was told by doctors in the asylums that although they categorized insanity under the divisions of dementia, mania, alcoholic, or melancholic, they themselves knew that the distinctions were not well understood. She tried to assign hand signs to these divisions. The hand of the melancholic, for example, had soft mounts, flabby palms, club-tipped fingers, and all the major lines of the palm broken.[38]

This kind of asylum work was more similar to the institutional work of medical doctor Charlotte Wolff (chapter 10) than it was to the Ellis family's work or West End encounters choreographed by Edward Heron-Allen, Cheiro, or Keiro. Katharine St. Hill was desperate for palm reading to become part of medicine. But in so many instances, medical doctors were already doing just that: the Langdon Down family, for instance, were certainly looking at the hands and palms of patients in their asylum. Yet palmistry's route into clinical medicine was troubled, to say the least. It emerged from, and within, a chilling time of asylums and crude psychiatry, of offensive hypotheses about race, evolution, and physical anthropology, and when degeneration anxieties were high among the British (most of all among the people coming up with those very ideas). This was a warped time when eugenics could be construed as welfare humanitarianism, even a kindness.

Inspecting the hands of "idiots" and "imbeciles" was already a long palmistry tradition, crossing directly into doctors' clinical domains. In drawings, ink impressions, and wax or plaster casts, palmists and doctors shared knowledge of the hand shape and palm lines of the mentally ill, and especially the institutionalized. Cheiro illustrated phases of insanity in the hand: melancholy and religious madness, hands that show a suicidal tendency, and the hand of "the natural madman." St. Hill's colleague Ina Oxenford printed the hand of "a mad subject," in her life-sized *Life Studies in Palmistry*, alongside those of "a professor of history" and "a mesmerist." A generation earlier, in 1864, Richard Beamish drew the hand of "a male idiot" in a Cork asylum. It became one of Katharine St. Hill's specialties (fig. 9.1).[39]

Granted access to asylums, St. Hill was moved and troubled. In those terrible institutions, human hands descended to their "lowest form," she wrote. "Nothing is right, nothing is normal." And as a diagnostician, it was uninteresting because everything about the hand was so obvious, from rudimentary thumbs to clubbed fingers. Some had no thumbs at all. Visits to asylums were deeply uncomfortable for her, let alone the patients. "I went into the padded room and examined the hands of raving maniacs while the keepers held them down . . . but the idiots upset me almost more." This was at the height of British eugenics, when rights to birth and death were often deemed not individual but public business. She told a journalist in 1922: "I really think those poor creatures ought not to be allowed to live." Yet she imagined saying so to be a kindness, witnessing in her view "the greatest cruelty imaginable to keep alive." And it all cost the public purse. Her expression of eugenics was explicit and clear: "The production of idiocy should be checked," by which she meant not just their own reproduction, but also their birth in the first place by "generally half-witted mothers."[40]

St. Hill thought asylums for "idiots" and the "imbecile"—children and adults with major mental and physical disabilities—the saddest places in the world. Human—even animal—senses were all but erased: "They had no mind and no senses, no taste, no sight and no smell, no use of their limbs . . . and their only expression was a pitiful howl like a little wild beast, and they looked perfectly miserable." She thought it far more merciful to give the asylum's most disabled a peaceful end with a painless drug and let them go to Paradise. At least that was a signal that she thought these people human, not animal, their souls going to Heaven. But in the longstanding theological debate about animal souls, Katharine St. Hill probably subscribed to the idea that nonhuman animals would also be saved. It so happens that animal welfare was Katharine St. Hill's other passion. She even said that such asylums would be better used by retired work animals—horses and donkeys from the East End—who would at least be capable of "both happiness and gratitude."[41]

At the beginning of her career, St. Hill was trapped by the constraints of an occult tradition, as well as its popularity, and by

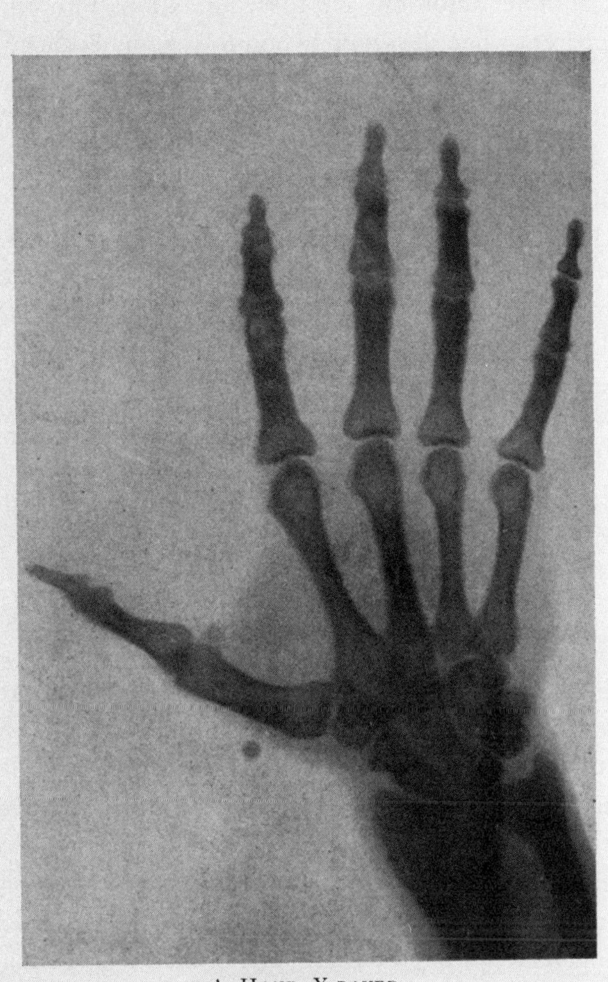

7.5 Medicalized hands. Frontispiece to Ina Oxenford, *Characteristic Hands* (F. N. Fowler, 1912). Copy in author's possession.

the strongest association between palmistry and Gipsy fortune-telling. Both her ambitions and these limitations and confusions were apparent when she set up in South Wales in 1891 at a "Grand Fancy Fair" in Pontypridd. Of the many stalls of the "bazaar," Professor Katharine St. Hill, as she was billed, inhabited the "Gipsy Tent." She attracted a great number of people for her character reading by palmistry. St. Hill's career *could* have duplicated that of O. Hashnu Hara, capitalizing on the occult and the hermetic and the spiritual. She *could* have duplicated Ida Ellis's capacity to make money and a career in lay health and hygiene. Instead, she grafted her interest and knowledge, and indeed her reputation and income, onto the more orthodox medical world of her father and brothers, of hospitals and asylums. By the end of her career, she was a clinical researcher of sorts.[42]

Not all members of her Chirological Society followed through on her medical palmistry, but some did. Ina Oxenford, FCS, opened her book *Characteristic Hands* (1912) with an X-ray image (fig. 7.5). It aligned palmistry with the latest diagnostic technology, but it obliterated all the fleshy specialties of the palmist. Gone were the lines "like hieroglyphics of old." The powers of the seer were being replaced by the technology that could see through the hand, revealing only its skeletal materiality. It is a peculiar and counterintuitive image to open a book of palmistry, a symbol of Katharine St. Hill's life with the language of the hand, from enchantment to disenchantment. Beginning with the thriving world of the occult, palmistry for Katharine St. Hill did end up a sister to medicine, through Reason, not Intuition, garnering status and significance, or at least so she thought. In truth, others did too. Within her own lifetime, reading signs on the hand was already part of the clinical and research practice of orthodox physicians. Probably that was what Katharine St. Hill had wanted to be all along.

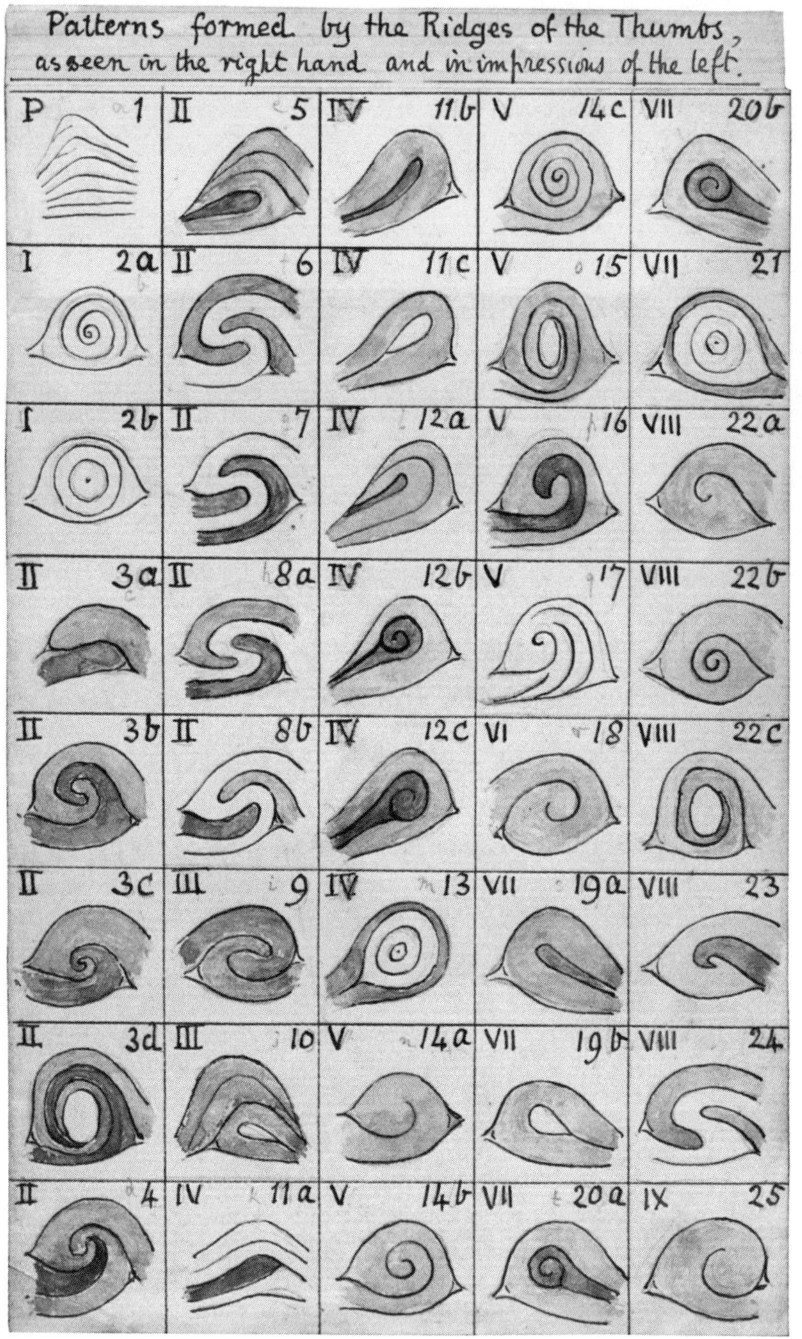

8.1 A new geometry of the hand. "Reconstruction of Ridges . . . the original experiments," n.d. Galton Papers, University College London, GALTON/2/9/6/2/4/14. © University of London, image supplied by UCL Library Services, Special Collections.

PART III

DECODING *the* HAND *in* TWENTIETH-CENTURY BIOSCIENCES

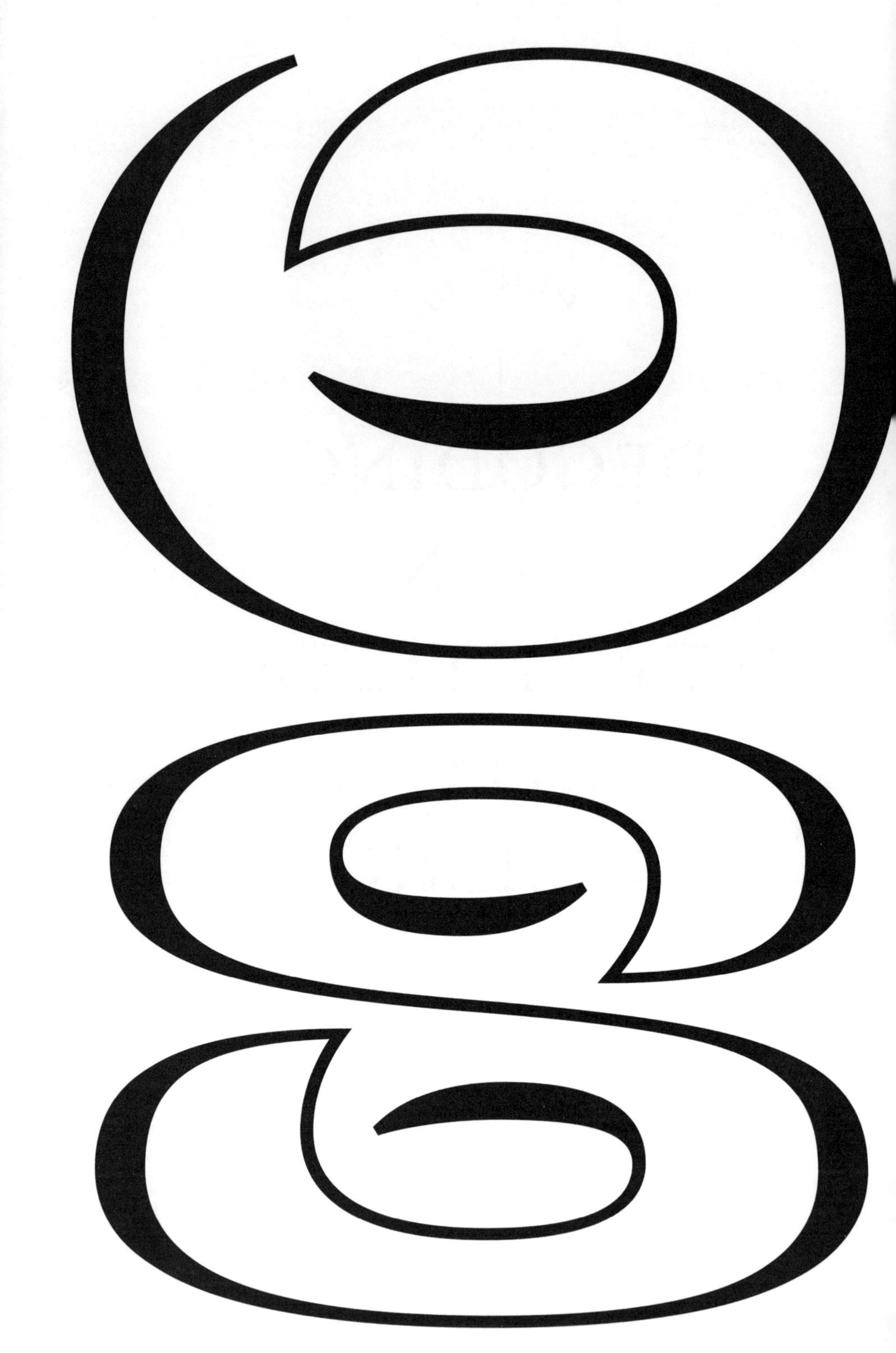

EIGHT

Fingerprints

FRANCIS GALTON'S DOCTRINE OF SIGNATURES

Francis Galton—statistician, explorer, maker of gadgets—was fascinated by the physical appearance of both individuals and types. In the 1870s, he pursued a new technique that he called "composite portraiture," layering multiple photographs of faces of many individuals of one group—a family, a race, or a prison population—on top of one another. In this manner, he created an image that represented no particular person, "but portrays an imaginary figure possessing the average features of any given group of men." One of the late nineteenth century's key theorists of heredity, Galton wondered how, precisely, faces might be both individual and familial, personalized and composite. He delighted in his new photographic invention, its paraphernalia, and its possible applications, although this particular idea sank. Still, Galton retained an active interest in faces, an incessant observer of human difference, a physiognomist of sorts. He quite liked his own face, commissioning multiple portraits, though that was a favored pastime of any gentleman-scholar of his milieu. Galton also had his silhouette drawn and cut, after the manner that Lavater had popularized, and regularly attempted to draw profiles himself, of both ordinary people and those in his

London circle (fig. 8.2). Physiognomy and portraiture often went together. Indeed, as his own long life was fading, Galton bothered to instruct that a death mask be cast of his own visage.[1]

Faces and profiles might be singular and recognizable, but were they really the most reliable part of the human body from which one could gather data about individuals and populations, and rigorously study patterns and correlations? For that matter, wondered Galton, were the skulls around which facial features were wrapped the best index of human types, skulls that had been so incessantly measured by comparative anatomists and physical anthropologists? And was the "mug shot" invented by his French competitor, Alphonse Bertillon, really the most refined means of individual identification? Galton thought not, and turned emphatically from faces to hands.

At the tips of fingers and thumbs, Galton found minute papillary ridges that make patterns of such wondrous variety that for decades the poly-mathematician could not look away. He took thousands upon thousands of prints, providing the kind of data that thrilled a brilliant statistician. Could those patterns present a precise and reliably measurable set of markers that might correlate to different human types? Might they even mark out an individual? Through painstaking research over the 1890s, Galton reached two conclusions: first, that fingerprints were "absolutely unchangeable throughout life"; and second, that there was an infinite variety of patterns making each set of prints singular. An inventor of systems and gadgets as well as ideas, Galton was always looking to the future, as avidly as any chiromancer. But not even he could have predicted how widely and deeply his intricate research on fingerprints would be applied. He had confirmed two qualities of the hand that enabled biometric states across the world to function, fingerprint technologies that have been rebooted in our own time with implications as spectacular and dazzling as they can be mysterious and dark.[2]

Francis Galton was no palmist or scientific chirologist. Nonetheless, he was most certainly an interpreter of the hand, as focused on those fingerprints as palmists were on crease lines, if differently so. Contemporary consumers of his fingerprint studies made the

8.2 After the manner of Lavater, who popularized silhouettes, Galton had his own profile crafted, and constantly drew the profiles of others. Silhouette of Francis Galton by Francis Smyth Baden-Powell, 1890s. © National Portrait Gallery, London, NPG D425.

connection: "It is a 'far cry' from the Gipsies to Mr Francis Galton," one journalist wrote, "but there is something in common between them." Galton anticipated the connection. He noted that palms and soles had two kinds of marks: the creases or folds of the skin, which interest the palmist ("these are no more significant than the creases in old clothes"); and the so-called papillary ridges, by far the most numerous, and for him the most telling. Yet Galton was dissembling when he insisted that studying fingerprints was altogether different to palmistry or chiromantic hand reading. For one thing, fingertips are part of the hand. For another, none of the medical palmists and the so-called astral palmists of his own generation looked only at the creases of the palm, as we have seen. And for a third, a later Galton professor at University College London, geneticist Lionel Penrose, was to show how wrong his intellectual predecessor was to trivialize those creases. With these developments in mind, what if we consider Francis Galton as a reader of hands not just in the anatomical tradition of Purkinje and Bell, but also in the chiromantic tradition of Richard Saunders and Edward Heron-Allen, and even in the physiognomic tradition of Richard Beamish and Ida Ellis? While Galton's methods, learning, and mathematical sophistication were far superior, when we ask *why* he looked at the hand, the answer aligns: in order to find signs of identity, health, criminality, insanity, race, individuals, types, and even the future, now a probability more than a prognostication. In any case, it is insufficient *not* to understand Francis Galton as a reader of hands, simply because his records and papers are filled with palm prints, including his own.[3]

With regard to the craft of printing from the hands, also, Galton was of a piece with Cheiro, Keiro, Katharine St. Hill, Edward Heron-Allen, and the Ellis family. They all experimented with inks, waxes, plasters, photographs, and microscopes, Galton most obsessively of all: how to maneuver the hands of the subject for the best print, how to magnify to the highest quality, how to most efficiently read the prints, and how to choose the best paper. Yet in the latter lies the great symbolic difference. Galton's own palm was printed on grid paper: the paper of choice for mathematicians (fig. 8.3). The triangles and quadrangles that chiromancers

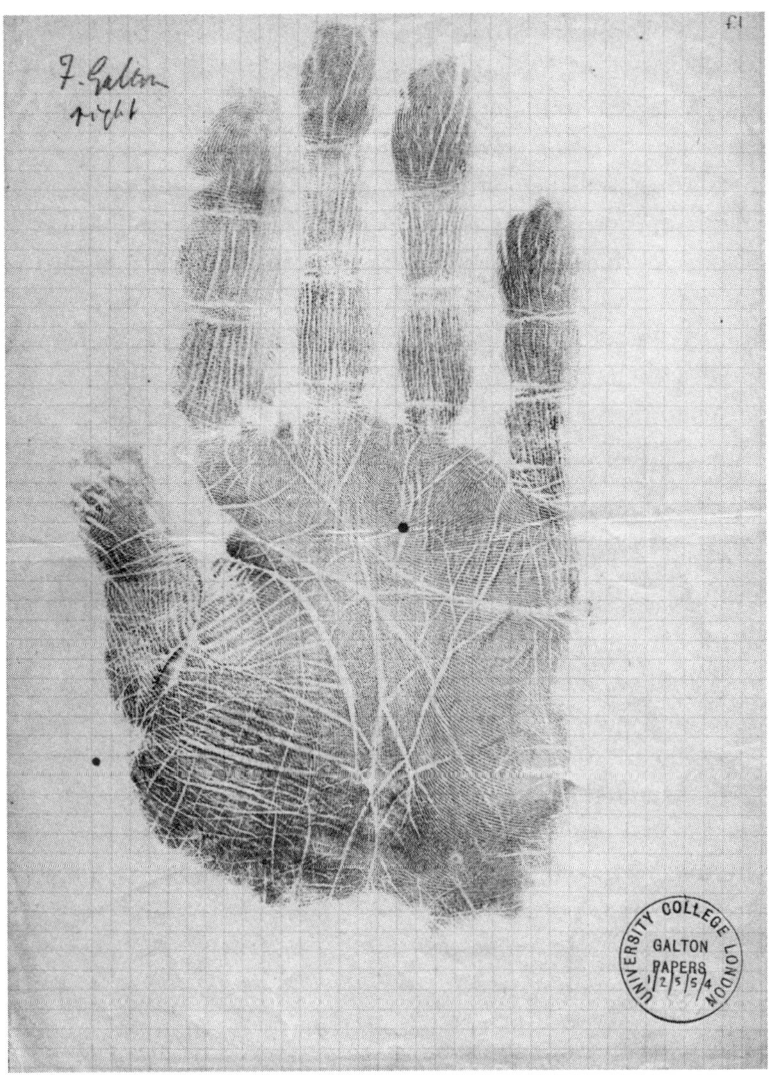

8.3 Francis Galton's right handprint, 1902–1903. Galton Papers, University College London, GALTON/1/2/5/5/4. © University of London, image supplied by UCL Library Services, Special Collections.

had for centuries isolated were now minute patterns—arches, whorls, curves—that could be measured far more precisely, and were reliably correlated by one of modernity's most influential statisticians. Galton had the means to turn massive data into numerals that could be plotted against lines, means, and bell curves, calculating correlations as well as probabilities with more chance of accuracy than any chiromantic prognostication and many medical prognoses. Francis Galton measured the hand in a way it had never been before. He found identities in the skin that decoded and disclosed the secrets of identity. It turned out to be a doctrine of signatures for the modern world.

THE HAND AND HUMAN TYPES

Galton's was the age of measurement, of anthropometry, of biometrics. In France, the so-called Bertillon system of identification was based on five bodily measurements applied within policing and court systems: head length and breadth, length of the middle finger, length of the left foot, and length of the cubit. Alphonse Bertillon added photography, now known as the mug shot of the face, another legacy of Lavater's eighteenth-century physiognomy. These records were collated into a system for police to identify those they had arrested and who were being charged and tried. Yet in truth, everyone was measuring human bodies. The esteemed British Association for the Advancement of Science created its own Anthropometry Section, and, typically indefatigable, Francis Galton thought (knew) he could improve on it. He created an Anthropometry Laboratory attached to the International Health Exhibition in South Kensington in 1884. Some of the four million people who flocked to the exhibition participated in a kind of citizen-science project that would boost his anthropometric data, eagerly presenting themselves to be measured, paying 3d for the pleasure, and for a summary of their own physical signs. The Laboratory gathered, anonymously, height, weight, arm span, color sense, and quality of breathing power, hearing, and sight. The hand was measured through its strength of pull and squeeze, and quickness of blow. A postal assessment of personal bodily efficiency could also be acquired, paralleling the cottage industry in remote health set up by the Ellis family in Blackpool. Anyone could fill in forms with their own measurements and post them to the Anthropometric Laboratory for assessment, and inclusion in the national count, if the data were deemed satisfactory. Yet the incentive for the Ellis family postal health service was embarrassingly commercial and self-interested compared to the lofty national ambitions of anthropometry for the biometric state that England aspired to be. In good mathematical hands, individual data could be rendered into complex statistics "to discover the efficiency of the nation as a whole and in its several parts, and the direction in which it is changing, whether for better or worse." Everything can be statistically compared, Galton delighted in saying: "schools, occupations, residences, races, &c." He was a force of nature, driving a system that could measure "types" or subpopulations and at the same time deliver interesting information to the individual. Techniques to know "each and all" were the special power of what later came to be called "biopolitics." It was also the special power, so to say, of eugenics, the term that Galton invented in 1883.[4]

Several millennia before Galton set up his laboratory to measure British humans, Aristotle had also wondered about the purpose of studying human physiognomy: whether conclusions could or should be drawn about the generic or the particular; a human type, on the one hand, or a specific individual, on the other. "Aristotle" instructed carefully on how widely the reading of a bodily sign can and should legitimately apply: "In physiognomy we try to infer from bodily signs the character of this or that particular person, and not the characters of the whole human race." And yet there was a typological ambition in Aristotelian physiognomy too, to know how an Egyptian or a Thracian, for example, presented physically. By Galton's time, versions of racial physiognomy had become entrenched, and he built actively on them, as we shall see. Yet "race" was not the only nineteenth-century division of humanity, especially when it came to hands and their measurement.[5]

Another was class, the great new human group of Galton's industrialized era. And although industrialization was a matter of machines, it was still the phenomenon of *manual* labor that constituted the "working class." Karl Marx and Friedrich Engels knew as much, but they had different takes on the hand. While Marx focused critically and theoretically on Adam Smith's "invisible hand," the eighteenth-century economist's metaphor for a self-regulating free market, his colleague Engels put his own mind and pen to the actual biological hand that labored. "Labour created man himself," declared Engels in 1876. And by that he meant in every literal sense that the hand was a working instrument that created humankind. After reading Darwin on the hands of the first humans to stand upright, Engels saw therein the key to freedom of a species: "This was *the decisive step in the transition from ape to man . . . the hand had become free* and could henceforth attain ever greater dexterity; the greater flexibility thus acquired was inherited and increased from generation to generation." For any comparative anatomist or evolutionary biologist, this was a familiar claim, but Engels made the capacity of the human body to morph and change over a long human history part of his socialist theory. In *The Part Played by Labour in the Transition from Ape to Man*, he rendered the human hand part of his dialectics of nature, his search for the material drivers of history. Manual labor could, and did, literally mark particular humans; repetitive use of tools actually shaped hands permanently. Economy imprinted nature.[6]

Just this idea was pursued systematically by Galton's French alter ego, the biometrician Alphonse Bertillon, who also wondered whether acquired hand marks could reliably identify a human type, based on their manual trade. In Galton's collection of papers at University College London is an album of photographs of human hands that was purported evidence for just this: *Signes Caractéristiques des Professions Manuelles*. The album was put together by the Service d'Identification de la Préfecture de Police, in Paris, in 1891. It contains photographs of any number of manual laborers and artisans: a cook, a machinist, a jeweler, a boilermaker, a coachman, a coppersmith, even *un pianiste*. Each entry displayed a photograph of the entire person with their tool of trade, an extension of their body, of their very embodied selves (fig. 8.4). Each

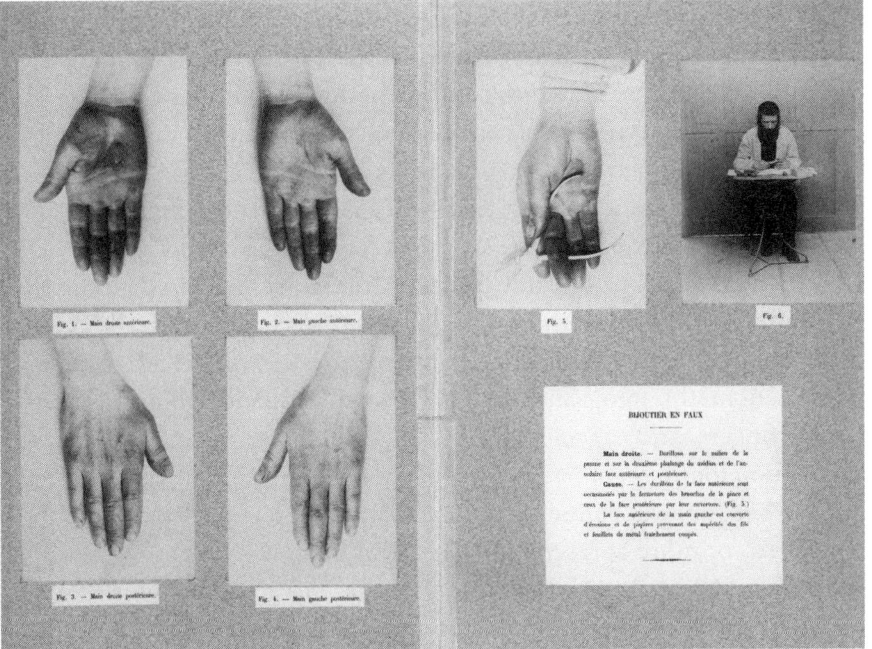

8.4 *Signes Caractéristiques des Professions Manuelles: Photographies et observations recueillies sous la direction de M. Alphonse Bertillon*. Par M. Felix Geoffray, attaché au Service d'Identification de la Préfecture de Police, Paris, 1891. Galton Papers, University College London, GALTON/2/9/7. © University of London, image supplied by UCL Library Services, Special Collections.

also reproduced a close photograph of their hands, left and right, front and back; the telltale part of their body that was literally marked by their work. The photographs record visual evidence of precisely how hands held a particular instrument or tool, and what mark was left. Engels could not have asked for a more precise demonstration and record of the hand as a product of labor.[7]

Such telltale marks on the hand interested Galton, but the equation that labor created biology was the wrong way round for the scientist: nature—biology—came first and was his real business. And in any case, measuring heads or laboring hands, let alone handwriting after the style of Bertillon's "graphology," was too imprecise and crude for Galton. He turned instead to finer patterns of fingerprints and thumbprints.

Galton did try to correlate those patterns with manual labor or its absence. Could different classes be discerned reliably by marks in the fingertips? He set to work gathering individuals who represented class types. In his own Bloomsbury backyard were many "persons of much culture," and he readily took the fingerprints of university students in Arts and Science schools. Through intermediaries, he took a sequence of prints from English field laborers. And personally he took the prints of "the worst idiots in the London district," like Katharine St. Hill visiting asylums with the aid and agreement of medical superintendents. And yet, for all his apparent methodological care, for all his skill in identifying patterns on fingers, and capacity to translate those patterns to numbers, Galton found no innate mark of class in the fingerprint, as Bertillon had found acquired in the hand of the laborer. The experiment failed, or more correctly, produced an unambiguously negative result: "No indications of temperament, character, or ability are to be found in finger marks," he concluded, redeploying old phrenological and physiognomic terms.[8]

Galton tried again to correlate particular human types to fingerprint patterns, this time as a study of race, another way of considering the marker of a *group* of humans to which an individual might belong. For generations, natural scientists across Europe had measured and compared lengths of bones, arms and legs, hands and feet, and especially skulls, trying to fix what had come to be known as "race." The manner of doing so had become increasingly precise over the nineteenth century, and leaders in the field tried to standardize various biometric methods. Galton thought his layered portrait photography method might create an "average" face of a Jew or an Englishman (as if the former could not be the latter). But fingerprints were a new possibility.[9]

In his early fingerprint studies, Galton experimented with the identification of four races, in his terms, English, pure Welsh, Hebrew, and Negro. To a lesser extent he worked with some data from Basques in the French Pyrenees. As a young man in the 1840s, Francis Galton had traveled adventurously through North Africa, the Middle East, and Southwest Africa.[10] Now firmly London based, he deployed agents, contacts, and intermediaries to gather data. These came from schools in the Welsh-speaking mountains; from London's Hebrew schools; and from agents in contact with the Royal Niger Company. The study proceeded by comparing one single mark, the arch on the right forefinger, and yet, despite great statistical care, patience, and caution, his conclusion was again emphatically negative, and disappointing (to him): "There is no *peculiar* pattern which characterises persons of any of the above races." As with class, there was no positive result. Few in the 1890s could have executed the same statistical work. Still, Galton thought there might be some benefit in pursuing studies in vastly differentiated and isolated groups: "The most hopeful direction in which this inquiry admits of being pursued is among the Hill tribes of India, Australian blacks, and other diverse and so-called aboriginal races. The field of ethnology is large, and it would be unwise as yet to neglect the chance of somewhere finding characteristic patterns." As we shall see in the next chapter, a small but dedicated group of anatomists, mainly based in the United States, pursued just that.[11]

A SINGULAR IDENTITY

Francis Galton had failed to correlate fingerprint markings with either class or race; his negative results were clear. Undeterred, he pursued another question through his statistical work: whether fingerprints were singular, whether they marked *individual* identity, even if they did not mark a type. Key figures in the physiognomic canon had wondered intermittently about the individuality of hands, palms, and fingers. As we have seen, Lavater had asserted that particular hands can only belong to particular people. "This is easy enough to verify," he had claimed nearly a century earlier. "Choose a hand for an example, compare it with a thousand others, and amid this number there will not be a single one that could be substituted for the first." Galton was to do just this, literally comparing one hand to a thousand others, but his mathematical and methodological capacity was of another order altogether to Lavater's crude directive "to compare."[12]

The first matter was to determine whether marks on the hand, including the papillary ridges of the fingers, changed over any given lifetime. Some anatomists were already claiming that creases stayed the same from embryonic life to death. Palmists, however, generally thought they did change. We have seen how Katharine St. Hill premised aspects of her medical palmistry on lines that appeared and disappeared, and how an infant's neonatal lines resembled those of the parents and then were replaced by an individuated pattern. Over the early 1890s, Galton's research began to show the permanence of the papillary ridges at the ends of fingers and thumbs (fig. 8.5).

In India, fingerprinting had been in operation for a generation. As early as 1859, William Herschel, then in the Indian Civil Service, had begun taking fingerprints and had the peculiar opportunity to repeat prints from the same person over several decades. His interest was inspired initially by handprints as a way to authorize contracts, both a seal and a signature (fig. 8.6). The first Galton Chair of Eugenics at UCL, Karl Pearson, reproduced this contract in his *Life* of Galton, noting that the signatory, Rajyadhar Konai, confirmed the contract at Herschel's request with a "signature" that was a handprint. Herschel was in the habit of sending fingerprint samples to Galton, who made 296 points of comparison between the sets and found them to be identical over time: "Not one failed to exist in both impressions of the same set." Galton was also sent a packet of prints taken from people in Bengal in 1878 and again in 1892. Another impression had been taken in sealing wax in 1873, a fingerprint repeated in 1890 when the subject was over eighty years old. Repeatedly, these time series revealed identical patterns. The papillary ridges showed an "extraordinary persistence" throughout life, and he began confidently to claim that he had shown that the minutiae are "independent of age and growth." Galton started to see that this single measure was a vast improvement on Bertillon's composite system, and certainly on anything that could be claimed from the shape of heads and faces, or even photographs. This is precisely what Galton told a commission of inquiry on anthropometry, charged with comparing and evaluating Bertillon's system vis-à-vis his own, a committee

8.5 Fingerprint portraiture. Just one of thousands of fingerprints magnified and mounted in a photograph album and codified by the number 4236. Galton Papers, University College London, GALTON/2/9/6/3/1/23. © University of London, image supplied by UCL Library Services, Special Collections.

to whom Galton also put into evidence any number of prints of palms and of complete hands, including his own. It was fingerprints, however, which were permanent and unchanging. Even after burns, Galton showed the committee, the original marks reappear exactly as before.[13]

How did he know this? What constituted these 296 points of comparison? Galton followed up the work of Bohemian anatomist

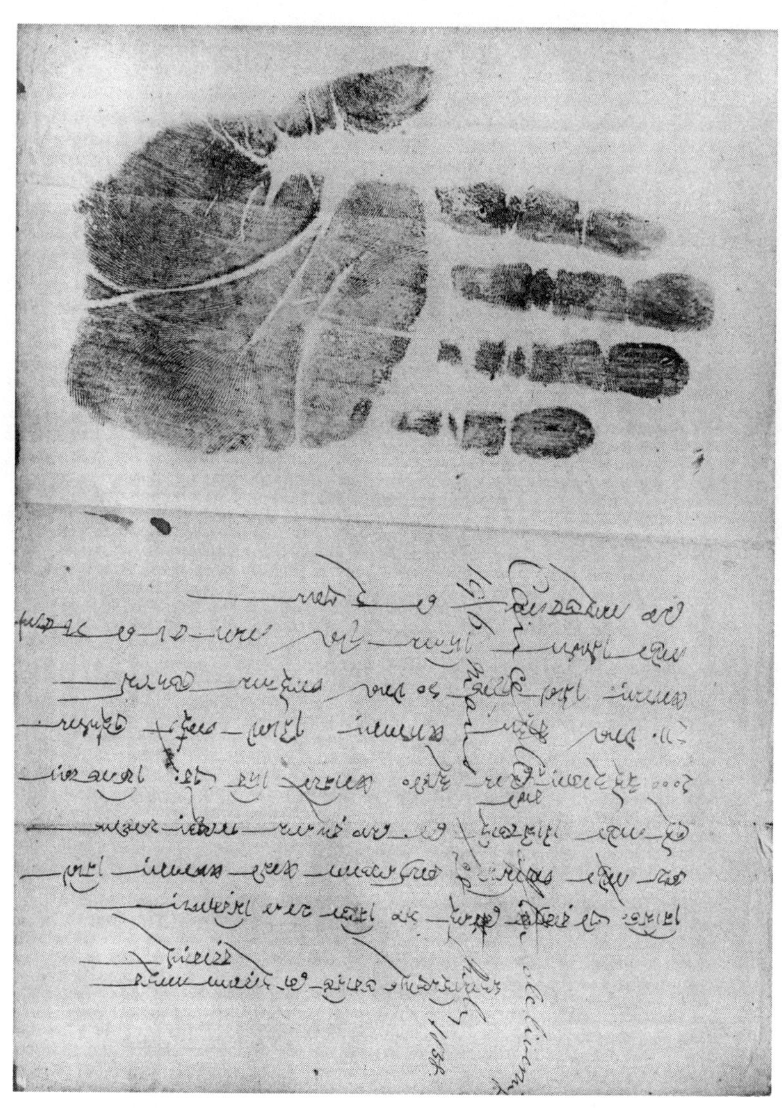

8.6 Rajyadhar Konai's contract made at Hooghly, 1858, with the sign of the palm. Karl Pearson wrote that Konai "signed" the document at Sir William Herschel's request, "with an imprint of his right hand as an identifiable sign manual." Reprinted in Karl Pearson, *Life, Letters, and Labours of Francis Galton*, vol. 3a (Cambridge University Press, 1930), 146. Reproduced with permission from the National Library of Australia.

Jan Purkinje, translating his thesis from its original Latin, paying attention to "the arrangement of the minute furrows of the organs of touch," or, perhaps "depressed lines" was a better translation, Galton self-corrected in his manuscript. Purkinje was a classifier after Galton's heart, who translated the early anatomist's terms for nine identified patterns as transverse curves, central longitudinal streak, oblique streak, oblique sinus, amygdalus (almond-shaped), spiral, ellipse or elliptic, circle or circular whorl, and double whorl. Over countless hours of analysis, Galton simplified Purkinje's classification into three "elementary divisions," which he named primaries (or arches), whorls, and loops (fig. 8.1). But since Galton's business was statistics and specifically correlations, he had to be able to plot what he found, and so assigned these primary patterns particular numbers. He translated—his word—the patterns into the sign of 1 (for arches), 2, 3 (for whorls), and 4, 5, and 6 (for loops). Whorls and loops were subcategorized, requiring a double enumeration in order to signal the varying axes that accompanied a sloping pattern. Over time, Galton refined his geometry of the hand, standardizing his measurements from an imaginary line drawn from the tip of the forefinger to the base of the little finger. These he translated into another numerical sequence. Normal slopes away from that line he expressed by odd numerals, 1, 3, 5; abnormal slopes by even numerals, 2, 4, 6. The whorls themselves were now subcategorized into circles, ellipses, and spirals "both simple and compound," determined by the direction of closeness of their twist. The loops could be "feathered" or "bent." Each of these variations was expressed algebraically by *p* or *q* or *l* or *m*. These data yielded 2,890 different finger marks, sufficient for his statistical work. This was a geometry of the hand of an altogether different order to the recognition of a triangle or a quadrangle by medieval ecclesiastics.[14]

Having failed to correlate fingerprints with groups or types of humans, Francis Galton now showed that fingerprints were not just permanent, they were singular. Given the number of variations on a single print, he was confident that "the chance of two fingerprints being identical is less than 1 in 64,000,000,000."

One fingerprint may hold some error, or similarity with the then 1,600,000,000 people on the planet, but assessing the prints of three or four fingers for each person eliminated that error entirely. If, then, two fingerprints are compared and found to coincide exactly, "it is practically certain that they are prints of the same finger of the same person." Even identical twins had fingerprints that were differentiated, he showed with minute observations of the hands of sisters Dorothy and Muriel Brooke, unceremoniously displayed to the Royal Society in June 1895.[15]

Francis Galton had verified a new bodily signature. "Each person thus possesses a formula which is, as it were, a personal name, that may be read from his fingerprints." And quickly, fingerprints *became* identity. One anatomist declared that "they may serve the authorities in the identification of a human body, living or dead." And others started to think about the long history through which humans make their singular mark, beginning with ancient stenciled hands on rock faces. Throughout history, man has given his hand as his seal, wrote one journalist appraising Galton's new "wonder." This signature, unlike Paracelsus's signatures from all those centuries before, far from being hidden, was an explicit, readable, declarable sign of the person, even a substitute for the person, their will and their authority, as was the handprint in Herschel's Indian contract. It was not just *like* a seal, it *was* a signature. And indeed, Galton occasionally recorded prints in sealing wax rather than with ink, as did some palmists. Going one step further, he had a seal cut from his own right ring-finger print, and it went on to be used on the name cards at the Annual Galton Laboratory Dinner.[16]

Some palmists had announced the singularity of each human's palm print but had no way of substantiating the claim beyond their own anecdotal evidence. Yet, as historians of their own practice, they also enjoyed declaring the significance of the hand as a longstanding imprint of singular authority, like a seal. Heron-Allen and Cheiro, we have seen, had their celebrity handprints signed and dated, an unarguable correlation of the print of the hand with the person. And indeed, just like these palmists, Galton took and kept, and sometimes even published, the prints of the famous.

Many were in his own circle, including Gladstone (again!), Zola, Wallace, Herbert Spencer, and his special relatives, the Darwins.[17]

Yet it was in the *craft* of printing and reading that Galton most directly followed a palmistry tradition. His aspirations were always to *apply* his mathematical work, and to that end he was particular with his instructions, like so many palmists writing handbooks and textbooks, literal manuals. Galton was ridiculously precise about how to make the best impressions of both whole hands and fingerprints. An older method of printing using dye or watercolor was inadequate, because the dye fills the furrows as well as the ridges. He preferred a plate of copper blackened with printer's ink, thinly spread with a printer's roller. One typescript in his papers offers detailed "Directions for Making Palm and Sole Prints." Perhaps they were Galton's own, but they may have come from US anatomist H. H. Wilder. It was precise about the ink, the best being the tubes for use with the Edison Mimeograph, or "mimeograph ink." The fingers should be separated out, with "considerable pressure of the hollow of the palm until the subject is conscious of a complete contact with the surface." The removal of the hand should be rapid. Once the print was secured satisfactorily, its designation was critical. The hands needed to be identified, by initials or some form of numbering that corresponded to a list retained by the operator. The sex and nationality should also be recorded, with instruction that by nationality was meant "the race, not the political relations" of each individual. The process needed to be achievable by everyone—the police, civil servants, officers of prisons—and necessarily standardized. Galton took the time to create his own perfect and perfectly printed fingers as the title page to his 1892 book.[18]

Various well-known palmists, contemporaries of Galton's, had experimented with different papers and inks, and with methods of magnification. Some had used microscopes through which they could observe hand marks or prints of them. Cheiro had investigated palms and fingerprints through a microscope with Mark Twain, for example. Ina Oxenford, colleague of Katharine St. Hill's, was especially determined to stress the significance of the minute: "the eighth of an inch difference on a phalange or a finger in one hand giving a clue which he can follow up until a whole history is revealed." Just like Galton, tiny differences could reveal archetypes, or types, or singular identities. Galton's own suggestion was a reading-glass of the sort used by watchmakers, with a six-inch focus. Only with adequate magnification could ridges be easily and accurately counted. And like any chirologist, Galton instructed practice and repetition to perfect the art and craft of taking and reading fingerprints. "The learner should mix water-colour, fill a pen with it, and dot the deltas of his own fingers and those of such friends as will allow him, and trace the courses of the divergent ridges until the outline of the pattern they enclose is sufficiently apparent." He had himself refined, through practice, the best way to read; don't try to read the prints of the right hand and then the left. This is unnecessarily fatiguing for the eye. Rather, "read the prints straight through from left to right, like words in a line." Reading this new language of the hand was itself a manual exercise. Galton instructed not just on how to print, involving two sets of hands—the subjects' and the printers'—one over the other. How hard to press, how to release, how to touch. How to see, through a lens or a microscope, and how to use photography to enlarge the imprints.[19]

In the end, Galton's method of identification was vastly more sophisticated than palmistry. Yet, looked at another way, the *identity* that was Galton's endpoint—the biological singularity of that person—was infinitely less complex than the "soul" sought by early modern chiromancers, the "character" sought by physiognomists and phrenologists, or the personality sought by psychoanalytically trained hand readers.

IDENTIFICATION: CRIMINAL HANDS

Having correlated fingerprints to individual identity, Galton moved swiftly to the application of this idea, from identity to identification, realizing its potential with the criminal justice system.

In France, Bertillon's measurement of the human body always had forensic ambitions and applications. He tried to bring handwriting into the courts, for example, famously in the case of treason against Alfred Dreyfus, when the writer of a critical memo needed identification. Bertillon claimed that his graphological system was based on a calculus of probability, incorrectly as it turned out. And yet hands, not just Galton's fingerprints, continued to be considered as forensic identification. The US anatomist Wilder, for example, investigated the "longer and more conspicuous markings of the palms" for identification. This would "brand the possessor with an indelible mark," with the advantage that it could be seen clearly on a print in a courtroom. The initial application of Galton's fingerprints was not so much forensic as a means to identify a habitual criminal. Was the imprisoned person who they said they were? Or were they, in fact, a repeat offender? This was the direct concern of another committee appointed by the secretary of state, which reported in 1894. Was the system of registration and identification used in France superior to the new possibility of fingerprint identification espoused by Galton? Indefatigable, Galton presented all kinds of evidence, and the committee accompanied him to Pentonville Prison to witness the prints taken of more than one hundred prisoners. Galton was excited by "the real simplicity of the method," far more applicable and simpler than his previous work on racial distinctions and heredity, he was happy to say publicly. And he was confident: the criminal application should have a great future. In fact, it was already underway, with sixty-three English prisons taking prints by trained warders that were forwarded to a central bureau for expert classification. Since the fingerprint was an "automatic sign-manual subject to no fault of observation or clerical error, and trustworthy throughout life," it could be applied not only to criminal identification, but also to check fraudulent re-enlistments in the army, or for identifying pensioners. Galton persuaded prisons and the police to create large datasets, on cards and in filing cabinets, where real identities were stored away.[20]

The 1890s was the decade of great new strides in forensics. Palm prints as well as fingerprints entered the policing and justice systems, accompanied by the fictionalizing of just that, the creation of Sherlock Holmes, whose close eye never missed a thumbprint or a handprint. Sherlock Holmes was a Galtonian figure. "You have an extraordinary genius for minutiae," Dr. Watson notices at the beginning of Conan Doyle's novel *The Sign of Four*. And Sherlock Holmes obsessively catalogued his work: "Here is my monograph upon the tracing of footsteps, with some remarks upon the uses of plaster of Paris as a preserver of impresses. Here, too, is a curious little work upon the influence of a trade upon the form of the hand, with lithotypes of the hands of slaters, sailors, corkcutters, compositors, weavers, and diamond-polishers." Conan Doyle also thought of fingerprints as signatures, the great sign of identity.[21]

This world of signs, hands, identity, criminality, and identification intersected with palmistry. Palm readers routinely looked for marks on the hand that told of criminality or of the likelihood of murder, or of a terrible inclination to murder that could be guarded against. The celebrity palmists all published either a named or generic "hand of a murderer." In the seventeenth century, Richard Saunders had instructed that the character ♀ seen on the mount of the thumb signifies a man who will murder his own wife. In the modern era, palmists put themselves forward as forensically talented, a new kind of policing resource, now that handprints, palm prints, and fingerprints were being scrutinized. Edward Heron-Allen drew Jack the Ripper's hand in 1888, not from life—the murderer was never found—but from information publicly available on the character, the motivation, the type, and the "temperament." Like a hand-drawn facial composite that would assist police in identifying a criminal, Heron-Allen drew a "hand-portrait" that could be used to identify a suspect as the Whitechapel murderer; since the hand shape and "nerve lines" could not be modified, its "minutest details" could not be altered. The second finger would be the longest, "denoting the morbid, saturnine temperament," and in the center of the head line would be "a great loop or island, indicative always of cerebral disorder, monomania, heresy and the like." If a suspect showed such a hand, it could be counted on as a sure sign that he could commit those crimes, and even had committed them. It was a profiling technique.[22]

Cheiro's most famous criminal hand was that of Chicago serial murderer Dr. Henry Meyer, whose hand he first read in 1893.

Years later, Meyer was in Sing Sing, convicted of murder, and with three failed appeals at commutation of his sentence, death by the electric chair was imminent. Meyer asked to see Cheiro, who confirmed that he saw no sign on the hand that death was close:

> To me it was torture to see that poor wretch before me, to feel his cold clammy hands touching mine, and see his hollow eyes hungry for a word of comfort. Although I could hardly believe what I saw, I pointed out that his line of life showed no signs of any break, and so I left him, giving the hope that some miracle could still happen that would save him from the dreaded "chair."[23]

Of course, in this account Cheiro's prognostication was vindicated, and Meyer's execution commuted at the eleventh hour (fig. 9.2). Closer to Galton's London, and to Scotland Yard, the new home of the metropolitan police, Cheiro recounted another murder. The bloodstained hands of both the victim and the murderer were left on doors, and from this evidence, he deduced that they were related, or so he claimed. Vindicated again, the dead man's relative was arrested. In this way, if Galton the fingerprint expert took prints of the whole palm, the reverse was also true: Cheiro the palmist reproduced impressions of "the tips of the fingers," and noted how this was increasingly being used by Scotland Yard to track down criminals. He understood this work to be in the same mold as his own: reading the body for signs of identity and character, including criminality. Cheiro even reminisced that the British police scoffed "when Monsieur Bertillon and the French police" had first started investigating the body. Now it was clear that reading bodily signs, including signs of the hand, "could also be made invaluable in foreseeing tendencies towards insanity."[24]

Katharine St. Hill wanted to make a special study of the palms of criminals and murderers, but had difficulty accessing prisons in Britain, claiming it was easier in Paris, where various officers were much more sympathetic to investigators. Yet she also watched the uptake of fingerprinting by Scotland Yard. In another reverse gesture between fingerprinters and palmists, she read the hands of Inspector Maurice Moser of the Criminal Investigation Department of Scotland Yard, and reproduced them in her *Hands of Celebrities*, itself signaling a new archetype, the detective. Like most published chirologists, she also looked for and passed on information about "bad hands," which she classified as a form of insanity. It was complicated, because just as there were different kinds of murderers, so there were different hand signs of a murderer. There was even a "star of assassination." Evil is a deformity of the mind, she instructed, bringing psychology into her palmists' diagnostics. And although it might not be possible to exterminate in the individual, in the race it is, "with each generation in turn," signaling Galton's own eugenics of improvement. By the early twentieth century, forensic hand analysis was becoming well known. In Ida Ellis's court case in 1904, for example, her defending lawyer asked one of the policemen: "In this very court, are not finger prints of criminals taken down?"[25] Yes, the policeman replied, "because no people's fingers or faces are alike." He was confusing and conflating old physiognomy with new Galtonian hand reading.[26]

Galton's fingerprints are everywhere. Symbolically so, as we travel over global borders and as we open our devices daily. But also literally. I found Galton's own original fingerprints all over the archives, the historical records in and around hands and identity. One charming set revealed itself as I opened a folder in the metropolitan police files in the UK National Archives in Kew, London. There was an envelope marked "Sir Francis Galton's Fingerprints" (fig. 8.7). Clearly nobody had opened it for decades, and it had been stored and forgotten. But the set of fingerprints of the famous man once meant a lot. Galton himself had sent them to the sculptor Dame Katharine Furse in 1903, whose husband, the Chelsea artist Charles Wellington Furse, was just then putting the final touches to a Galton portrait. Katharine Furse would often carve the frames for her husband's work, but Galton had a special request. Could she carve his fingerprints into the frame? Here they were, in two dimensions at least, but could she reproduce that in

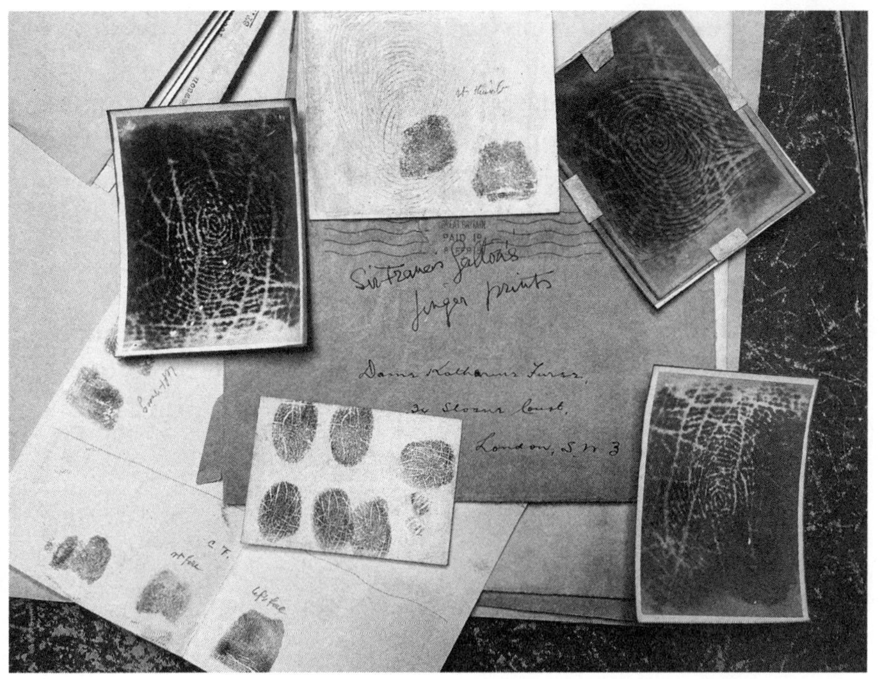

8.7 Francis Galton's fingerprints, from portraiture to police, eugenics to identity documents. Katharine Furse correspondence with New Scotland Yard, TNA, Mepo 2/9979. Photograph by author.

the timber? The request was the strangest mix of signature and portraiture, bringing the face and the hand together yet again, but this time in a manner as unique as the fingerprints themselves. There could not be a more singular representation of the human known as Sir Francis Galton.

This particular set of Galton fingerprints traveled in another direction too, circling back to their other, criminological, use. In 1936, Katharine Furse was selling her house and distributing various collectibles. She offered the celebrity fingerprints to New Scotland Yard. Was there a police museum that might value this part of Sir Francis Galton, like a saintly relic? Alternatively, she suggested, the Eugenics Society might be interested, "but they probably have them already." And on through the twentieth century, Galton and his fingerprints gathered significance. When Katharine Furse sought a visa for the United States in 1951, the tips of her own fingers were rolled and printed by a young officer at the American Embassy. He chatted with her as this was done: since his job was to take fingerprints all day, every day, he often thought about the man who had invented it all. He wished he could have met Sir Francis Galton, considering himself a kind of disciple. Katharine Furse was able to disclose that not only had Galton sent his own prints to her, but she herself had been measured as a child in Francis Galton's Anthropometry Laboratory in Kensington in 1884. Anthropometry, policing, eugenics, identification and identity documents: these were the applications of Francis Galton's doctrine of signatures, now located at the tips of every human hand.[27]

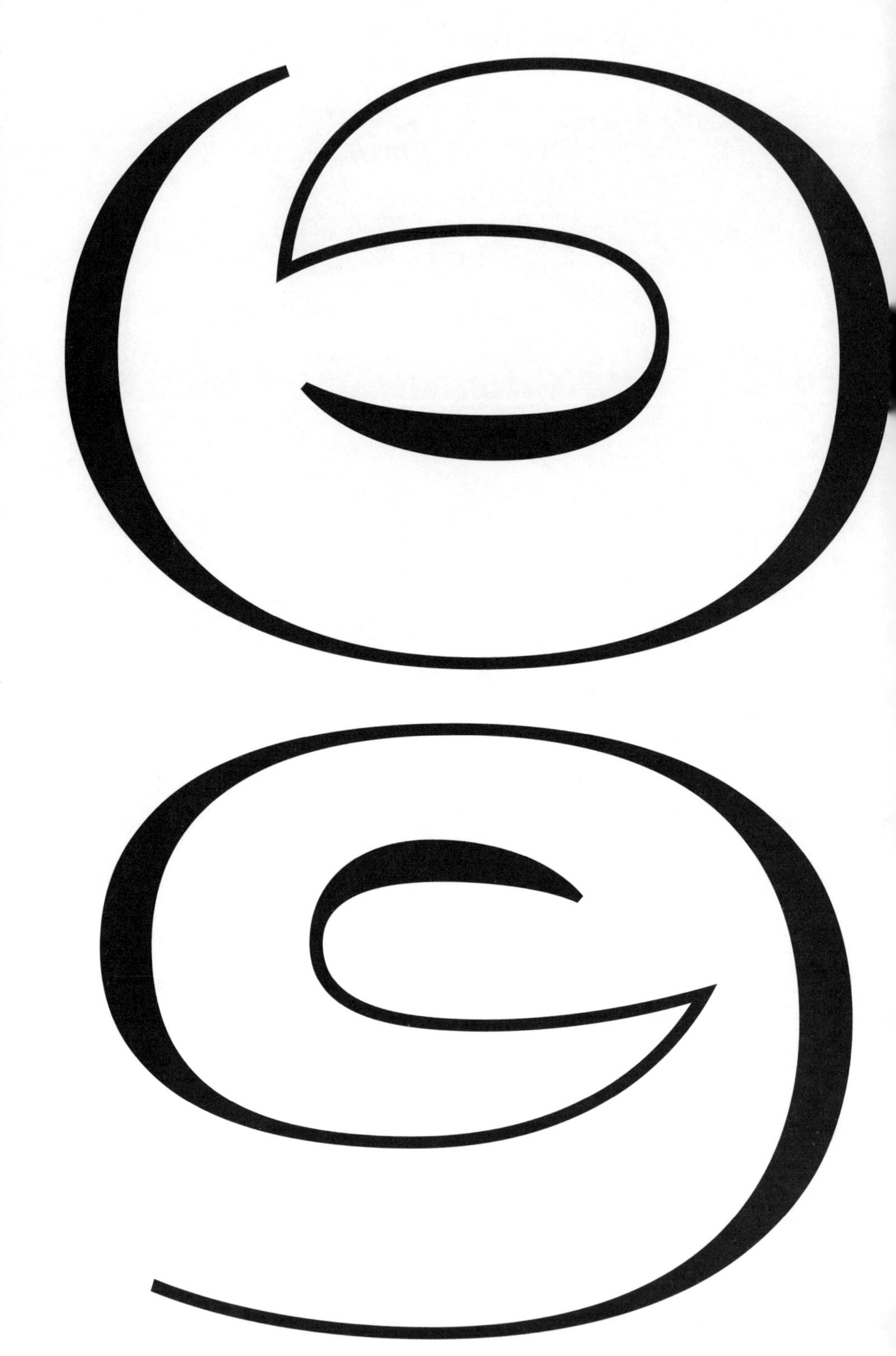

NINE

Anatomists of Race and the Simian Line

In the early twentieth century, clinicians were increasingly looking not just at the hands of their patients, but specifically at their palms. A medical family had encouraged them to do so, the specialists whose name was subsequently lent to Down syndrome. In the 1860s, John Langdon Down appraised the faces of his asylum patients and named the syndrome "Mongolism." One of his assistants, George Shuttleworth, also looked at their hand shape: "hands broad; thumb and little finger short; the latter often curved towards ring finger. Fingers taper at ends." The son, Reginald Langdon Down, turned the hand over. After the style of palmists, he presented palm *prints* rather than photographs to a medical meeting on "Mongolian Imbecility" in 1909 and noted a transverse line right across one or both palms. Many of his patients had two main palmar folds, not the more usual three; the head and heart lines joined. This was so common, he remarked, that it could be an aid to diagnosis. He had been aware of this for some time, admitting a patient in March 1907, recording "Hands, no head line."[1]

It had been noted before. Richard Beamish included a plate showing the hand tracing of a "congenital idiot" in his 1864 *Psychonomy of the Hand* (fig. 9.1). It showed clearly the joined heart and head lines, crossing the whole palm.[2] The French physical anthropologist Paul Broca is another instance. He had published

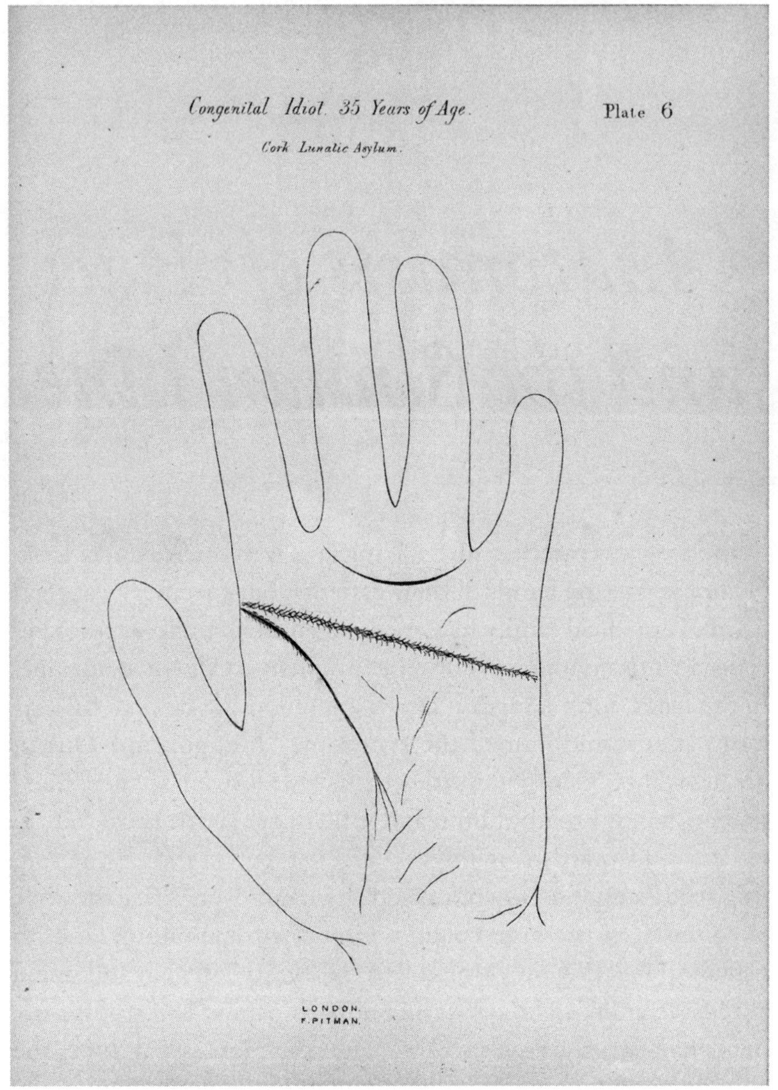

9.1 Now known as the "transverse line," this distinctive palm crease was also known as the "simian line," later the "four finger crease," and by French anatomists as "*le seul pli*," the single fold. "Congenital Idiot, 35 years of age," in Richard Beamish, *The Psychonomy of the Hand: or, the hand an index of mental development* (Frederick Pitman, 1865), plate 6. Copy in author's possession.

on such palm creases in the late 1870s, and made explicit links between simian and human palms, "*la disposition-simienne du pli palmaire.*" "The simian line" came to be a term as enduring, and as problematic, as "Mongolism."[3]

Over three generations, the Langdon Down family worked in asylums, caring for people with what came to be known as Down syndrome, first in the public Earlswood asylum in Surrey, which John Langdon Down reformed and transformed, and then in their own private and highly regarded Normansfield Hospital in Middlesex. Subsequent generations took on Normansfield, including Reginald Langdon Down, whose son John happened to be born there in 1905, with the syndrome named for his family. For all their personal and professional care and connection, the Down family belonged to an era and a milieu in which human difference was being thought through on multiple axes that mixed and confused race, ethnicity, and disability. The Downs were fully enmeshed.[4]

Hands—both their back and their front, their dorsal and anterior sides—were one telling part of the human body that could easily be studied and compared by all manner of clinicians, psychiatrists, primatologists, and physical anthropologists. Together, these specialists made up the third and fourth generations of comparative anatomists. But what anatomies were they comparing? In the first instance, they compared different types of humans in the present. This was anatomy in which living humans, unborn humans (the fetus), and dead humans (the cadaver as well as skeletal remains) were studied, dissected, measured, calibrated, and compared with each other across the world. In the second instance, such modern human anatomies were compared with pre-humans paleoanthropologically, as skeletal remains from a distant past. In the third instance, they compared human and so-called "subhuman" primates in the present. The hands of all manner of primates had long been a key organ in traditional comparative anatomy, the stakes increasingly high the more the comparisons were made within an evolutionary framework that posited the "transmutation" or the evolution of species. In the early twentieth century, the living hand offered fleshy, ridged skin, crease lines, and patterns to compare across species, across time, across sexes, and across the range of human being that generations of medico-scientists by then firmly categorized as "race."

This biomedical and evolutionary interest in hands, palms, and human difference took several turns in the early twentieth century. In the case of physician, psychiatrist, and Adlerian analyst Francis Crookshank (1873–1933), it was mixed with the many psychologies and philosophies of mind then taking shape about how the body could and should be read. He was adamant that "markings on the palm and even on the sole are of the utmost significance." Crookshank contributed a notorious study that folded together psychiatry, primate evolution, and comparative anatomy of race: *The Mongol in Our Midst*. Patterns on the palm were one of his telltale homologies; physical traits and characters present in different species, even in widely separated taxons, that suggested common descent. This was his evidence for an idiosyncratic thesis on what everyone in his generation called, after Down, "Mongoloids" or, worse to our ears, "Mongoloid imbeciles" or "Mongoloid idiots," the latter terms signaling as much technical psychiatric information then as they do offense now. His anatomical, evolutionary, and clinical research brought together a physical anthropology of race, with observations of people with Down syndrome, and incorporated both into research on subhuman primates. In doing so, he looked closely at the palm.[5]

At the same time in the United States, the zoologically trained anatomist Harris Hawthorne Wilder (1864–1928) studied palmar and plantar skin patterns that became a new medical subdiscipline. "Dermatoglyphics" took a generation to enter British medical and anatomical use and was most fully developed by another US anatomist, Harold Cummins (1894–1976), who codified the term at a meeting of the American Association of Anatomists in 1926, *derma*, skin + *glyphe*, carve. Even though his was a mid-twentieth-century career, Cummins carried forward nineteenth-century-style medical ambitions, vocabularies, and categories, inflected with statistical work from first-generation population genetics. His ambition was less to identify *individuals* (after the manner of Galton) and more to identify *types* of humans, by race or region or disability and through the largest samples and most disparate communities he could find.[6]

For each of these medical men, the hand, the palm, and even the sole were analyzed as part of comparative anatomy and evolutionary physical anthropology that typically included racial theorizing. But it was all too elaborate, too presumptive, and (as population geneticists were already showing by the 1930s) simply too inaccurate. Nonetheless, in the form of "dermatoglyphics"—invented in the United States and taken up in Britain—reading the hand became fully and technically part of the discipline of human anatomy and subsequently human genetics.

FRANCIS CROOKSHANK: THE PALMISTRY OF POLYGENY

Francis Graham Crookshank lived, worked, and thought in the rich but confused space between clinical medicine, psychology, and philosophies of mind. He learned these many ways of thinking in Bloomsbury, a mile or so from Cheiro's West End rooms and from Katharine St. Hill's home in Knightsbridge. In 1895, the year she was writing *Hands of Celebrities* and Cheiro published his *Language of the Hand*, young Francis Crookshank graduated with his medical degree from University College London, with clinical work undertaken at University College Hospital. Earning a gold medal from the prestigious medical institution, it was an auspicious start. He was early drawn to psychiatry, working in Northampton County Asylum and during the war serving as medical director of the English Military Hospital at Caen, where he treated soldiers with "unusual cerebral symptoms," the new disease entity named *encephalitis lethargica*. Medicine of the mind and body was a sudden focus during and after that war, a range of new therapies arising alongside these new diagnoses. Crookshank became a Fellow of the Royal College of Physicians in 1920, but the more philosophical and psychological he became, the more his golden start in medicine started to tarnish. He had "a curious literary ability, writing medical essays," one obituarist offered, "but straying into philosophical studies of medicine, as well as psychology and psychotherapeutics." An ungenerous obituary, but true nonetheless.[7]

Crookshank brought psychology to the hand, inspired by Viennese doctor and analyst Alfred Adler, the "Master" to whom Crookshank cast himself as "Pupil." And when he came to consider hands and palms in his study of Mongolism, he based even that project on Adler's signature "Individual Psychology," on the principles of "the functional and structural unity of the organism, or individual." It was a heady time for the thoughtful, and Crookshank was steeped in scholarship on language, meaning, and the somatic. A semiotics of the hand, so clear if complex to Paracelsus centuries earlier, also enchanted Francis Crookshank, fascinated by minds and mentality, by human interiors and how they manifested: the secrets that the body disclosed.[8]

Basic clinical medicine was (and indeed remains) a practice of decoding bodily signs, of interpreting this line, that mark, as a sign of this or that disease. Symptoms "are at once appreciated by the sufferer and often by the observer," explained Crookshank in one of his many tracts on the philosophy of diagnostics, while "physical signs" are deliberately sought by the clinician and are not necessarily comprehended by the sufferer. Every physician knew this. At a more abstract level, however—Crookshank would say at a more genuinely scientific level—he lobbied hard for medicine's uptake of a "Theory of Signs." Medical practice had descended to being simply an art or (possibly worse) a calling, he thought, no longer a science because those in the field paid scant attention to first principles. One of those principles was how, in diagnostics, the actual disease should be considered in relation to the word attached to it. In his view, physicians too easily imagined a disease word to be a real thing, the more or less random diagnostic neologism *encephalitis lethargica* his constant instance. Like contemporary linguists, he wanted to bring to medicine a clear awareness of the difference between the signifier and the signified.[9]

And yet Crookshank knew that the Swiss linguist Ferdinand de Saussure's thesis that the signifier and the signified were *randomly* connected was insufficient. In medicine, the bodily sign (say the simian line) at the very least had a correlation to the signified (say the symptoms of "Mongolism"), and possibly a causal one. Another linguist, the American Charles Peirce, had thought about a triangular relationship between a symbol, a thought, and a referent, and Crookshank thought this triple distinction far more plausible and useful. He tried to explain "Names, Notions and Happenings" in a lecture delivered to the Royal College of Physicians, but he was literally lecturing *at* the room of doctors, who looked back blankly. Still, two clever Cambridge linguist dons spoke Crookshank's language, and sought his contribution for an important book that brought Saussure and Peirce, from the previous generation, up to date. And so, the arcane *Meaning of Meaning* by the linguists Ogden and Richards included an essay by clinician Francis Crookshank, "The importance of a theory of signs and a critique of language in the study of medicine." Only linguists and anthropologists ever read it, however, and Crookshank despaired at his profession's hopeless inability to comprehend a proper theory of signs. He turned bitter about his brethren, who would say that "only 'mad doctors' in these scientific times dabble in Philosophy." He committed suicide in his home in London in October 1933.[10]

In his lifetime and afterward, Crookshank came to be best known for a book that was both notorious and successful, *The Mongol in Our Midst* (first edition, 1924). And although the subtitle was about the face—"A Study of Man and his Three Faces"—the anatomy of the hand was also featured. It turns out that Francis Crookshank in Wimpole Street, Cheiro in Bond Street, and Katharine St. Hill in Park Row, Knightsbridge, all shared simultaneously a precise interest in how to read signs of the palm.

Crookshank was most interested in those hands that held the sign of Mongolism. His peculiar thesis drew from the Downs' work, especially from the nineteenth-century father who had introduced the term "Mongolism" in the first place. In this regard, Crookshank's book is marginally better than its title sounds, since he insisted that mongoloids were not all "idiots" or "imbeciles" in asylums but were around us and with us all the time, "passing" as normal, a term he used himself. And so, "mongols" were "in our midst." On the other hand, the book is far worse even than its shocking title, since Crookshank's thesis was that newborns with Down syndrome were atavistic evolutionary remnants. This was a development of John Langdon Down's original idea about "retrogression of ethnic type." That had not been received particularly well, however, and even Langdon Down's student Shuttleworth said in 1909 that he could not go so far as to adopt this theory. By the 1920s, decades of high anxiety about British "racial" degeneration had accumulated, igniting a popular as well as scientific eugenic concern precisely for those "in our midst" who might silently reproduce undesirable traits. This offered Crookshank a more receptive context. For him, Mongolism opened up not only theories on evolutionary human origin and on the relationships among current "races of mankind," but also between humankind and particular species of existing apes.[11]

Crookshank derived his theory from the famous early evolutionary book *Vestiges of the Natural History of Creation* (1844). In a section on the growth of the human fetus and how that reveals and repeats ancient developments and transitions, the author had written that the different races of humans in his present, though all of one stock, embodied those successive stages of development. "The leading characters, in short, of the various races of mankind, are simply representations of particular stages in the development of the highest or Caucasian type." And so, in this logic and for Crookshank, "the Mongolian is an arrested infant newly born." What sparked Francis Crookshank's particular attention was a sentence that followed: "It is found that parents too nearly related tend to produce offspring of the Mongolian type,—that is, persons who in maturity still are a kind of children." All those decades before—before Darwin's *Origin of Species*, before eugenics, before degeneration anxieties among the British higher classes—the author of *Vestiges*, Scottish Robert Chambers, had not just made an interesting "ethnic" classification (Langdon Down's term), but had even attached it to "suitableness and unsuitableness of marriages." Crookshank read *Vestiges* back through Langdon Down's "ethnic" classification of idiots and imbeciles, turning them both into a much harder racial biology of evolutionary descent. Crookshank interpreted Chambers's monogenist argument—and in fact Darwin's as well—to be evidence for what he called the "polyphyletic descent" of humans. Different groups of humans have different common ancestors whom they each share with different apes: the so-called white human race shared a common ancestor with the chimpanzee, the black human race with the gorilla, and the yellow human race with the orangutan. It was a polygenesis of sorts, suggesting different species of human. In Crookshank's hands, both Mongolians the Central Asian people and Mongoloids with the syndrome provided physical evidence that substantiated this theory.[12]

In the first instance, Crookshank suggested that Mongolians look like orangutans because both human and nonhuman primates share a common ancestor, just as whites look like chimpanzees because those two primates share another common ancestor. In the second instance, and crucially, he argued that Mongo*loids* don't just "look like" Mongols (Langdon Down's nineteenth-century proposition), but actually are phylogenetically linked to that race of humans known as Mongolians. In the third instance, and even more shockingly, he triangulated this apparent similarity between Mongolians, Mongoloids, and orangutans. In his asylum work especially, he noted physical similarities between "the Mongolian and Malayan races on the one hand, and the orangutans on the other, as well as between the imbecile mongoloids of our asylums and these Bornean and Sumatran apes." In short, this physician was constantly trying to expand the dyad of comparative anatomy (human race + subhuman primates) to a triangulated comparison: "normal" human race (Mongolians) + "abnormal humans" (Mongoloids) + subhuman primates (Orangs). He correlated a range of specific homologies to substantiate a common ancestor. For Crookshank it was axiomatic, after Darwin's principle, that "homologies prove descent." And so, he drew physical connections between Mongolians, Mongoloids and, very specifically, the great ape called the orang. Partly based on the evidence of palm lines and hand shape, he developed a thesis about three kinds of humans, related in the present to three kinds of apes, each with a separate, not common evolutionary ancestry. Palmistry met polygeny, or what Crookshank presented as the polyphyletic scheme of human descent.[13]

Crookshank had long been puzzled as to why his medical colleagues rarely sought to connect Mongolism with insights and research from physical anthropology. One of the reasons was that this kind of homology-seeking comparative anatomy was old-fashioned for the 1920s and 1930s. Certainly it was old-school anthropology, the cutting edge just then shifting away from bodies and bones toward culture, ritual, kinship, and belief. The method of substantiating evolutionary descent by looking for physical homologies between humans and apes derived from the intellectual world of mid-Victorian Thomas Henry Huxley, not the 1920s world of cultural anthropologist Bronislaw Malinowski, for example (whose supplementary essay on ritual language happened to sit alongside Crookshank's in *The Meaning of Meaning*). Still, there *were* old-school comparative anatomists leading various disciplines, especially among paleoanthropologists, who put together piece by piece a fossil record that told the human origin story. Sir Arthur Keith was one, an anatomist in charge of the Hunterian Museum of the Royal College of Surgeons. He lectured in 1901 on "the Anatomy of Palmistry." Grafton Elliot Smith was another, more interested in brains and heads than hands, but certainly a researcher who paid attention to anthropoids' opposable thumbs. Lamarckian anatomist Frederic Wood Jones was a third. He had recently written not just *Arboreal Man* (1916), with another round of attention to that famous opposable thumb, but even more specifically *The Principles of Anatomy as Seen in the Hand* (1920). They were all extending a century of comparative anatomy, primatology, and research on species' homologies and on "vestigial structures," residual organs from humanity's long evolutionary journey. These had been theorized and substantiated by Charles Darwin, by Thomas Henry Huxley, and by a hundred other Victorian naturalists interested in similarities and differences, especially among human and subhuman primates. None, however, had nominated the peculiar "orangoid homologies" in the manner of Francis Crookshank.[14]

A library could be written in response to Crookshank's thesis on the three faces of man, but our question is: How did hands and palms enter his evidence and his claims? As regards humans, Crookshank considered the correlations between palm lines, hand shape, and "psychical styles and patterns" to be far stronger than most in the medical world believed. But for such a learned man, he didn't help his case by making wild statements about what he called common sense. He dipped back into old-fashioned physiognomy, for example: "The correlation between certain palm lines and lines of conduct during life is as indisputable as is that between a big nose and pushfulness, or obstinacy and a jutting chin." And worse, so far as reasoning goes, was his *ex post facto* rationalization if one encountered the opposite: that could be explained away by Adlerian psychology (Crookshank's favored school), "which teaches us how often over-compensation is induced by an organ-inferiority." Moving from physiognomy to palmistry—itself common sense "since untold ages"—he thought it simply inadequate for such reading of bodies to be cast aside, especially with flippant gibes about the fortune teller. To do so was naive; similar to dismissive comments about "dream-books in the servant's hall," when in fact everyone was reading and benefiting from Freud; or sneering commentary about astrology when it was plain that epidemiologists were correlating epidemics with cosmic cycles. Crookshank was irritated also by a "turgidly magisterial" article on palmistry that had recently appeared in the *Encyclopedia Britannica*, which put palm reading "beyond reason." If that was the case, he asked: What are we to do with empirical, clinical, asylum, and consulting-room work in which correlations between hands and conditions had long been evident? Can we be so sure that the signs of the palm have no predictive meaning? And why can that hypothesis, at the very least, not be put forward by reasonable and reasoning medical men?[15]

And so he did. Like the papillary ridges so well known to Scotland Yard, Crookshank thought that all palm lines deserve and require anthropological study. Following the nineteenth-century French chirologist D'Arpentigny, he first turned to hand shape and form, actively wondering just why this had received so little attention from anatomists and physical anthropologists who were busy measuring and comparing every other part of the human body. He described a "typical" Mongolian hand, meaning those of Central Asian people, and started to draw his own comparisons and correlations, not with other "races" but with various medical conditions that could be diagnosed. Similar hands are seen in people with Cushing's hypopiturianism, for example. And so, also, for the Mongoloid hand in which the same qualities were exaggerated: "Lax, plump, stubby, square-palmed, short-thumbed, and with a wabbly little finger." Drawing strongly on Wood Jones's primatology, Crookshank explained a "digital formula" common in orangutans: how the fingers are measured against one another. There was a pattern, apparently: "3, 4, 2, 5, 1." Reading a little like Cheiro's numerology, Crookshank identified this metric in his asylum patients as well: the thumb shorter than the fifth finger, shorter than the middle finger. And then the message: "a definitely simian feature."[16]

Turning the hand over, to look at the palm, he found much more. The three main palm lines—the life line, the heart line, the head line—are normally present in the last weeks of fetal life, soon to be joined by many more detailed lines in the early weeks of life. But some infants are born with only two, not three, main lines: the line of life is present, but only one other crosses the palm from side to side. In identifying the simian line, Crookshank drew from two authorities, in addition to Reginald Langdon Down. The first was nineteenth-century Richard Beamish, who had seen such a palm "on the hand of an idiot at Cork." The second is even more interesting. Twenty years earlier, Crookshank had met and discussed hands with a famous palmist. Cheiro had shown him the handprint of Dr. Meyer, who had been executed in Chicago for murder. The single fold was apparent (fig. 9.2).[17]

Just like Cheiro in his many books, Crookshank illustrated the simian line with two clear plates, using photographs now, not ink prints. One was of a "normal" human palm, and the other the hand of an "imbecile mongoloid" showing the simian line or "orangoid line." That they came from the photographic collections of his colleagues, Dr. Primrose and Dr. Brushfield, indicates how interesting palm lines now were to clinicians. Crookshank speculated phylogenetically on these hands; that the more usual three palm

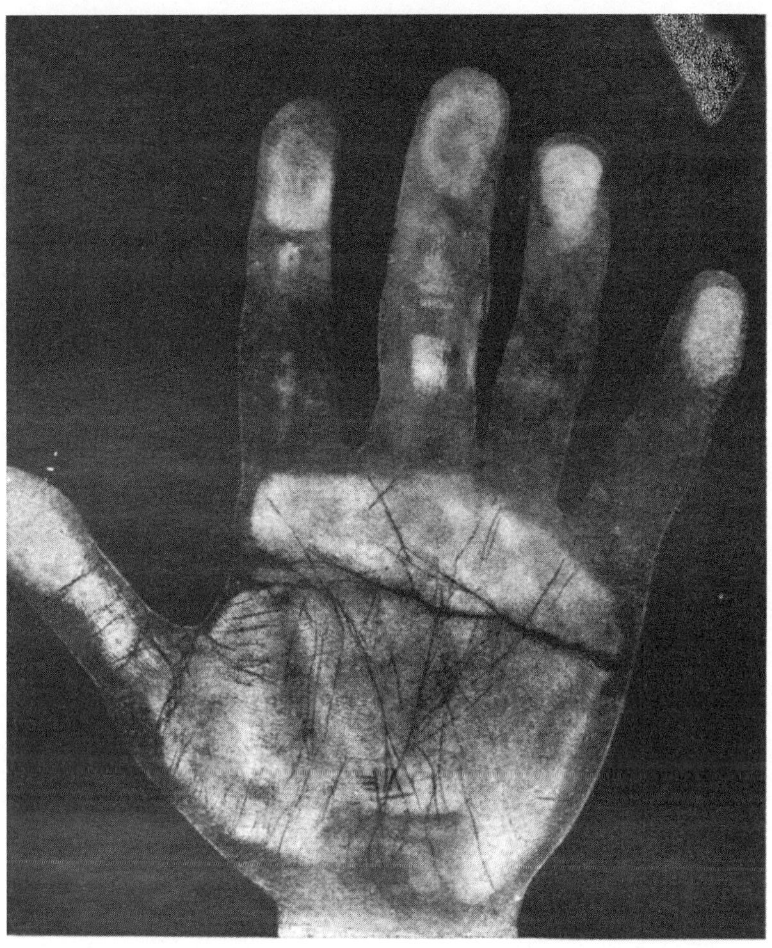

THE RIGHT HAND OF DR. MEYER, CONVICTED OF MURDER IN NEW YORK

Note the centre horizontal line (line of mentality) across the palm in an abnormal position, rising toward fourth finger

9.2 The murderer's hand discussed by Cheiro and Crookshank. From Cheiro, *Cheiro's Language of the Hand* (Herbert Jenkins, 1900). Copy in author's possession.

lines in humans had evolved over time from an earlier two palm-line hand. The single transverse line—the "simian line"—was thus primitive, an "atavistic" sign that appeared in some humans, as in some subhuman primates. When seen on the newborn, the simian line, what the French sometimes called *le seul pli*, the single fold, held "predictive meaning."[18] Reginald Langdon Down had shown in 1909 that it occurred frequently in one or both hands of his patients. But Crookshank considered "simian" to be more than a qualifying adjective, applying it literally and directly. His own patients were, he wrote, "even more simian than Mongolian." Indeed, he invented a peculiar reverse nomenclature for the simian line, calling it occasionally a "Mongolian or Orangoid" line (fig. 9.3).[19]

To substantiate such claims, Crookshank turned to the palms of anthropoid apes themselves, building on contemporary primatology that showed the palms of apes were indeed characterized by one or more continuous transverse lines. Crookshank claimed, erroneously, that the line "most certainly" did not occur among gorillas and chimpanzees. Drawing from Wood Jones, again, it *was* present in the Rhesus monkey, but most reliably apparent on the palm of the orang. In 1924, C. F. Sonntag, an anatomist employed by the Zoological Society (and mentored by Wood Jones), had shown clearly that the orang had the single transverse line, and sometimes even two in parallel, both running right across the palm. Crookshank's conclusion, then, was that in humans the line was a "stigma of mongolism" *because* it was also a definite homology with the "normal orang-utan." His homologies thus correlated "Mongols," "Mongoloids," and orangutans. Offering yet another in the long list of biological and medical classification of primates, this time based on the hand, he nominated this group as "Type M." And then he created a "Type N" hand—Negro—in which sometimes a third transverse line "nearer the fingers than that called the Line of Heart and not as a rule running right across Palmists know this as the Girdle of Venus." This girdle of Venus corresponded to Sonntag's illustration of the gorilla's palm. And finally, "Type W" was the hand of the artist, the primitive "Aryan" and the "schizoid." These "white" hands were "chimpanzoid," showing a different characteristic set of palm lines altogether,

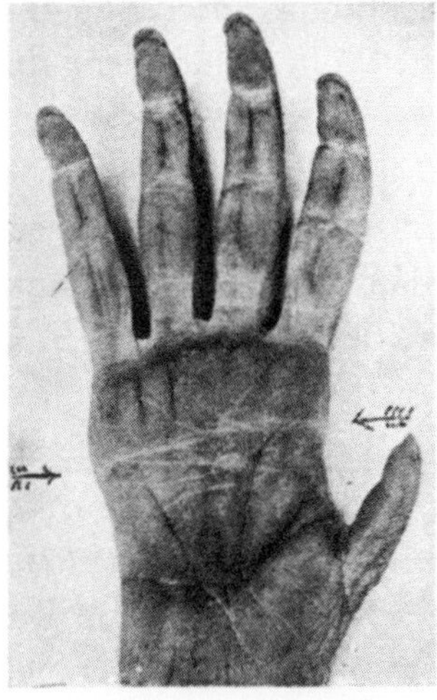

Fig. 1. Hand of an Orang-utan.
(Photograph by Dr Primrose.)

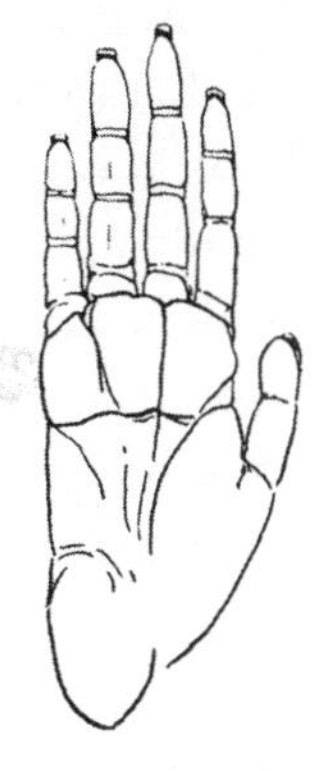

Fig. 2.
Hand of a Rhesus Monkey.

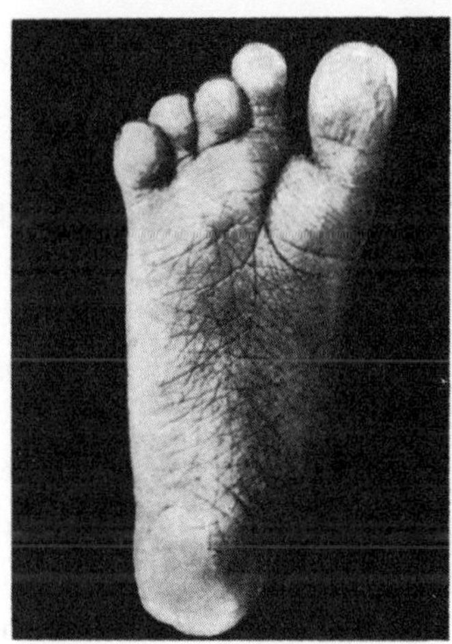

Fig. 3. Foot of an Imbecile Mongoloid.
(Dr Brushfield's case.)

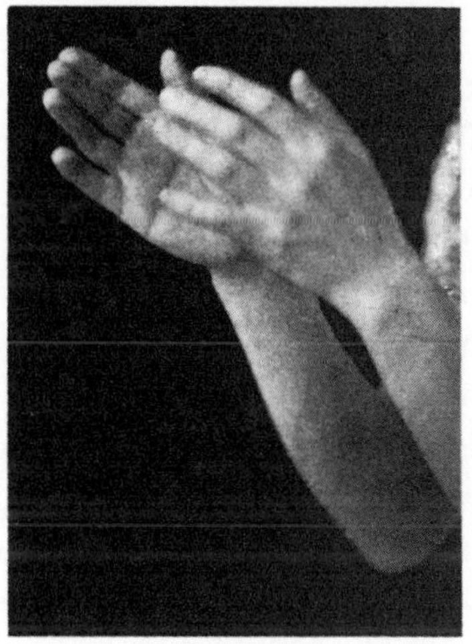

Fig. 4. Hands of a Russian Lady of
Tartar ancestry.

HANDS AND A FOOT.

9.3 Crookshank's illustrative evidence to show similarity between the orangutan, "mongoloid" humans, and "normal" humans with Mongol—Central Asian—ancestry. F. G. Crookshank, *The Mongol in Our Midst* (Kegan, Paul, Trench, Trübner, 1924), plate XVI. Copy in author's possession.

apparently held in common with chimps. Soon enough, medical doctor and palm reader Charlotte Wolff would verify that it was gorilla palms that were most humanlike.[20]

Crookshank was probably most interested in how the simian line occasionally appeared on the hands of the apparently "normal," certainly among Mongolian people. In the "normal" of other races too, the single transverse line could best be read as a sign of a subordinated racial ancestry (read geographically Central Asian). He saw it in the hand of a Jewish doctor, for example, "who confessed to me that he was of East Baltic and probably Khazar ancestry." Yet he had never seen this sign on the hand of any Sephardic Jew. Nor, he said, had he seen it in the hand of a Hindu, or of any "pure Aryan." While it was common in France, it was almost entirely limited to those obviously "mongoloid." It also appeared among "Nordics," but in that case typically took obtrusive form, in people passing as normal.[21]

Crookshank claimed that every person he had encountered with the single line displayed some sign of mongolism, however slight; they were all "in *some degree* atavistic or, at least, orangoid." A fellow physician with a simian line quite reasonably took offense, and Crookshank was quick to clarify his thesis. The simian line was a sign of atavism, he insisted, but not necessarily one of imbecility. His colleague, in other words, was one of the "mongols in our midst." But that hardly improved matters. Although the sign was evident on the palms of "quite intelligent people," they were still likely to show slight traits of being "different." Indeed, in some cultures this difference was positively received, he offered. Chinese and Japanese artists were "oddly reticent" to discuss the special line, not because it was a bad sign, but because it was a "sacred line." For those who wished to verify, he challenged, it was clearly visible in Chinese art and sculpture held by the British Museum. Crookshank looked to European art for evidence as well, since portraitists were also keen anatomical observers. In a final attempt to appease those "in our midst" with the simian line, but who were disinclined to being linked ancestrally to Mongolians, Mongoloids, or orangutans, Crookshank gestured to the work of Renaissance Milanese master Bernadino Luini. In one of his paintings, hanging now in Budapest, we can see for ourselves that even the Virgin Mary had a simian line.[22]

DERMATOGLYPHICS: THE RACIALIZED ANATOMY OF PALMS AND SOLES

At the same time in the United States, the scientific study of creases and lines on palms and feet was pursued afresh by a slightly different breed of anatomist. Harris Hawthorne Wilder, professor of anatomy and zoology at Smith College, Massachusetts, was looking at fingers, palms, and soles, including his own. This was a refinement of anthropometry, in this case the systematic measurement of palmar and plantar ridges. Via Francis Galton's work, American anatomists and physical anthropologists made fingers, palms, and soles their own, the study of dermatoglyphics later circling back to University College London, when Lionel Penrose and his laboratory of first-generation human and medical geneticists started to look again at hands (chapter 11).

It is noteworthy that Wilder joined a long tradition of scientific palmists in studying physiognomic faces too, and he occasionally used the term "physiognomy" to describe his method. Wilder also occasionally wrote of dermatoglyphics as "scientific palmistry," but the small suite of specialists who followed rarely conceded any palmistry genealogy for their work, and would routinely, if disingenuously, insist they studied *ridges*, not palm *lines*. These anatomists tended to evade everything "chiro": chiromancy, chirology, chirognomy, Cheiro, or Keiro. Yet this writing or etching on the skin—these "glyphs"—were still signs that needed to be decoded, an apparently new palmar and plantar language. We need only look at how these anatomists communicated meaning between themselves to see the enduring power of the palm. At their first meeting, Wilder the master held out his hand to a young apprentice, Harold Cummins, who was to spend his life studying hands at Tulane University, New Orleans. The gesture was not to shake hands, but to inspect them. "Notice how the hypothenar pattern resembles that of the monkeys," Wilder said of his own palm. And they sealed their commitment to a new brand of scientific palmistry with an exchange of handprints. As a salutation to the master, and to his inherited knowledge, but also like any number of contemporary palmists, Cummins went on to reproduce this print of the right hand of Harris Hawthorne Wilder as an authorizing frontispiece to his textbooks on fingerprints, palms, and soles.[23]

Wilder's and Cummins's anatomy of the palm was similar to Crookshank's, simultaneously comparative anatomy, physical anthropology, and primatology. It was as useful for the geneticist as for the morphologist or ethnologist.[24] This anatomy of palms and hands had a Galtonian origin that then crossed into the evolutionary history of *Homo sapiens*. Next-generation Cummins took "pedigree" studies a step further, approaching dermatoglyphics not just as a search for phenotypes of race, but also into the earliest moments of human and medical genetics; its population or statistical expression in the 1930s and 1940s. A formal and lifelong anatomist, he linked the older nineteenth-century comparative disciplines with the new twentieth-century fields. And although Cummins understood his dermatoglyphics in the genealogy of Francis Galton's statistical observations, it was also clearly prefaced by the long history of physiognomic and chiromantic sign-reading and mark-making (fig. 9.4).[25]

Wherever palms could be analyzed, Wilder and then Cummins followed the comparative research possibilities. Neither was a clinician, like Crookshank, but anatomists and pathologists did have opportunity to examine normal and abnormal, anatomical and pathological, bodies-in-death, as well as in life. Wilder, for example, looked closely at living conjoined twins, taking prints of their hands and feet. Cummins followed through with studies of "monsters," the technical, if already old-fashioned, anatomists' term for the grossly physically deformed. In 1923, he published a paper on epidermal ridges "in a human Acephalic monster." His growing interest in palms directed him toward anatomies of the disabled. And over time, it was Harold Cummins who refined the recognition of a set of dermatoglyphic patterns on the palm and finger that signaled mongolism, beyond the simian line. That came to be understood as his most important intervention.[26]

TRACING THE PEDIGREE 251

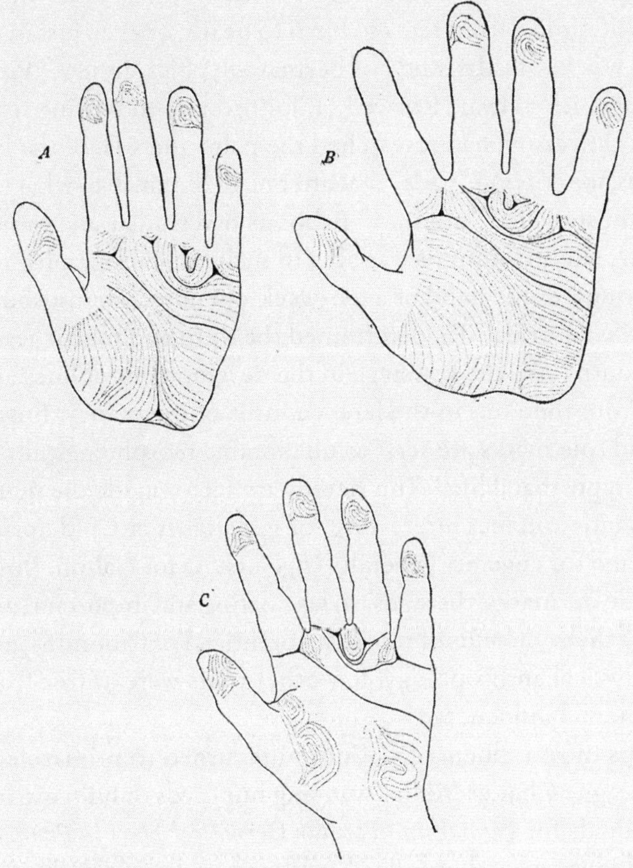

Fig. 85. Configuration of the epidermic ridges (friction-skin) upon the palm of the human hand.

a. Negrito, female. Island of Luzon. 18 years old. H. H. W. Coll. No. 479.

b. Maricopa Indian from Arizona. Male, 55 years old. H. H. W. Coll. No. 467.

c. American white, Portland, Me. Male, 12 years old. H. H. W. Coll. No. 409.

9.4 Comparative anatomy of human hands, showing "epidermic ridges." H. H. Wilder, *The Pedigree of the Human Race* (George G. Harrap, 1926), 251. Courtesy the Wellcome Collection, London.

Human embryology was another focus through which these anatomists sought to clarify just when palm creases normally formed in human fetuses, and whether they changed before birth. The ontogenetic history—the development of an organism within its own "lifetime," from fetus to birth to death, or even just in fetal life—was key to the study of dermatoglyphics. In 1899, Pittsburgh eugenics stalwart Roswell H. Johnson, then writing from Harvard University, had researched the palm and sole of the human fetus, identifying "pads" that after birth diminish to what (he said) palmists called "mounts." Baboons had similar anatomies. Fifty years later, Cummins was able to include photographs, not just drawings, of the hand of a ten-week-old human fetus, showing such "volar pads." He determined the third and fourth gestational month to be the key stage in the development of lines and ridges. From then on, in the later months of pregnancy, finger, palm, and sole marks are set, "as unalterable morphologically as they are in postnatal life." This hereditary factor made the markings of hands and feet interesting for geneticists at Cold Spring Harbor, and for eugenics generally (fig. 9.5). As for Galton, this is what made the marks, the signs, so fascinating and important, and what gave them all kinds of research potential. For Cummins, and within physical anthropology, few other traits were as free from environmental influences.[27]

Perhaps most influentially, Cummins turned to primatology with colleague Charles Midlo, studying hundreds of humans and other animals for *Palmar and Plantar Dermatoglyphics in Primates* (1942). This was what they called "comparative dermatoglyphics," in which the palmar and plantar prints were taken, photographed, drawn, and analyzed from the tree shrew, the lemur, the night monkey, from New World monkeys and Old World monkeys, from a particular orang named "Pongo," and from the palms of *Homo sapiens*. As early as 1904, Wilder had researched "racial differences in palms and sole configuration" and how such differences might be seen on the palm and sole prints of Japanese and Chinese people. This research too was taken up by Cummins, who applied the study of dermatoglyphics to physical anthropology and studies

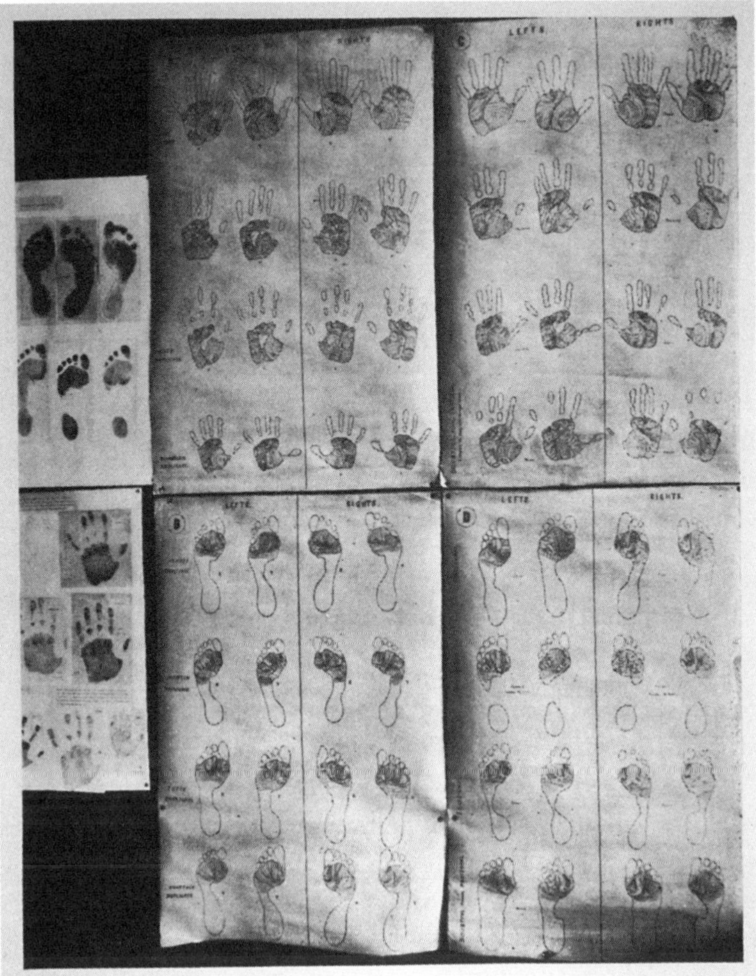

9.5 Inheritance of palms and sole prints, 1921, Cold Spring Harbor. E.06 Eug-2, Exhibits Book, Second International Exhibition of Eugenics, 91. Courtesy Cold Spring Harbor Laboratory Archives, New York.

of human population differences; that is, to the biological study of race.[28]

As we have seen, many claims were being made in this period based on skeletal remains, but researchers of skin clearly needed flesh, not bones. This did not necessarily need to be that of the living. Early in his career at Tulane University, New Orleans, Cummins dissected the body of an African American man, keeping the skin from one sole and preserving it between two sheets of glass. A later geneticist praised his foresight, including the fact that he had included skin from the toe. Race was never incidental in such work, not least in Cummins's shameful recollection that many cadavers in life had been social "dregs," but in death finally contributed socially. If anything a worse statement, he labeled them "passive teachers," whom we should therefore respect. Sometimes, however, race was precisely Cummins's biological point, especially when he transferred his dermatoglyphic research to the living. At a high point of physical anthropology, Cummins went on to study palm folds among "Eskimos" and among South African "Bushmen"; among "Negroes of West Africa" and "Indians of Southern Mexico," among the "Siamese" and "the Jews," among "peoples of the Near East" and in a "Dutch family series"; among "Rwala Bedouins" and "Spanish Americans." Colleagues across the world, taken with the possibility of dermatoglyphics, added to this list, a study of "three Veddahs," for example, and "six natives of Tierra del Fuego." This was second-generation tracking of "race" via hands. We have seen that Galton had tried, and statistically failed. But Wilder had pestered Galton in the years around 1900 with his own research on race, palms, and soles, offering him prints of "several rare and interesting human races," taken from the recent St. Louis Anthropological Exhibit. He gingerly trusted Galton with duplicate sets, and was excited: "I presume that those of the Central Africa pygmies are the first prints ever taken" (fig. 9.6). He followed up with prints from Ainus, Patagonians, and Indians of the Southwest United States. These, added to the rest of his collection, covered all the extremes of human types, he boasted, but like Galton, he was not sure that the data were supporting his hypothesis. Wilder tried again in the more elaborate "sculpture of the palms and soles," and thought it more promising. But still he could only conclude a minor positive result from an investigation of Maya Indians, Whites, African Americans, and Chinese. He did identify a pattern in the Maya (he claimed), "a high frequency of palmar thenar/first interdigital patterns and

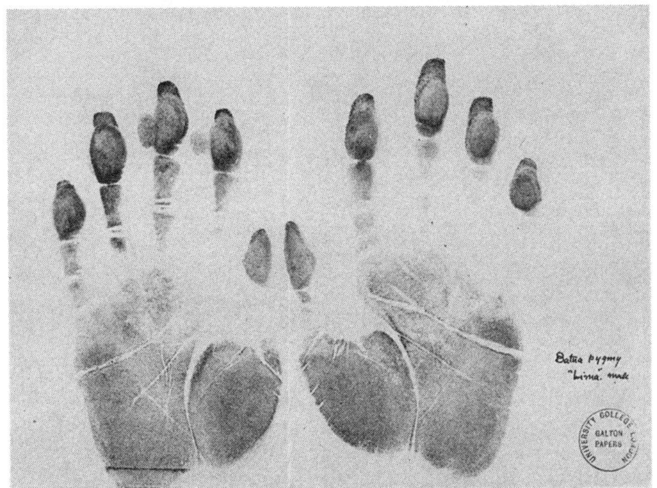

9.6 H. H. Wilder's handprint of an Echuya Batwa man, sent to Francis Galton, November 28, 1904. Francis Galton Papers, University College London, GALTON/2/9/6/2/2. By permission, UCL Library Services Special Collections. Courtesy the Wellcome Collection, London.

infrequent hypothenar patterns." And then he confided to the master, Francis Galton: "I am rather losing my faith in the value of the markings as racial criteria," though he still held some hope that with many more samples, there may be positive results.[29]

Wilder's protégé, Cummins, carried forward just that hope. In doing so, Cummins knew that he was the latest in a long line of comparative anatomists and physical anthropologists who had tried to classify humans into groups on any number of criteria; physical, cultural, linguistic, and geographic. Nineteenth-century physical anthropologists used multiple characters ranging from skin color, eye color, hair form, to nose form—modern physiognomy. In the twentieth century, blood groups had been used to try to identify patterns that would mark one group off, physically, clearly, and objectively, from another. But as Cummins well knew, the sets of characters would often conflict, not align, in this science of race, and had long done so. It was a problem. Indeed, this long record was already being tallied by other biologists and mathematicians—population geneticists—as the impossible biology of "race." Cummins continued to pursue it, however, seeking to add to the list of measurable "racial" characters toward an (increasingly elusive) fixity. His life's work was the scientific study of increasingly fine traits in populations, "types" and "races" of humans, as told by the hands. While he was well aware of Galton's and then Wilder's negative results, he nonetheless persisted, seeking to add to the data, and with his own work concluded that "racial differences in dermatoglyphics are real." By this he meant not that individuals could be thus identified as x or y "race," but that group differences were beginning to be reliably and statistically apparent at a population level.[30]

In Cummins's view, dermatoglyphic analysis recommended itself to such racial classification because it was now standardized methodologically, and it was objective; the trait could not be environmentally modified, since the marks remained the same from birth. He rightly indicated that up until his time it was fingerprints that had been the most studied, but palmar patterns should be systematically added, he thought, because they also remained consistent. They were "age stable." Cummins also knew that while skin ridges and patterns were heritable and therefore dependent—somehow—on genes, cellular-level knowledge was as yet insufficient. Pending developments in genetics, it was only possible to compare races at a group or population level and "on the basis of phenotypes." Palmar or plantar marks were additional characters that could and should be added to the long-established mix, helping show multi-trait trends that might make the group distinctive. Racial characteristics were certainly part of what he still called the human "constitution"; so were sex differences, or right- and left-handedness. For him, before cytogenetics, it was only statistical analysis of large samples that could reveal correlations and trends. In fact, he was clear that firm "diagnosis" of race, sex, criminality, or predisposition to disease in an individual was not possible solely through observation of dermatoglyphics. For example, there was no convincing correlation between dermatoglyphics and blood groups, and only a general trend in ridge breadth and palmar patterns that correlated with sex, that became clearer in some racial groups at any kind of statistical significance. Still, working from his own research and that published by others, Cummins persisted with the intention of gathering enough data to make hand, sole, and palm marks, a suite of phenotypes that together might biologically distinguish one group of humans from another. He studied, for example (these are his terms): European Americans, Germans, New Mexicans of Spanish descent, Jews, Eskimos, North American Indians, Indians of Mexico and Middle America, Chinese, Japanese, Gilyaks, Oroks, Siamese, and Negroes (East African Negroes for fingers and Liberian Negroes for other features).[31]

Cummins was not researching "race" naively, at least in his postwar work aware of the word's "ill repute." He used the term as biologically equivalent to "breed," of dogs or cattle: groups of humans with common characteristics, "traceable to inheritance." He well knew that it was methodologically insufficient to categorize "whites" as a race, and then switch to the category of nationality. But his was a thin critique, to say the least, with a biological, not a political motivation. He was not a critic, in other words, of how his own research on race might be formed by, let alone contribute to, *racism*. The problem for Cummins was the obvious one of dividing people into groups. Italians, for example, now one nationality, were made up of two races, apparently: "Alpines in the north, and Mediterraneans in the south." And, while any number of biologists of his generation were abandoning the search for biological race, Cummins never sought to overturn the field. Rather, he persisted in its apparent refinement, developing the well-worn—overworn—classificatory triad: "White, Yellow-brown and Black."[32]

The increasingly specific groups whose hands he studied were most often geographically selected for their isolation or for their self-defined specificity, sometimes religious. The Rwala Bedouins were interesting to him, for example, because "these Arabs represent a relatively pure racial group, isolated by virtue of their nomadic life and the hostility of the desert." He collaborated in this instance with William M. Shanklin, professor of histology at the American University in Beirut, who in the early 1930s had done work on blood groups, in which a high group O seemed to distinguish the Rwala from other "Near East peoples." This appeared to be confirmed by a corresponding different cephalic index as well. Could dermatoglyphics be a third phenotype that would confirm a biology of race? Shanklin himself took the palm prints of 231 Rwala men from five different camps. The palmar main lines were studied, with close investigation of where they terminated: this was part of Cummins's method, his "main-line index." They concluded that "hypothenar patterns" were more frequent in the Rwala left hand, a similar finding to studies of Chinese and of "various collections of American Indians," and in clear contrast to "Europeans," who have a right-hand trend. The finger and palm prints of Rwala men were compared with Lebanese males (Shanklin's students), New Orleans Jews (in Cummins's circle), and "Eskimos" and "North-American Indians" (previously studied by Cummins). Cummins did indicate that the distribution of similar and dissimilar dermatoglyphic patterns might well offer clues to the continuing mapping of "ancient lines of dispersal," of early human migration. Who might be revealed as related to whom, by a correlation of blood group, dermatoglyphic, and cephalic indicators? He speculated that a geographic pattern in density in humans in the present (dense palm patterns in Central Asia and least dense in southern Africa) might substantiate a 1930s thesis about the origins of humans in Central Asia.[33]

In his textbook on the field, Cummins dealt with the simian line and mongolism in a chapter called "Constitution." This encompassed "all the structural, physiological and psychological traits of an individual," partly inherited, partly acquired within environmental contexts. Palm and sole marks were part of the "structural constitution," heritable, though possibly also open to prenatal environmental influences. If "race" was difficult to diagnose dermatoglyphically, was correlation with particular diseases clear enough to serve as a diagnosis? If so, this would be of the "utmost import." Some early and small studies of neurofibromatosis warranted further investigation; what he called the heritable skin disease psoriasis showed similar early results that could be positive. More statistically significant samples in German studies of schizophrenia and dermatoglyphics had been confirmed several times, though he thought racial distinctions were complicating these conclusions. Others had studied epilepsy and general feeblemindedness, as expressed in the hand. As with "imbecility" and "idiocy," statistically significant correlations had not yet been shown. It was all different, he claimed, with "Mongolism." There, the distinctions in fingerprints and palmar configurations were "outstanding." Particular patterns—not just the simian line—were doubled compared to "normal groups." The "main-line index" was unprecedentedly high. As for the simian line, it was not exclusively a sign of mongolism, as many already knew. "Occasional in normal persons, [it] occurs quite frequently in mongoloid idiots." The simian line was now one mark which, when combined with a suite of others, *could* signify Mongolism, confirmed on the cusp of the shift of designation to "Down syndrome," and on the cusp of cytogenetics. A palm line was now, formally, a phenotype within human genetics.[34]

"New Palm Study is Real Science," announced a *Science News Letter* in 1932, with a photograph of Cummins at his craft. Scientists don't tell fortunes, or predict marriages, or numbers of children, but they may now be able to "reveal your race" and confirm whether one is naturally right- or left-handed. They are also able to confirm that you are different from every other person in the world. It is an unsurprising account, a predictable angle in popular journalism. And yet, the connection with traditional palmistry, and especially those aspirational "medical palmists" like Katharine St. Hill, is closer than Wilder or Cummins would ever allow.[35]

Anatomists and physical anthropologists who read dermatoglyphs sometimes deployed traditional palmists' vocabulary as their own, nominating the heart line, life line, and head line. Crookshank, for example, described the simian line thus: "instead of a distinct line of life and a distinct line of head, one transverse line only." But Calcutta physical anthropologist S. S. Sarkar was quick to point out the error, as any Indian familiar with Vedic palmistry would. "This is an obvious printing mistake, it should be line of heart." Sarkar's summary of work on the simian line, published in *Zeitschrift für Morphologie und Anthropologie*, effortlessly deployed "head and heart lines" as if they were conventional anatomical terms. He was trying to sort out why some palm students in the medical field understood the single transverse line to be a substitute for the head and heart lines. As late as 1961, Sarkar concluded that "the position of the true simian crease has not been satisfactorily defined." In other words, much of medical palm reading was ill judged. The simian crease, according to Sarkar, needed to be more precisely researched and defined.[36]

Cummins wondered occasionally about "character" and "temperament," the traditional domains of physiognomy as well as of palmistry. And while he was simply not going to attach these fields to his own specialty, he did consider "dactylomancy," which he dignified as an ancient method of fortune-telling from the fingerprint. Either he or his co-author Charles Midlo—they would not confess who—had had a reading taken by such a "fingerprint expert." It indicated "the hand of the thinker." The whorls showed "a degree of tenacity, stamina . . . on the other hand you should have double these virtues if you had two whorls (in the thumbs)." Predictions followed: "You will be shorter-lived than your co-workers who have all whorls." It was only dactylomancy, not the "pseudo-science of palmistry," that they conceded merits attention as genuine scientific inquiry. Perhaps via Galton, and by association, fingers were more credible than palms. They were told that "the fingerprint tells the truth." Cummins thought as much, and indeed a correlation between "dermatoglyphics and the character-temperament constitution may be ultimately demonstrated," sounding far too like Albert or Ida Ellis. It remains to be explored, he conceded. It was. The study of dermatoglyphics became a minor specialty that lasted into the 1980s—at least—well into the era of cytogenetics, and with enough devoted experts to form professional associations, who gathered as specialist panels at the world's conferences on human genetics.[37]

For all the easy if affectionate mocking of palm readers like Heron-Allen, St. Hill, or Cheiro, it is surely anatomists' study of "glyphs" on the fingers, palms, and soles that constitutes the strangest effort to decode the hand. The link between the diagnosis of Down syndrome, palm lines, and subhuman primates did not go away. By the time Cummins retired, it had become more rather than less formally scientific, as chromosomal diagnoses began to emerge, and perhaps stranger still, the study of dermatoglyphics had a long twentieth-century afterlife in marginal physical anthropology. Both the simian line and more general palmar and plantar ridges and patterns continued to be studied within medical and anatomical traditions, often as a sign of trisomy 21, and even more often, within physical anthropology, as a sign of race.[38]

TEN

The Hand That Speaks

CHARLOTTE WOLFF

When Dr. Charlotte Wolff (1897–1986) held the hand of Aldous Huxley in the 1930s, she decoded its signs in the manner of any palmist (and many physicians), whether from the 1630s, 1730s, or 1830s, assessing its shape, feel, color, creases, and marks. Yet Wolff declared herself to be neither a chiromancer nor a clairvoyant, instead reading those signs as a medically trained psychoanalyst. If the traditional palm readers we have met routinely announced a *language* of the hand, they did so in a period when the founding range of analytic psychologists were also comprehending just that: how bodies speak. "In anger, anxiety, sorrow or any other emotion," Austrian analyst Alfred Adler wrote, "the body always speaks; and each individual's body speaks in a language of its own." Charlotte Wolff's idiosyncratic twentieth-century life shows us that some interpreters of the language of the hand did not emerge from the occult tradition at all, but rather out of the great tradition of mind-body medicine that intensified across Europe in the interwar years: analytic psychology in Austria and Germany, developmental psychology in France, and experimental psychology in England.[1]

Charlotte Wolff was also, for a time, a primatologist. Having analyzed the hands of Aldous Huxley and his wife, Maria, Wolff climbed into the cage that was Peter the chimpanzee's home in the London Zoo. There she took an ink print of his hand and revealed *his* past, his interior, and his character as well (fig. 10.1). Seemingly a strange twist in the story of the modern hand, in fact Charlotte Wolff was extending the long tradition of comparative anatomy and primatology of extremities and its vital endowments. Indeed, she nominated her own intellectual genealogy as much in the natural history and comparative anatomy of the nineteenth century as in German and French psychologies or physiologies of the twentieth. For her, reading signs of the hand in the 1930s was a direct extension of Charles Bell's *The Hand* from the 1830s, vastly improved by a century of medical and scientific thought on primate comparisons, the link between intelligence and hands in evolutionary biology, and the great surge of insight into the language of the body. Charles Bell's eloquent hand now spoke in a new dialect.

A MEDICAL LIFE WITH HANDS

Wolff is not a well-known twentieth-century figure, and for those who do remember her, a life reading hands has been of less moment than her life loving women. Historians of sexuality and of sexology have intermittently noted Wolff's accomplished writing, including *Love Between Women*, published in London in 1971. Long-lived, she experienced both 1920s gay culture at the height of Weimar Berlin and the culture of sexual liberation in 1970s and 1980s London. Wolff was one of very few personally to connect interwar German sexology *and* its late twentieth-century reassessment by historians. She herself wrote an important biography of foundational German sexologist Magnus Hirschfeld, with whom she had worked, publishing it the year she died, 1986. In doing so, Wolff was remembering a period when she worked as a physician in sexual health and in dermatology, from 1925 an obstetric doctor with the state health insurance institution, the *Krankenkasse* in Berlin. There, she organized birth control clinics and combined this with a psychological practice.[2]

Educated as a medical doctor in Berlin after the First World

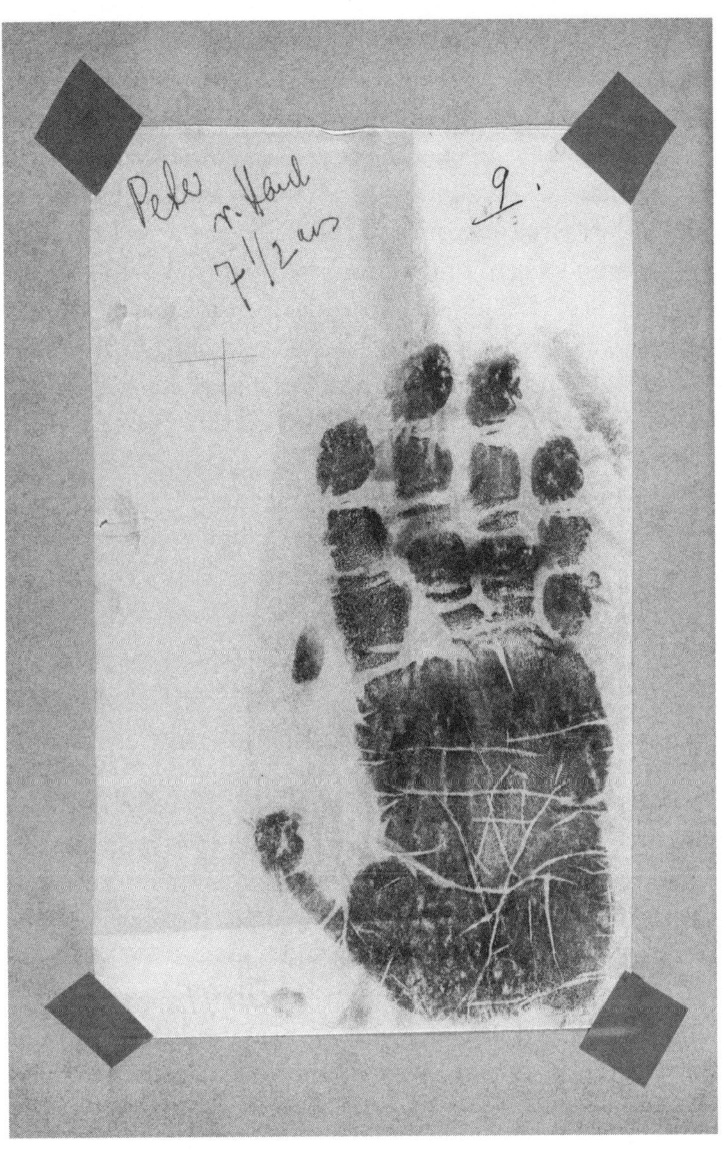

10.1 Print of the hand of Peter the chimpanzee, 7½ years, taken by Dr. Charlotte Wolff. In Album of Extremities of Chimpanzees and Orang Utans. British Psychological Society Archive, Charlotte Wolff Papers, the Wellcome Collection, London, PSY/WOL/3/2. By kind permission, the British Psychological Society.

War, Wolff's appreciation of the human body was expansive from the beginning. A life with hands emerged from a swirl of German and Austrian psychology, philosophy, and medicine on the mind's relation to the body. In 1919, she attended lectures by the philosopher of language Ludwig Klages, who had developed a theory of psychology focused on expression, including the analysis of handwriting, "graphology," or *Handschriftendeutung*. Wolff also studied literature and philosophy at universities in Freiburg and Tübingen. There she was taught by philosophers Edmund Husserl and Martin Heidegger, just then developing "phenomenology," how the phenomena of the external world are lived subjectively and experientially, necessarily through one's body. From their lectures, she began to think about gesture and meaning. Hands and touch are everywhere in the philosophy of embodiment that is phenomenology: hands shaking, hands reaching, hands in contact; perhaps one's hand touching another's, or perhaps touching one's own other hand. Through them, she was instructed philosophically on touch, on sense, and on sensation, even as she learned the physiology of skin in her intermittent work in dermatology.[3]

Wolff's great interest in how the body speaks led her to analytic psychology too. By her own definition she became a "wild" analyst, an ill-fitting one: "I did not belong to the Freudian or the Jungian or the Adlerian schools." In many ways it is unsurprising that psychoanalysis occasionally produced readers of hands, since the interface between minds and bodies was its core business: "Inasmuch as the living body contains the secret of life," mused Carl Jung, sounding a little like Paracelsus, "it is an intelligence." Jung, as it happens, was famously interested in the occult, but it was his psychology of embodiment that alerted him to a self-described "psycho-chirologist," Julius Spier, in Zurich. Not medically trained, Spier nonetheless held chirology courses for physicians in Berlin from 1929, and it was at one such meeting that inquiring young doctor Charlotte Wolff became curious.[4]

Wolff's original work with the body and the mind was German inspired. Yet as Weimar Germany became Nazi Germany, as a Jew her professional and personal life became first compromised, then difficult, and then threatened. As early as 1931, the growing nationalist movement made birth control work less viable, by her own account, and the head of the *Krankenkasse* removed her to a position within an Institute for Electro-Physical Therapy. We may well note here in passing the extent to which a medical career could cross over substantively with the kind of vernacular health practiced by the palmists in this story. Like Katharine St. Hill, Wolff was interested in handwriting and its character reveal; like Cheiro, she learned from, and even *was*, an electrical therapist; like Ida and Albert Ellis, she actively taught and promoted birth control methods. There was more common ground between regular and irregular medicine and health than is often recognized, and well into the twentieth century. Yet Charlotte Wolff's life course was more challenging than any of the palmists we have met, even though her class bearing and obvious learning would have been read by any decent chiromancer or physiognomist as signaling good fortune. Instead, she was caught on the dark side of physiognomy—its manifestation as Nazi racial science.[5]

In May 1933, Wolff was removed from her medical work entirely, and she fled Germany for Paris and Sanary-sur-Mer, where she lived and work until October 1936. In so doing, she left the world of German-speaking philosophy and psychology behind, and entered a Francophone universe of art, bodies, signs, and meaning. She was adopted by the modernist intelligentsia in Paris and Sanary, especially by the Surrealists, who were already enamored with the occult and indeed with the hand. In Charlotte Wolff, they found their perfect hand analyst, and artistic circles appreciated her as a kind of seer, and understandably so. She was, after all, *reading* their palms. Wolff herself, however, routinely distinguished her practice from any occult revival and from that of "gypsies and fortune-tellers." She actively distanced herself not only from the outdated esotericism of the previous avant-garde generation, but also from modern "scientific palmists" like Katharine St. Hill, who themselves insisted that hand reading should be pursued without recourse to fortune-telling. Further, not only did Wolff consider her hand reading *not* palmistry and certainly not clairvoyance, her "enchanted" spiritual life lay elsewhere altogether. A Jew, Wolff was also drawn to the Society of Friends, first in Paris, then in London. "The need for religion is inbuilt in the human make-up," she wrote in her autobiography, and while a nonbeliever, she sought a "religion without a church," and found it with other Quakers.[6]

In the tradition of celebrity palmists like Cheiro, Wolff read the hands of the famous and then published, first in a 1935 issue of the surrealist journal *Le Minotaure*. Alongside pieces by Salvador Dalí and André Breton, Dr. Lotte Wolff offered "Les Révélations psychiques de la main," including the prints and her analyses of André Gide, Maurice Ravel, and Marcel Duchamp. At the same time, she was connecting with medical and psychological colleagues in France. She actively absorbed a French tradition, beginning with the work of Paris-based Romanian Nicolae Vaschide, who a generation earlier had been taken not just with hands, but also with chiromancy. A pupil of the foundational French psychologist of intelligence, Alfred Binet, Vaschide's *Essai sur la psychologie de la main* (1909) was built on a theory of emotions more than on external signs. Wolff long considered him a pioneer, his thesis offering a new way of thinking about the hand. This was a tension-relaxation theory of emotions that had repercussions on "the whole psychophysical organism of man," including the hand. She also absorbed the psychiatry and philosophy of Gilbert Robin and Eugène Minkowski, the latter a psychopathologist, phenomenologist, and one of the founders of the new journal *L'Évolution psychiatrique*. Indeed, she took and read prints of his hands, and those of a range of medical doctors, specifically comparing their hands with philosophers. These sit in the Wellcome Collection in London in folders alongside the hands of artists, of simians, and of *Mongoliens*.[7]

Wolff's Parisian years were highly productive, yet by 1936 France was starting to look more dangerous than hospitable for any Jewish German psychotherapist-refugee, let alone one constantly in and out of lesbian love. Early that year, Hitler's troops violated articles of the treaties of Versailles and Lucarno by entering a demilitarized Rhineland, with no response from Paris or London. The French right, *Action française*, was increasingly actively antisemitic, as were members of the Paris Municipal Council and even the Parliament, despite or because of the election of *Front populaire* and its new leader, Jewish Léon Blum. In Sanary, Wolff had fortuitously met

Aldous and Maria Huxley, with whom she remained close friends. And it was through their influence that she fled Paris, seeking and finding refuge in London. Yet her difficulties were repeated there: as a German refugee, she was unable to practice medicine. Out of necessity, Charlotte Wolff traded on the icon of the seer, turning to palm reading—or "hand reading," as she always had it—for money. The social round of palmistry performance in Sanary, in Paris, and later in London was dissatisfying, even "humiliating" for Charlotte Wolff, but it proved a necessary and relatively easy income for a constrained German Jewish refugee. It was her good fortune on one level, and her burden on another.[8]

It was the Huxleys who urged her to write a book in English on the human hand, Aldous pressing his links with publishers Chatto and Windus. He cajoled his own circle of interwar cultural luminaries to have their hands impressed as well, and Wolff's *Studies in Hand Reading* was published in 1936, revealing the palms and personalities of Bernard Shaw, Virginia Woolf, T. S. Eliot, John Gielgud, Ottoline Morrell, and more. Aldous Huxley, already famous for *Brave New World*, strongly authorized the book not only by presenting his own palm to be analyzed comparatively with his then equally famous scientist-brother Julian, but also by offering an insightful introduction. Probably through Huxley connections again, she was invited to be one of a handful of subjects through whom initial BBC television demonstration footage was produced in 1937. Now kept in the BBC archives, we are still able to see and hear Charlotte Wolff read hands. Denied any right to practice as a doctor, but incorporated into the very beginning of the most British of institutions, the BBC, London was home until the end of her long life: "I never, of course, went back to Germany."[9]

PSYCHO-CHIROLOGY: THE HANDS OF *ANALYSANDS*

In the English-speaking world, Charlotte Wolff's *Studies in Hand Reading* was received and reviewed in a predictable way, with little acknowledgment of her immersion in medicine, philosophy, phenomenology, and clinical psychopathology. Perhaps understandably, journalists and reviewers read it as part of the familiar tradition of esoteric publishing: the books by the Ellis family, by Edward Heron-Allen, and by Katharine St. Hill. One reviewed it as upstaging and updating Cheiro, who suddenly seemed old-fashioned and unscientific. Another understood that Aldous Huxley "thinks the whole business more than a joke," but still made too many jokes about bazaars and Gipsy tents. It was often Huxley's introduction more than Wolff's book that was noticed. *The Morning Post* summarized his call for a large-scale investigation into the scientific basis of palmistry. "One does not have to know anything of palmistry . . . to realise the hands are often as expressive as faces." He also called for a study of hospital patients to research correlations between hand features and mental afflictions such as mania, paranoia, and dementia praecox, or physical diseases such as cancer, tuberculosis, or diabetes. *The Listener* was more respectful to the author herself. Dr. Charlotte Woolf [*sic*] "probably comes nearer than anybody to reducing palmistry to a science." The *Spectator* reviewer saw it as a short cut to self-knowledge, while the *New York Times* simply praised: "She is very scientific, very modern, very brilliant."[10]

One journalist had had his hands read by Wolff in Paris, his secrets disclosed: "She unlocks your inmost secrets, gives advice which is so sound you wish you had thought it yourself." He detailed the technical and bodily processes:

> She smears both your hands with grease, presses them palm downwards on paper, sprinkles prints with black powder, studies resulting "portraits" closely, also your hands themselves. She then talks, very intensely in French and German; discusses your gland-system, blood-circulation, faculty of character (very frankly!), profession . . . For nearly an hour she talked. At the end she was exhausted. I felt that I had obeyed the old Greek injunction "Know thyself."

This was a therapeutic and sensual practice and experience, an intense combination of the visual, the optic, the tactile, the oral, and the aural. Wolff herself privileged the sense of touch, authorizing this sensory expertise not only through phenomenology taught to her by Heidegger, but also through her favored medical antecedent, Charles Bell. A century earlier, she explained, he had nominated touch as a special human sense, separating humans from animals. The soft pads at the ends of human fingers were "'inlets to the knowledge of matter,'" she quoted him. Just this was caught in an arresting portrait of Charlotte Wolff by the interwar period's most adventurous photographer, Man Ray, who visually captured a sense of touch-knowledge more powerfully than philosopher Heidegger or anatomist Bell ever did in their texts (fig. 10.2).[11]

Man Ray's portrait also captures the extent to which an expert eye needed to be refined for hand reading, keenly assessing creases, shapes, ridges, lines, or whorls, the magical word that Francis Galton had applied to finger patterns. Hand reading was also visual, perhaps primarily so. It was, after all, a form of reading. Touch, sight, and insight is what Man Ray collectively portrayed in his portrait, and in a reverse gesture of intimacy, Wolff took Man Ray's palm print. She often called such prints "portraits" of her analysands.[12]

Following the work of D'Arpentigny, from whom so many English palmists derived their styles, Wolff read far more than crease lines. Her practical method took in the form of the hand (which would give a general impression of physical constitution, emotional potential, and innate gifts); the nails and the hand's physical qualities (which would indicate heredity and health); and the parts of the hand, for example, length and shape of fingers, radial, ulnar, and middle zones, which might show the force of the will or, in a new vocabulary, "the relative strength of ego and id." With all that assessed, only then could the lines be read accurately. From creases, mounds, and crosses an astute reader could derive knowledge of "degenerative traits," including "simian patterns of the papillary ridges"; "the strength or weakness of the super-ego"; and "the degree of nervous stability and resistance."[13]

In this way, Wolff "diagnosed" first an individual's constitution, their physique and physiological tendencies. Second, she diagnosed their "temperament," the phrenologically remnant term

10.2 Man Ray, *Admirateur des mains, Dr. Charlotte Wolff*, 1932. Solarized gelatin silver ferrotype print. © MAN RAY 2015 TRUST / ADAGP-ARS – 2024. Image: Telimage, Paris.

that she now explained physiologically: "It largely depends upon the endocrine glands and the autonomic nervous system, but its essence is the speed and intensity of nervous and emotional reactions." Third, she classified people's "mentality," or different qualities of intelligence, and fourth, their "vocation," "the capacities for which the individual is gifted by nature." This, also, was a convention of traditional palmistry, fully claimed by the Ellis family, for example, as their special expertise. All this together enabled diagnosis of illness or conditions, of endocrine disorders or heart problems. She always insisted that it was only as medical prognosis that hands told a future.[14]

Wolff typically called her method "hand psychology," after Julius Spier, but insisted that it was more systematic than his own approach, since as a medical researcher she could analyze correlations between hand traits and illnesses in her patients. She did so first under the supervision of psychiatrist Henri Wallon, leading a French school of developmental psychology. Wallon was intrigued by psycho-motor systems in successive stages of child development, and Wolff began intermittent clinical research at his asylum at Livry Gargan, newly alert to psycho-motor reactions and "the origin of certain hand traits." She was able to study the hands of 400 patients, blindly comparing her hand diagnoses with his clinical diagnoses, and, not unlike Katharine St. Hill or Reginald Langdon Down, she documented correlations between hand shape, palm creases, and psychiatric symptoms. The results were positive enough for Wallon to extend her an invitation to contribute an article on hand interpretation for the *Encyclopédie Française*. This medical psychopathology made Wolff think afresh about the complex relations between the mind, hands, touch, affect, and sensation. What is imprinted in the hands of the human fetus, newborn, child, and adult? She always assessed Wallon's work as highly significant on her later practice, and complementary to her inclination toward German-acquired Gestalt psychology as well as her comprehension of how an *id*, an *ego*, and a *superego* might literally manifest.[15]

Reading palms and hands was always and ultimately, for Wolff, psychological. Interpreting multiple features of the hand facilitated a therapist's capacity to "reach a real structure, the *gestalt*, of an individual, that which decisively differentiates him from every other individual." It was a learned extension of the confessional therapeutic encounter that is discernible even in traditional palmistry, as we have seen. Katharine St. Hill, for example, understood palmistry to be underwritten by an insight that demanded honesty from the practitioner: "Nothing is hidden from you, the human soul is unmasked to your eyes, its motives, principles, and springs of action revealed." In this regard, a half generation earlier, St. Hill had instructed her students to conduct themselves more like doctors than members of the clergy, since that was how they were perceived. "In the same way that people will let a medical man know all that concerns their bodies, which they would often conceal from their nearest friends, so will they let you tell them all that concerns their minds, their feelings, and intellect." Clients will soon learn that "it is useless to be annoyed or ashamed, or to try to conceal the truth from you." Cheiro, too, congratulated himself on releasing Oscar Wilde's dangerously contained emotions and inexpertly imagined himself as palmist-therapist. With far superior learning, training, and substance, Charlotte Wolff *was* a hand-reading psychotherapist. Her idiosyncratic life spent looking at and thinking through hands shows us how a new kind of human interior was revealed. This was part of the history of a modern search for a self. Indeed, Charlotte Wolff is an underappreciated theorist, her life account—*On the Way to Myself*—far exceeding traditional autobiography in its learned presentation of interwar European philosophies of minds, bodies, and interior lives. This trajectory of physiognomy—seeing human souls through their bodies—is part of the history of analytic psychology, of psychiatry, and of the twentieth-century clinical therapeutic encounter. With Virginia Woolf, for example, a face-to-face reading of the hand was supplemented with long talks sitting back-to-back, "as in a psychoanalytic session."[16]

Naturally, Wolff was alert to traditional palmistry, and would sometimes retrieve and reauthorize parts of palmist lore, even astrological nomenclature. She would not dismiss as superstition the common claim among palmists that the length and shape of the thumb "are connected with the power of personality in general and with will power in particular." But we have also seen that many anatomists thought just that. She sometimes explicitly tested such chiromantic belief, verifying it or discarding it, according to her own clinical results. While Wolff distanced herself from palmists, her work was more akin than she often allowed. Put another way, we need to see palmistry such as that practiced by Katharine St. Hill on a continuum with Charlotte Wolff's clinical medical and psychological work.[17]

Unlike a "medical palmist," however, Wolff's grafting over time of hand reading onto analytic and developmental psychology was not a journey via esotericism or spiritualism or the occult. Nor was it a journey, in the more orthodox medical sphere, via mesmerism, hypnosis, or the history of the trance. In fact, far from her own work with hands being framed as a kind of magic or an unexplained marvel, she reserved that term for her own astonished response to the work of others in the medical domain, who on any other measure were more conventional than she was. For example, she watched the hypnosis performed by her sometime research supervisor Charles Earl, medical superintendent of the Monyhull Colony in Birmingham. Wolff well knew that hypnosis was an accepted medium of therapy, yet when she observed how Earl instantly created a trance state in his young patients, she exclaimed: "The magic worked: the boy was asleep in a 'trance.'" This therapy, indeed, she thought as effective as some healings at Lourdes.[18]

Yet it was nothing like Wolff's own clinical work. She repudiated any sense that her hand reading was a form of thaumaturgy, or the working of wonders and miracles. It was not palmistry or fortune-telling or clairvoyance, and it was not akin to hypnosis. There was no supernatural or hermetic occult in Charlotte Wolff's psycho-chirology or hand psychology, and for this Jewish hand reader, no remnant of the Kabbalah either. We need to understand Charlotte Wolff's hand reading more squarely as a modern extension of the nineteenth-century physiology of the hand, of senses, and of the mind.

ANATOMY AND PHYSIOLOGY: READING HANDS IN ASYLUMS AND HOSPITALS

There was a basic anatomical tradition behind Wolff's hand reading: the crease lines could be reduced to the effect of the bending of the fingers, *Vierfingerfurche*, she called it. She sometimes ascribed her own intellectual genealogy in the Galtonian tradition that began with Purkinje, followed by the work of anatomist Breslau on the papillary ridges of monkeys, and Broca and Alix in the French physical anthropology tradition. She was well versed in scholarship on the papillary ridge—"like an engraver's hatching"—that had been closely researched in the criminological tradition of fingerprint work and as dermatoglyphics. Nonetheless, she thought important psychological implications had been neglected, sacrificed to the identification imperative. Her own clinical research work addressed that gap.[19]

When her refugee circumstances allowed, Wolff researched hands in many different hospitals, asylums, and even schools, in both France and England: at Livry Gargan with Wallon, at St. Lawrence's Hospital Caterham with medical officer Henry Rollin, part of London County Council's mental health services, at St. Bernard's Hospital Southall, and at the Willesden General Hospital, where she undertook research in the endocrine clinic. In looking at hand morphology and its correlation to disease, syndromes, and symptoms, she was hardly alone in the psychological and psychiatric domain. She drew on Crookshank's work, on Cummins's work, and on Lionel Penrose's work. Supported by Cyril Burt, professor of psychology and measurer of intelligence at University College London, she followed his lead in the measurement of "mental capacity," undertaking a wartime study there on hands, character, and mentality. In 1945, she presented percentage incidence of hand features in "mental defectives and in 100 control cases," concluding particular correlations between "mental deficiency," hand structure, and crease lines: ankylosis of the fifth finger; simian-like insertion of the thumb; lack of one crease on the fifth finger; abnormal digital formula; syndactylism; concave primary nails; the absence of the long longitudinal line; the too-short upper transverse line. All these were "completely absent in the hands of the control group."[20]

While Wolff retained traditional palmistry's astrological nomenclature in her earliest book, *Studies in Hand Reading*, over time it all became more anatomical. In her mature work—*The Human Hand* (1942), *A Psychology of Gesture* (1945), and *The Hand in Psychological Diagnosis* (1951)—the four main lines were the thenar line (the life line), the upper transverse line (the heart line), the lower transverse line (the head line), and the long longitudinal line (the line of destiny). She began to use the term *dermatoglyphics* as well, after Harold Cummins, who reviewed *The Human Hand* in 1944 for *The American Journal of Psychology*. Some might be inclined to dismiss it as "cheiromantic drivel," he patronized his learned medical peer, but it repaid more serious consideration. He was partial to Wolff's use of Vaschide's *images motoriques* regarding motor sensibility. He also noted another principle that drove her work: "The author advances a complementary theory of tactile images; of imprints in the hand, as in the brain, of repeated cutaneous experiences." This was the long life of the sympathetic hand of which Glanvill had written in the 1600s. Yet here it was less about correspondences than a physiology that appreciated the hand as "the visible part of the brain."[21]

The turn-of-the-century generation had been inclined to link minds and bodies via speculation on the nervous system, sometimes even a revival of ideas about electricity and magnetism. In the interwar years, however, Wolff was intrigued by a newly named endocrine system, then cutting-edge physiology. She frequently correlated hand characteristics to endocrine disorder; indeed, late in life she considered her particular theory confirmed: "We all show in our hands *an individual endocrine formula*." Wolff had studied correlations between endocrine deficiencies and hand traits first in Paris with Gilbert Robin at the Dispensaire pour les Enfants Difficiles et Arriéres, and then with Samuel Leonard Simpson at the Willesden General Hospital. She built on medical hypotheses regarding links between endocrine glands and personality. Studies such as *The Glands Regulating Personality* (1921) by Louis Berman interested her, although she knew that his ideas were considered extreme. "Gland folk-lore," the distinguished experimental physiologist Lancelot Hogben dismissed. However, the reputable physiologist S. L. Simpson was also inclined to study mental and emotional implications of endocrine disorders, and outlined connected personality traits in his book *Major Endocrine Disorders* (1948), including discussion of correlations with hand traits. Endocrinologist August Werner, too, had identified nine "endocrine hand types," including hypothyroid hands, hands in cretins, hands in eunuchism, hands in the Babinski-Frohlich syndrome, and hands in Addison's disease. These were all studied by endocrinologists as well as by Wolff. She went on to present her own classification of endocrinological hand types: pituitary hands; thyroid hands; gonadal (or euchanoid) hands; and adrenal hands. She then hypothesized correlations between endocrine disfunction, hand type, and "temperament and mentality," followed by a personality description, a mix of phrenology and psychoanalysis. The hyperpituitary hand, for example, is oversized, often with very large moons in the nails, with rigid palms and fingers. People with those hands are moody, with quick tempers, and tendency to depression. Obsessional and paranoid tendences are often noted. It was all perhaps most like Richard Beamish's revival of complexions, humors, and temperament in the hand (fig. 10.3).[22]

Charlotte Wolff developed knowledge of the simian line through the mental health research she undertook in Paris, London, Birmingham, and Surrey hospitals. In her method of reading all and any hands, she would sometimes look for "atavistic features of crease-lines," after Crookshank, noting the "simian-like direction" of any crease. Those with simian lines will often lack the long longitudinal line, the life line, or it will be imperfectly developed, she instructed, falling back like other doctors onto palmists' terminology. Wolff sometimes called the simian line a "stigmata," recalling chiromancy's long Christian tradition from early modernity, up to and including Charles Bell's natural theology. But for her, the "stigma" was a medical, not a divine sign. That sign might mean emotional disturbance, it might mean "schizophrenia simplex," it might mean dementia or propensity toward dementia,

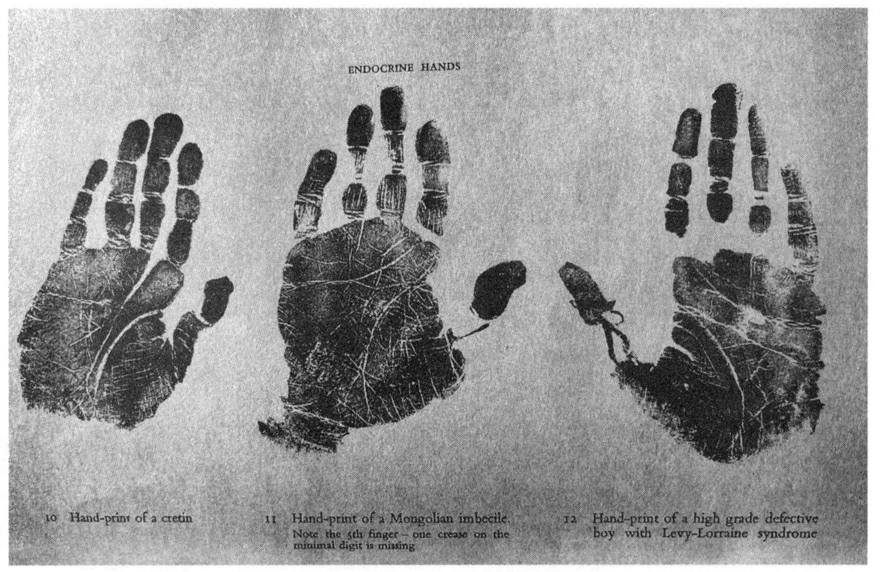

10.3 Endocrine hands. From Charlotte Wolff, *The Hand in Psychological Diagnosis* (Methuen, 1951), plates 10–12. Copy in author's possession.

and it might mean "mongolian imbecility," or any number of endocrinological conditions. One patient's hand had a simian transverse line, as well as a "typically simian" thumb shape. Another boy diagnosed with then-named schizophrenia showed "the extreme of the long sensitive form of hand," as well as three transverse lines, the middle one a "simian line." Printing and analyzing the hands of a thirteen-year-old "with a dull schizothyme temperament and a subnormal intelligence," she noted that his thumb was stiff, clumsy, and immobile, and the digits "lifeless." This contributed to the minimal number of creases on the hand, with the simian-like line prominent. Wolff described this particular hand as "inarticulate." It was the opposite of the eloquent hand conceptualized by Charles Bell.[23]

Charlotte Wolff almost always noted Bell's famous comparative study from 1833, setting her own psychoanalytic language to his formulation of the eloquent hand, its special capacity to communicate, to "speak." She understood Bell to have perceived "the whole psychological significance of the hand a century ago." As we have seen, Bell had then claimed the hand to be "the ready instrument of the mind," and Wolff assessed this as an accurate early perception of the great connection between the hand and individualized intelligence.[24]

There was, in other words, an evolutionary biology behind Wolff's psychology of the hand that drew her toward comparative primatology. In explaining the link between the hand and intelligence, she drew attention, like hundreds before her, to the evolutionary significance of the hand that was freed from locomotion. Yet uniquely, she connected this explicitly to palm lines. The special mobility of the human thumb and index finger was linked not just to human intelligence, but also, in her own thesis, to the formation of particular flexure lines. And so, the properties of certain crease lines "are likely to be indicative of intelligence also." The index finger too was unique in humans, signaling the capacity to indicate, to discriminate between objects, pointing to something, or pointing out something, and this too formed lines. Most humans' two main transverse lines are linked to the independent movement of the index finger correlated with intelligence. Indeed, "the quality and design of all creases [are] due to the movements of the first two fingers." Apes by contrast cannot move the index finger separately and so lack the capacity to make this meaning. Significantly, also, their incapacity to move the index finger independently is precisely what creates the four-finger crease in apes (the simian line), she explained. It was within this tradition of comparative anatomy—also Harold Cummins's disciplinary home—that Wolff sought to throw some of her own light on evolutionary history and biology by examining the palms of apes and monkeys.[25]

PRIMATOLOGY: READING HANDS IN LONDON ZOO

Over her years in London, any number of the extended Huxley family offered their hands to be read and their palms to be printed by Charlotte Wolff. But one of them—Aldous's brother Julian Huxley—happened to be secretary of the London Zoological

Society and lived at that point inside the zoo's grounds. A gracious zoological host, he secured Charlotte Wolff good access to his monkeys and apes, allowed her to take hundreds of handprints, footprints, and in some cases, finger- and thumbprints. He instructed the superintendent of the zoo and keeper of the primates to assist her to the fullest, and she was advised on which monkeys and apes were safe, and which were not.[26]

Julian Huxley just then had his own biological and evolutionary interest in the hand as an evolved fin, work that exceeded palmistry and was different to his brother's inclination to more esoteric wonders. He was an expert in his famous grandfather's own comparative anatomy of primates and the evolutionary biology of functions of hands and feet, fingers and thumbs. Indeed, Wolff routinely nominated the great mid-nineteenth-century anatomist Thomas Henry Huxley to have propelled her own interest in the hand, even calling him "the father of these studies." One of her research questions was Huxley's own: What do comparative studies of a range of human and ape extremities tell us about their anatomical similarities and their evolutionary connection? Yet if Huxley had pursued this through close examination of dead and dry skeletal remains, Wolff did so through research on the fleshy hands and feet of monkeys and apes who were very much alive.[27]

Like a generation of occult palmists, as well as Galtonian fingerprint technicians, Wolff's was a practice, a *craft*, of printing, even, perhaps especially when she had to deal with nonhuman primates. She used different materials to Galton. The animal's hands or feet were smeared with brilliantine, the men's hair oil, then pressed on paper that was placed over an india-rubber pad. The print was then sprinkled with powdered copper dioxide and a fixative applied, commonly that used to fix charcoal drawing. In this way, she took hundreds of prints herself, but usually a keeper took the all-important gorilla prints (fig. 10.4). Wolff then arranged them into albums, and these are an intimate and underappreciated record of individual animal palms, soles, and souls at the London Zoo.[28]

Just as Wolff sought an individual psychology in humans—their gestalt—so she found individuals in the apes with whom she worked, beginning with their zoo-bestowed pet names. She

10.4 Moina, the lowland gorilla, having her prints taken by the keeper under instruction from Dr. Charlotte Wolff, London Zoo, July 9, 1937. Courtesy Smith Archive / Alamy Stock Photo.

took prints of the hands and feet of "Daisy" and "Girty," for example, both two years old, "Fifi," fourteen years old, and "Boobou," sixteen years old. They all showed simian lines, what she called, in actual simians, "the lower transverse crease line." And she also analyzed, as she would in any human, all the other intersecting features: where the thumb is set, the proportion of the cubital margin below the transverse crease to the radial margin of the hand, generally much longer in apes, though a structure she had also found in some humans; the ring finger longer than the index finger, a feature she found in about 50 percent of humans. Her freshest find was that the crease lines in simians changed with age. The older the chimpanzee, the more longitudinal crease lines appeared to cross the transverse lines.[29] Wolff also thought palm lines changed over time in humans.

Julian Huxley read Wolff's papers to the Zoological Society, and they were published in its *Proceedings*, meaning that she was momentarily a zoologist as well as an anatomist, psychotherapist, and hand psychologist. In one paper, Wolff printed and compared the hands and feet of every kind of simian from whom she could take prints: a gray lemur, a green monkey, a tantalus monkey, an olive baboon, a Guinea baboon, a marmoset, a squirrel monkey, a lar gibbon, twelve chimpanzees, and one orangutan, the only animal she named: Mary, aged ten years. She compared them to the paws of a hedgehog and a tree shrew, whose delicate prints she also took and read. In the gibbons, she found that "secondary flexure-lines cover the thenar and hallux-ball, and they also form a kind of communication with the principal crease-lines. This pattern resembles that of the human hand." She also found that the right and left hand lines were markedly different in the baboon, another similarity to humans. And while the crease lines in the foot of an aged chimpanzee showed no resemblance to "normal Man," it did show some resemblance "to those of an insane person." With others—especially comparative anatomist Hermann Klaatsch—she based her study on the fact that extremities of primates were organs of touch.[30]

Wolff's research on primates' hands recapitulated a minor line of nineteenth-century palmists' interest in gorillas and other apes. Richard Beamish had included the outline of a gorilla's hand, not long after the species became known to European naturalists. There were no live gorillas on the Continent at that point, but a cluster of dead ones, or parts of them, prized specimens in scientific collections. Beamish bestowed the gorilla the honor of plate 1 in his *Psychonomy of the Hand*, perhaps prime position, perhaps lowest. In the next generation, geneticist and biometrician Raphael Weldon took the fingerprints of a dead chimpanzee in 1894, sending them to Francis Galton, apologizing for their quality: "The creature had been dead for a day or two when I received him, and his fingers had become a little dried and crumpled." Later again, Dr. Henry Faulds, the Scottish missionary-doctor who had anticipated Galton's fingerprint work (and possibly inspired it by writing to Darwin, who passed the letter on to his cousin, Francis Galton), took a print of the right hand of Chloe, a juvenile gorilla very much alive at the London Zoo. And closer to Wolff's own time, psychologist Noel Jaquin displayed a gorilla print in his 1933 book on "the hand of man," which dealt with its "psychological, sexual, superstitious and medical aspects."[31]

All marks and shapes considered, Charlotte Wolff concluded that the gorilla hand resembles the human hand more closely than that of any other anthropoid ape. This included her reading of the crease lines that she pursued as a comparative anatomist, though not at the level of detail set out by Frederic Wood Jones in his *Principles of Anatomy as Seen in the Hand* (1920), which is intricate in its documentation of nerves, muscles, tendons, papillary ridges, and of course bones of the gorilla's hand. He did look at crease lines, but this was Wolff's special expertise. Sometimes she interpreted complex lines to indicate a higher development of the beast than the human. Comparing a gorilla print with that of a patient, for example, she declared that "the idiot appears to be the less differentiated and developed being."[32]

When Congo-freeborn gorilla Mok died in a Regent's Park cage in 1938, Charlotte Wolff obtained prints of his hands two days after his death (fig. I.4). Even the keeper had been unable to secure them safely from Mok in life. She reprinted Mok's hand in the *Proceedings of the Zoological Society of London*, in her books, and in a French follow-up article. Wolff also took fingerprints and thumbprints, considering the gorilla's thumb to be especially significant. Its patterns were complicated and similar to human prints, and they were alike on both thumbs. "This is of unusual psychological interest," she noted, citing Hamburg medical fingerprint researcher Arthur Kollmann's work on correlations between thumb papillary ridges and gyri of the brain. Indeed, Kollmann actually called the papillary ridges "gyri of the skin."[33]

At one level, this was all old-fashioned comparative anatomy. Yet her work always exceeded the purely anatomical through a return of psychological interest. The more she looked at apes' palms, the more psychological questions kept intruding and intriguing. In some ways, Charlotte Wolff's career with hands added up to a kind of primate psychobiology, including of humans. Apes have interior lives too. And so, when she took the print of seven-year-old Peter the chimpanzee, she read it thus:

> Peter, who is seven years and six months of age, has a hand with particularly few lines and an arrangement which is distinctly of the juvenile type. As the juvenile chimpanzee, which is not yet fully under the influence of the sexual emotions, is more intelligent that the adult one, the unusual intelligence of the chimpanzee Peter may perhaps be explained as a result of delayed emotional development.[34]

Charlotte Wolff was engaged in simian palm reading as well as simian analysis. Just like humans, she posited that "the mentality and emotivity of chimpanzees have a correlation with the lineal composition of their hands." Poor Peter was paraded as "the most intelligent chimpanzee in England," including on very early BBC television. In 1937, he was taken to the studio, where Wolff explained to the camera his special intelligence. She also displayed his handprints, decoding them for the British public, showing how they revealed Peter's particular interior and earlier life. In the end, Charlotte Wolff read the hand of Peter in the same manner and with the same method as she read the hands of Man Ray, Aldous Huxley, or Virginia Woolf, in the process turning the chimpanzee into a minor celebrity too.[35]

Over and over again, Wolff found herself having to explain to various medical superiors that her clinical research on hands was not palmistry. Well might we sympathize, since it is clear that of all the modes of hand reading discussed here, hers was indeed least aligned with popular fortune-telling. She certainly sought to unravel the soul, the mind, the person, the human interior, but in disciplinary and knowledge-practice terms that had little, or really nothing, to do with occult or hermetic knowledge. She might have been disclosing secrets of the body, but not via modern wonders. On the contrary, the provenance of her own hand expertise lay far more truly in comparative anatomy, evolutionary biology, and the links developed over the mid-nineteenth century between the hand, intelligence, and the mind. From there, late nineteenth-century psychologists and a generation of medically trained neurologists, on one trajectory, developed the psychoanalytic traditions that governed so much twentieth-century perception of the modern self, the modern soul, and the modern body that speaks. Far from being a "wonder" or a kind of magic, then, reading hands in this tradition was, perhaps unexpectedly, part of materialist evolutionary biology and medical diagnostics. Charlotte Wolff's primatological work was a twentieth-century extension of a long if specialized history of hand anatomy that morphed into detailed comparative anatomy of primates' palms, simian and human. There is, even now, an "evolutionary palmistry."[36]

In her clinical work, Wolff built on aspects of anatomy, physiology, neurology, and psychology, and well into the 1970s. In 1972, a new kind of transverse line was identified by pediatrician S. G. Purvis-Smith, subsequently called "the Sydney line." By then elderly Charlotte Wolff complained that this was what she had called "the Simian-Like Line" some forty years earlier. Indeed, over her long career, Dr. Charlotte Wolff had no hesitation in diagnosing individuals from the hand, serving as a kind of global consultant. Psychiatric researchers actively sought her hand expertise as an aid to their own diagnoses, especially regarding transverse lines. In reading a patient's prints sent to her from a US institute of psychiatry in 1973, for example, Wolff explained: "The simian line runs in families and is of genetic significance. Where there is no indication of mental deficiency, this trait points to emotional unbalance, generally in a schizoid personality." Sometimes she offered psychoanalytic interpretations: "These stigmata, which are characteristic of neurotic disturbances—some conflict between the ego and the id—confirm that the left is the hand of the subconscious mind and the primate emotions." One US psychiatric researcher told her of a group in Iowa that was trying to correlate flexion creases and PKU (phenylketonuria). In 1974, she sent comments on handprints sent to her by the Waisman Center on Mental Retardation and Human Development in Madison, Wisconsin, of three children of a mother with PKU. She noted, again, a simian-like line in one, an "endocrine hand" in another, speculating on hypothyroidism, and in the third, "the main creases imperfectly developed, which may explain some lack of motor and psychomotor coordination. According to the prints the child is NOT a mental defective." One of the prints also suggested the possibility of a "pre-psychotic condition."[37]

The history of the hand is, at one level, simply part of humans looking at their own anatomy. And so, we might well take Wolff at her own learned word when she claimed early nineteenth-century anatomist Charles Bell as an original hand reader. Yet counterintuitively, there is no necessary correlation between formal anatomy of the hand and disenchanted knowledge of it. Bell's human hand was every bit as enchanted, and marked by a Divine, as was the hand for medieval ecclesiastics, or Renaissance kabbalists, or for theosophical occult revivalists. The human hand was for him a sign—*the* sign—of humanity's special creation by God. Charlotte Wolff inherited Charles Bell's hand, and then she secularized it. She transformed Bell's divinely created palm into a set of somatic signs to be read through his anatomy, to be sure, but also through neurology, endocrinology, and any number of psy disciplines and therapies.

As a reader of hands, Charlotte Wolff was a mélange. Her interest in the human interior placed her in the tradition of physiognomy, from Richard Saunders to Ida Ellis, reading modern souls and selves through the body. She was herself the medical embodiment of would-be clinical researcher Katharine St. Hill. And clearly she extended the twentieth-century work of Francis Crookshank in reading the simian line, as well as the primatological implications developed by Cummins and other students of dermatoglyphics. For all of them, signs on the hand invited and required highly expert decoding. For Charlotte Wolff, a focus on, expertise in, and fascination with hands was a twentieth-century journey through unlikely ways of knowing the mind, the body, and the modern self, as well as primate histories and identities.

ELEVEN

The Hand in the Age of Human Genetics

LIONEL PENROSE

From the 1930s to his death in 1972, the distinguished geneticist, medical researcher, and mathematician Lionel Penrose looked at hands incessantly, decoding their signs with ingenuity, with a fascination for their patterns, and with the conceptual tools of a new scientific era. The twentieth century might have been the age of "psy," but it was also the age of genetics. Penrose entered the field at the beginning of the rise of population genetics, after Mendel's work on patterns of inheritance had been rediscovered and absorbed by some brilliant minds into the Galtonian biometric tradition. This generation of quantitative geneticists was able to make new kinds of statistical inferences from large datasets on all manner of physical traits and their distributions. Another breed of biological researcher was looking at *how* hereditary information was transferred between generations, uncovering the genes that carry that information within chromosomes, the structure of DNA. In the mid-1950s, they settled the fact that in humans twenty-three pairs of chromosomes exist in each cell, normally. Over the twentieth century, then, Lionel Penrose straddled two great developments, first in the mathematics of population genetics and second in the study of cytogenetics, of cells and the chromosomes within.[1] His work on the hand was woven into both.

Penrose's population-level work extended the significance of Galton's fingerprints, and Galton's own interest in how fingerprint patterns were inherited, how they might signal traceable kinship. Without a viable understanding of the mechanism of inheritance, however, Galton had had no real means to research this. But he did have money, and left a large bequest to University College London in the hope that his work would continue. It did. Karl Pearson was the first Galton Professor of Eugenics, the biometrician who occasionally looked up from his work to read treasured palmistry books from the 1400s (fig. 1.5). Ronald Fisher was the second, the brilliant biostatistician who dealt with the future like a modern-day chiromancer: he offered entirely new ways to calculate probability and "maximum likelihood," yet failed to calculate a future in which his own eugenic sympathies would come undone. Lionel Penrose was the third Galton professor, the one who cleaved UCL genetics from eugenics, but not from Galton's original hand-based research. He worked over Galton's vast collection of fingerprints and palm prints and added sophisticated mathematics, population genetics, and ultimately cytogenetics to the field that had been far more crudely researched by anatomists: dermatoglyphics.[2]

Penrose brought another biomedical tradition to his work on the hand, inheritance, and genetics: his clinical and research work on Down syndrome. Around the time that Crookshank was expounding his *Mongol in Our Midst* thesis, Penrose secured a major research position to inquire into the inheritance of "mental defect." In so doing, he began to look for the simian line. From the early 1930s to his final talks on dermatoglyphics in the early 1970s, Penrose considered the transverse line in the palm to be diagnostically significant. He lived through, and was part of, great twentieth-century changes in Down syndrome research too. First, the discovery of trisomy 21 in 1959, an additional copy of part, or all, of chromosome 21: most humans have forty-six chromosomes, some have forty-seven. Second, the shift in terminology from "mongolism" to Down syndrome, in which Penrose was a key actor.

Lionel Penrose studied hands constantly, correlating not just the transverse crease to Down syndrome, but all manner of additional hand signs to other syndromes as well. Palm lines, dermatoglyphic ridges, and chromosomal information now formed one dialect in the global language of the hand (fig. 11.1). It was a disenchanted endpoint for chiromancy. "There are no mysterious incalculable forces that come into play," sociologist Weber diagnosed of the modern era generally; rather, one can "master all things by calculation." The hand was certainly mastered in this way, but it is one piece of twentieth-century good fortune that the master calculator was the modest Lionel Penrose, Quaker, pacifist, antiracist, and anti-eugenicist.[3]

LIONEL PENROSE

Son of a portraitist, perhaps Lionel Penrose was destined to become a modern physiognomist, looking for inner secrets manifest on the outer body. From the beginning, his inclination was both to patterns and to human interiority. He came to medicine, and eventually to human genetics, via an idiosyncratic combination of Cambridge mathematics and psychology in 1919–1921. These were difficult years in Britain, recovery from the Great War and the great influenza pandemic both taking their toll. Penrose had been a conscientious objector, following a Quaker upbringing and education, and had served in France with a Friends' Ambulance unit. In Cambridge, he was drawn to the same swirling world of psychological ideas that had swept Charlotte Wolff toward psychoanalysis, and Penrose worked with Frederic Bartlett, professor of experimental psychology. He then went to the Vienna Psychological Institute to work with new director Karl Bühler, who had made a name for himself with *Die Gestaltwahrnehmungen* [The Gestalt Perceptions, 1913] and then *Die geistige Entwicklung des Kindes* [The Mental Development of the Child, 1918]. Penrose met Freud in this high moment and core place of psychoanalysis. This foundation in

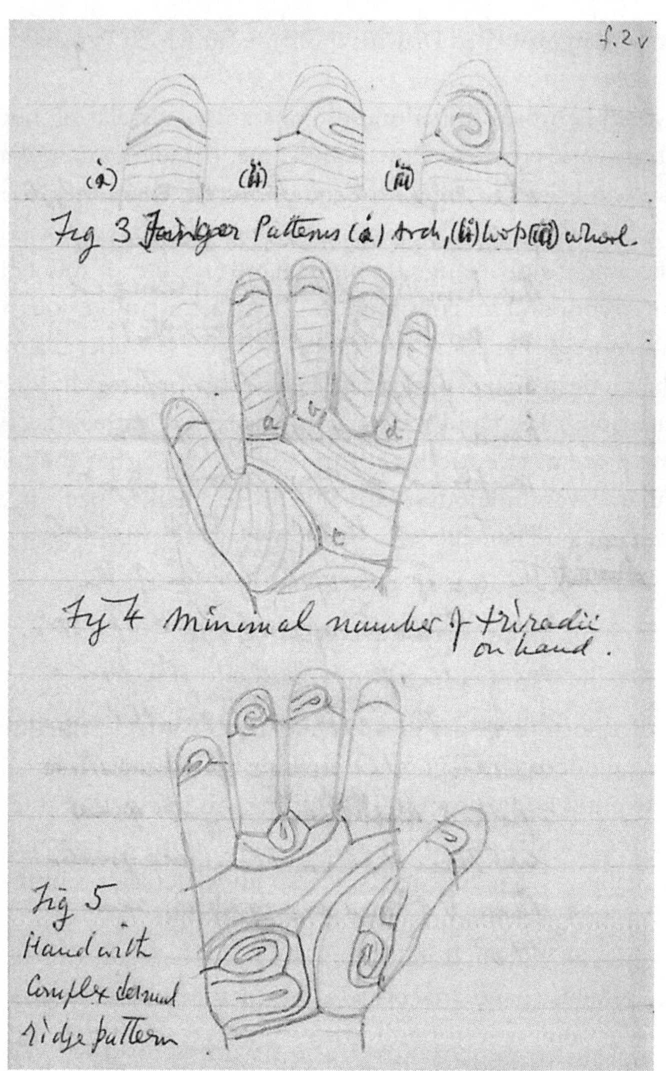

11.1 Lionel Penrose, notebook. Lionel S. Penrose Papers, University College London, Special Collections, n.d. PENROSE/2/42/2/1–2. By kind permission, Professor Shirley Hodgson and University College London Library Services, Special Collections.

the mind and in mathematics was extended by study of the body, after Penrose took another degree, this time in medicine back at Cambridge, with clinical work at St. Thomas's Hospital, London. By 1928—mathematics, the mind, and the body in place—he then took the turn that would be his research path for life: understanding the causes and correlations of mental disorder, of intellectual disability.[4]

This began with distinction when Penrose was appointed research medical officer at Colchester's Royal Eastern Counties Institution for Mental Defectives in 1931, a large residential institution that had originally opened in 1859 as the Eastern Counties Asylum for Idiots and Imbeciles. Penrose began research for an extensive clinical and genetic study of 1,280 residents. In 1933, he published *Mental Defect*, incorporating a chapter on physical examination, including the "stigmata of degeneration," by which he meant the physical signs that had been well documented in medical literature for generations; high and narrow palates, short minimal digits, malformed ears, and distinctive creases in the palm. This part of his work at Colchester was straightforward enough, if difficult for the residents: measuring and documenting. Far larger than anything Cummins or Wolff had undertaken, it was not dissimilar in scale and ambition to Galton's anthropometry. Yet if Galton had collected thousands of measurements on intrigued people who wandered voluntarily into the Anthropometry Laboratory at the London Exhibition in the 1880s, Penrose's Colchester Survey gathered data from a contained population who had little choice as to whether their palates, hands, and heads were to be measured. Penrose's parameters extended those of Galton's in two ways: first with more refined laboratory analysis, especially on blood grouping, one of the most recent innovations in the long quest to find reliable markers and patterns among human groups; and second, Lionel Penrose had far more sophisticated tools and models with which to understand inheritance, both biologically and statistically.[5]

Penrose benefited from Karl Pearson's biometry in the early part of the century, and especially from the later innovations of Ronald Fisher. It was Fisher who had taken the biological sciences by storm with his new mathematical theories and models of inheritance, the data-based population genetics that defined that generation's work. New statistical sense was made of larger and larger datasets of increasingly precise characters and patterns that were measurable. In this way, Penrose's entrée into research in the 1930s was undertaken at the exciting moment when mathematicians were turning Mendelian laws into probability, even toward the prediction of individual inheritance of certain characters. Galton, Pearson, Fisher, and Penrose were linked in an intergenerational world of curves, coefficients, correlations, and regressions. Fisher had worked up a method of "discriminant analysis" to deal with the problem of multiple criteria, which Penrose applied to his Colchester data. His equations included information on, for example, parental age, birth order, and incidence of "defect" among both the patients and their near relative. How can we measure the extent to which two or more variables are related to one another? The answer was a kind of fortune-telling: regression analysis of patterns of distributions enabled the prediction of a future trend.[6]

It also looked to the past, to inheritance. One ambition of the Colchester Survey was to test presumptions about "hereditary mental degeneracy." Lionel Penrose's task was clinical *and* genetic, making him an exemplary appointment, given his special combination of mathematical and medical training. This also made Ronald Fisher's statistical and theoretical breakthroughs key, since Penrose was not just gathering data from residents and patients, but also from their wider families, intergenerational patterns the foremost question. The research on families and relatives was novel; much later assessed as "formidable" by one of Penrose's many obituarists.[7]

In addition to this biometric tradition, Penrose sometimes put his research in the anatomical and fingerprint genealogy of Purkinje, Galton, Wilder, and Cummins. Once he assumed the Galton chair in 1945, Penrose was proximate to Galton's own voluminous and sprawling research records, and like Karl Pearson, he couldn't resist dipping back into it, asking new questions of it, and applying new statistical techniques to it. He literally used Galton's fingerprint records as his own data. At least once, Penrose was able to study the same actual hand. In 1890, Galton recorded the prints of the fingers of the right hand of an eight-year-old girl. In 1961, Penrose took prints from the same hand, now belonging to an old woman. Over those seventy years he was able to confirm, again, Galton's conclusion: "No change in details of pattern was seen." While Cummins said over and over that the object of inquiry in dermatoglyphics was the skin ridges, not palm creases, like Galton, Penrose was definitely interested in the latter, beginning with what he sometimes called the simian line crease, and sometimes the four-finger or transverse crease.[8]

Over the long years of the Second World War, Penrose was director of psychiatric research for Ontario, Canada, and returned to England to take up the Galton Professorship of Eugenics at University College London, which he held for twenty years, from 1945 to 1965. This stamped his biometric pedigree—Pearson, Fisher, Penrose—like a Galtonian chosen family. Or perhaps it was more like an intellectual bloodline that he couldn't escape: it was part of him, and he was part of it. Still, he *could* re-narrate the family history, and did so forcefully by changing the family name, from the Galton Professor of Eugenics to Galton Professor of Human Genetics. That biometric family had offspring too, less well-known geneticists who sometimes bridged the generations. Sarah Holt, for example, worked both with Ronald Fisher and with Lionel Penrose, and contributed independently to the UCL specialty, with studies of the fingerprints "of a small sample of the British population" (1949), a comparative quantitative study of the fingerprints of "mongolian imbeciles and normal individuals" (1951), and the genetics of dermal ridges, with a focus on "parent-child correlations" (1956).[9]

Being disinclined to eugenics did not mean that Penrose and his team stopped measuring humans. Alongside many other physical characters, he continued to measure hands and their engraved signs, a brilliant mathematician who knew what could be done with puzzling data built from multiple variables. He was a national and world leader in mid-twentieth-century genetics, president of the Genetical Society of Great Britain, 1955–1958, and president of the Third International Congress of Genetics held in Chicago in 1966. By then, genetics had transformed from mathematics to chromosomes, population genetics to cytogenetics, and a whole new era of possibilities opened up. DNA could be seen and studied, inheritance now investigated via the gene and its chemicals, as much as the patterns of distribution of particular physical characters across populations. It was Lionel Penrose's good fortune that his career spanned this great shift in the age of genetics, but as we shall see, his specific research on hand marks as *phenotypes* also spanned this profound shift, a twentieth-century constant.

In retirement after 1965, Penrose continued his work in medical genetics, increasingly focusing on the hand, as the director of the Kennedy-Galton Centre at Harperbury Hospital, near St. Albans, north of London, once the Middlesex Colony for the mentally disabled. His center specialized in clinical genetics and in the prenatal diagnosis of chromosomal abnormalities in the fetus. As in Colchester, this was an institutional community in which more than a thousand people spent their lives at any given point, the last gasp of the mass institutionalization of those with intellectual and physical disability. After Penrose's death in 1972, it declined, and was dismantled over the 1980s, part of the shift to care in the community. At its height, however, it was one community in which hands were raised for two purposes. The first was the teaching of Makaton signing, John Bulwer's language of the hand now fully owned by the hearing impaired. The second was Lionel Penrose's longstanding clinical and genetic research on the patterns of lines, ridges, creases, and shapes of hands that turned into a different language altogether.

DOWN SYNDROME: FROM PHENOTYPE TO GENOTYPE

From his early years with the Colchester Survey, Lionel Penrose was interested in Down syndrome, and although he often puzzled at the strange term "mongolism," which meant so little biologically, he also used it. Those who knew him said Penrose always inspected the hand first, "for the characteristic pattern." Well might we say that he looked at those palms with as much interest as Katharine St. Hill and Edward Heron-Allen, as Reginald Langdon Down and Francis Crookshank, if with different learning. All of them would see the palmar anomaly immediately, the single transverse crease. Penrose knew that the so-called simian line was common in this population, but so were some other patterns in the hand. In 1931, he showed that 27 percent of his patients had only one flexion crease on the little finger. And he listed these two hand signs among the seven identifying characters of Down syndrome, a combination of which should be observed to secure a diagnosis: intelligence quotient; cephalic index; eye folds; fissured tongue; conjunctivitis; the transverse palmar line; and one crease only on the minimal digit of either hand.[10]

Over the 1930s and 1940s, it was not yet clear whether Down syndrome was an inherited condition, or a genetic one, or neither. Even for Lionel Penrose, working so closely with Colchester families, it was mysterious, not least because so few people with the syndrome themselves reproduced. But there was a long tradition of speculation about its etiology. In the late nineteenth century, John Langdon Down's collaborator Shuttleworth had connected "mongolism" to maternal age, but this idea was left hanging for decades. In the next generation, at a meeting of the Royal Anthropological Institute in 1921, Francis Crookshank presented his racial theories to a distinguished audience that included Reginald Langdon Down, but at the meeting anatomist Grafton Elliot Smith insisted that "mongolism" was pathological, not some kind of racial atavism. He was taken with a recent theory of "interference with pre-natal growth in the seventh week of intra-uterine existence," and thought this occurred mainly "in the offspring of young or worn-out mothers." By the time Lionel Penrose took his research job with the Colchester Survey, clinicians speculated that Down syndrome *might* be heritable, but it might also have an intrauterine origin, caused by some unknown damage to the "germ cell," or it might be an endocrine-based disorder. Perhaps it had something to do with a phenomenon called a "chromosome." All these were possible. In 1932, Dutch geneticist Petrus Johannes Waardenburg was the first to suggest that Down syndrome "might be due to chromosomal nondisjunction," the failure of chromosomes to separate, and similar hypotheses were put forward later that decade by the US physician Adrien Bleyer, by the Swiss Guido Fanconi, and by the French Raymond Turpin. In 1939 Lionel Penrose also proposed that Down syndrome *could be* caused by a chromosomal anomaly, a duplication or deletion. "Such an abnormal chromosome could be transmitted from a parent to a small proportion of the offspring." Yet he did not specify the nondisjunction that would turn out to be correct.[11]

Despite this lack of clarity, almost all specialists who observed the common simian line were in some way either researching or presuming a heritable basis of Down syndrome. Charlotte Wolff's clinical observations crossed over into basic epidemiological and statistical work too. Although it was often claimed that the simian line was present in about 2 percent of humans, in her group she found it in fifty of 532 cases—about 10 percent. "Faulty heredity," Charlotte Wolff put it, convinced that a simian line in parents signaled higher chances of the reproduction of an impaired newborn. In 1948, biologist Doris Mann Snedeker agreed that there are trends in the palm patterns of "mongoloid imbeciles," and that these patterns are in place between twelve and fourteen weeks in the human fetus, remaining unchanged throughout life. The simian line occurs occasionally in normal persons, but she confirmed it as frequent in the "mongoloid"—still the term. The simian crease was only one element of palmar diagnosis, however. There were other loops and whorls on the palm that interested her, just like any good palmist, and just like Lionel Penrose. But what was the application of such knowledge? One was early genetic counseling, she said, presuming heritability. Along with others, she was interested in common patterns between siblings and across generations, and so sought the palm prints of both parents and siblings of the "mongoloids." A second application was to settle the lingering question about race and Down syndrome: "whether the dermatoglyphics of mongoloid imbeciles resemble those of normal Mongolian stocks." And a third was diagnostic, particularly in cases of doubtful diagnosis. Harold Cummins also applied his dermatoglyphics to inheritance within families. Interestingly, this included its application in cases of questioned paternity. All manner of researchers, then, from many disciplines, continued to look at the palms of those with Down syndrome, either presuming or speculating on its heritability, or wondering about a genetic cause, but one that was not necessarily heritable.[12]

In 1954, Penrose surveyed the literature on the etiology of Down syndrome and considered it likely to be multifactorial, possibly including a chromosomal abnormality. But that conclusion was about to change. His survey was undertaken on the cusp of the 1956 breakthrough that confirmed that humans normally hold forty-six chromosomes: twenty-three pairs. Following quickly in 1959, a French team—Jérôme Lejeune, Marthe Gautier, and Raymond Turpin—found something different in the Down patients they studied: forty-seven chromosomes, and they identified the nondisjunction as "trisomy 21." This great revolution in understanding the cause of Down syndrome, as in human genetic research more generally, heralded a wholly new era, discontinuous with the wild speculations of a Francis Crookshank and the great data-driven Colchester Survey on "mental defect." It was a paradigm shift.[13]

Another great paradigm shift was cultural, not biological; a 1961 letter to the *Lancet* and to the *American Journal of Human Genetics* that dismissed the term "mongolism." It was now deemed "embarrassing," not only to those with the syndrome, but also to the suite of East Asian researchers closely involved, the letter explained. As importantly, expressions that implied "a racial aspect of the condition" were simply incorrect. Lionel Penrose was a signatory, as were Jérôme Lejeune and Raymond Turpin. So, indeed, was Norman Langdon-Down, grandson of John Langdon Down, who had come up with "mongolism" in the first place, nephew of Reginald, who had looked at those palms in the early years of the twentieth century, and cousin of John, whose own chromosome 21 was affected.[14]

A curious and important fact: the French team who discovered trisomy 21 had previously worked on dermatoglyphics too. As recently as 1953, Lejeune and Turpin had published "Dermatoglyphic Study of the Palms of Mongolian Idiots and of Their Parents and Siblings." Just like Lionel Penrose, Parisian Lejeune had worked on palm prints of a group of ninety-three people with Down syndrome, their families, and two large control groups. And just like Charlotte Wolff, Francis Crookshank, and Harold Cummins, he had also turned to anthropoid apes to consider the simian line. His own question concerned possible accumulated gene changes over evolutionary time. And he speculated also about single-generation polygenic difference between human parent and child. Might that involve the chromosome? With the essential addition of Marthe Gautier's skill in tissue culturing, the team's attention turned from the phenotype on the hand to the genotype in the cell.[15]

While we might imagine that the switch from phenotype to genotype would make physical observations of something like a palm line seem antiquated or at best insufficient, in fact Lejeune, like Penrose, continued to be intrigued with those hand markings, even after the trisomy 21 discovery. Both men kept looking at the palm and its morphology, wondering about the correlations of its patterns to other chromosomal abnormalities. Through the 1960s, letters and drawings and ideas crisscrossed the English Channel, as the two distinguished specialists in Down syndrome and its genetics, one in Paris, the other in London, compared palm prints (fig. 11.2). There was no clear break, in other words, from the population genetics in which the phenotype on the hand was researched (correlations of patterns of ridges and creases) and the cytogeneticists in which the genotype was researched.

Lionel Penrose in London and Turpin and Lejeune in Paris were not alone in continuing to measure palms, fingers, and hands. Other researchers established new-generation standardizations of measurements, and not just for Down syndrome. A Swedish team published from a study of the fingerprints of 200 people diagnosed with then-named schizophrenia. This followed a Michigan-based study from the previous year in which fingerprints of 100 male "schizophrenics" were compared with a control series provided by Scotland Yard. A higher incidence in whorls in the fingers of male schizophrenics was confirmed, although it was difficult to conclude differences in the frequencies of palmar patterns or in "ridge disturbances." The psychiatric and criminological parameters of a study like this seem more like a Galtonian project from 1890 than the chromosomally aware and exciting 1960s. Penrose's

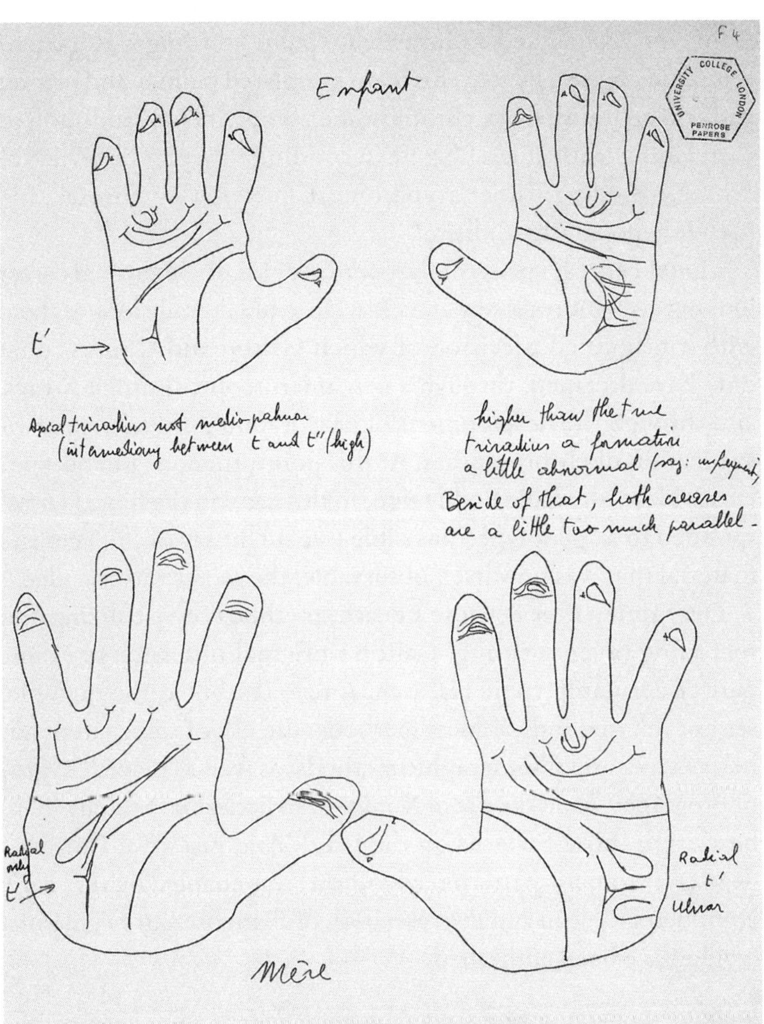

11.2 Phenotypes of the hand. The population geneticist (Penrose) and the cytogeneticist (Lejeune) continued to discuss palm lines, after the discovery of trisomy 21 in 1959. J. Lejeune to L. Penrose, February 27, 1962. Lionel S. Penrose Papers, University College London Special Collections, PENROSE/2/20/10/11. By kind permission, Professor Shirley Hodgson and University College London Library Services, Special Collections.

team, too, continued to investigate palm and finger ridge patterns and creases. By 1965, his team correlated palmar and plantar patterns to unusual sex chromosome complements, and showed correlations with at least eighteen conditions beyond Down syndrome, including Turner's syndrome, Klinefelter's syndrome, and pseudohypoparathyroidism.[16]

Lionel Penrose spent this last period of his distinguished career looking even more closely at hands. He could literally look at them with a magnified precision of which Galton and Cheiro could only have dreamed; through a new microscopy, through X-rays, and through the development of karyotyping, the science of visualization of chromosomes. At this point, then, a "phenotype" (readable, observable bodily sign, in this case on the hand) corresponded to a "genotype," invisible (we might say occult) genetic material that was now itself observable: the secret was decoded.[17]

Over time, after Penrose's death, methods of visualizing and recording (even printing, Galton's original obsession) became part of dermatoglyphic research. A 1976 textbook, for example, set out ink methods, inkless methods, the use of transparent adhesive tape, and photographic methods, as well as newer "hygrophotography," and the use of X-rays—"radiodermatography"—to best catch those patterns on the hand. And just after Penrose's death, "automatic pattern recognition" was enabled by very early computers: Galton and Penrose, both brilliant human calculators, would have been enthralled.[18]

LIONEL PENROSE'S GEOMETRY OF THE HAND

In twelfth-century Canterbury, priests and scribes looked for a triangle made by the intersection of the three main lines in the hand, the higher line, the middle line, and the oblique line. In his 1522 chiromancy, Johannes Indagine had crafted chapters "Of the Triangle" and "Of the Quadrangle." In the eighteenth century, the popular *Conjurer's Guide* explained that "the line of the brain usually called the liver line, reaches to the table line, making a triangle." For Lionel Penrose—mathematician, chess theorist, and artist-observer—the geometry of the hand, its lines and shapes, were just as compelling. He developed a more complex angle for medical hand readers, but one that in the end would have been entirely comprehensible to Indagine or Praetorius, Richard Saunders or Isaac Newton, so long as they had their measuring compass at the ready. Penrose called it the *atd* angle, and he was simply trying to standardize a clearly measurable set of lines in the palm. In "Fingerprints, Palms and Chromosomes," in 1963, *Nature* published one of Penrose's characteristically clear diagrams, showing a range of angles measured from the base of the index and little fingers—points *a* and *d*—to a point *t* in the palm. The different mean angles of the "axial triradius" ranged from 48° to 108°, correlated to "Normal," "Turner," "Mongolism," and "Trisomy 13" (fig. I.3). This was a measurement precise enough, but also easy enough, to be used in quantitative genetic analysis within Down syndrome–affected families.[19]

The geometry of such dermal configurations was applied diagnostically. In 1965, a Swedish team attempted a simplified scoring method of eight distinct patterns. One of the early scientific papers to use the new term "Down syndrome" (but still explaining it for readers as "mongolism"), Beckman, Gustavson, and Norring by then knew (from the Penrose lab work) that several other syndromes were caused by chromosomal aberrations, and that they also showed deviations in dermal configurations. It was known that people with Down syndrome show "in their fingerprint patterns, an increase in the frequency of ulnar loops especially on finger II and an increase of radial loops in fingers IV and V." They also had a "high *atd* angle." Flexion creases were also notable, especially the "transverse palmar crease." In the suite of characteristics, the "typical four finger line" was first, with scores allocated differently for males and females, and for the left and right hand. Penrose's *atd* angle was to be scored next, followed by presence of particular hypothenar patterns, interdigital patterns, an ulnar loop on finger IV, a radical loop on finger V, and the single crease on finger V, the little finger, which Penrose had early established (fig. 11.3). With equipment available at the Institute of Medical Genetics in Uppsala, they were able cytogenetically to check the

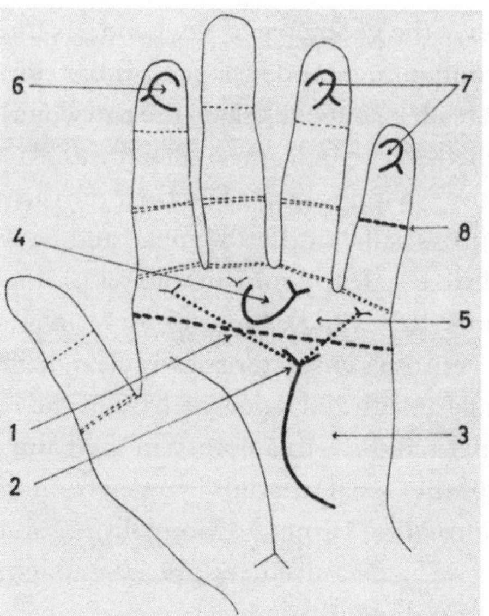

Fig. 1. Schematic picture showing the dermal traits used in the diagnosis of the Down syndrome.
1 = four finger line, 2 = atd angle, 3 = hypothenar pattern, 4 = pattern in interdigital area III, 5 = absence of pattern in interdigital area IV, 6 = ulnar loop on finger II, 7 = radial loops on fingers IV or V, and 8 = one crease on finger V.

11.3 Diagnosis by the hand: "the dermal traits used in the diagnosis of the Down syndrome." From L. Beckman, K. H. Gustavson, and A. Norring, "Dermal Configurations in the Diagnosis of the Down Syndrome: An Attempt at a Simplified Scoring Method," *Acta Genetica et Statistica Medica* 15, no. 1 (1965): 6. Copyright © 1965 Karger Publishers, Basel, Switzerland.

accuracy of their suggested index, a score that expressed logarithmically this suite of hand-based characters. It was announced as an accurate diagnostic method, either separately from or verified by chromosomal identification. The latter, they identified, was not always available, laboratory demand might be high, and expense might be exclusive, and so "a good solution would be to combine both dermatoglyphic and cytogenetic methods in the diagnostic laboratories." In saying so, they cited Penrose's 1963 paper in *Nature* in which he suggested that dermatoglyphic analysis might help cytologists distinguish just which chromosome was triplicated "in patients with trisomy for a small chromosome."[20]

Penrose applied more complex geometries to the hand too,

especially when he returned to think about Galton's loops and whorls. He started with the whorl pattern, in a geometric sense composed of two loops. Then it was patterns of parallel, but rarely straight lines on living things that fascinated him: the stripes on a zebra, the shells on some snails. We can sense his late-in-life excitement, because the shape of living things, he explained, "can seldom be described by simple geometrical concepts." His innovation was to apply topology, the geometry of relationships, to fingers and hands, the mathematics of spaces that are invariant under any continuous deformation. Penrose explained topology as "a branch of mathematics well-suited to the study of how parallel lines can cover complex surfaces." The ridges of the skin on the palmar surfaces of the hands and feet in primates he called "one of the most attractive anatomical systems of this kind." In the long tradition of measuring human anatomies, he certainly thought topology much needed, succeeding the crude arithmetic of comparative anatomy that had long measured the number or the length or the weight of a structure: vertebrae or teeth or limbs or fingers. As a geneticist, he appreciated working with such patterns because they were morphologically stable from the twelfth week of fetal life, and generally not influenced by age: adults and children could be accurately compared. This consistency made them useful for genetic studies, unlike a trait such as hair color. There were three basic questions in genetic studies, Penrose set out with characteristic clarity: what characters or measurements are to be used; by what method the material is to be summarized; and how the results are to be interpreted. He had long before decided that lines on the hand formed the character that would yield the most. And now topology helped its summary and interpretation. In 1965, Penrose set out "Dermatoglyphic Topology" for *Nature*. It was all certainly an extension of Galton's work, but the shapes Galton had called "deltas" were now known as "triradii," natural centers of the patterns of the lines that produced Galton's "whorls," "arches," and "loops." Yet there are regularities in the patterns, as well as "normal variations, which represent hereditary differences between members of separate populations, members of the same population, or members of the same family."[21]

The geometry of the hand endured and enchanted. Danuta Loesch, author of *Quantitative Dermatoglyphics*, published in the Oxford Monographs on Medical Genetics (1983), was associate professor of human genetics in Warsaw, and had undertaken close topological research with Lionel Penrose. They had planned a jointly authored book, sadly interrupted by Penrose's death. But the study she finished without him was dedicated to his memory and stands as a memorial of the twentieth-century geometry of the hand, detailing topology, the geometry of loops, and triradii. One language of the hand was now so geometrical and mathematical as to be incomprehensible to most, fascinating to a few (fig. 11.4).[22]

THE AFTERLIFE OF DERMATOGLYPHICS

In no sense did the discovery of forty-six chromosomes in each human cell, and forty-seven in some humans' cells, spell the end of phenotype-based dermatoglyphics, neither for Penrose, nor for Lejeune, nor for the modest network of researchers that had accumulated since Cummins had named the field in 1926. Indeed, the reverse was the case: building on the great upsurge in human genetics, researchers of dermatoglyphics started to organize themselves, forming a specialized committee at international conferences on human genetics, first in Copenhagen in 1956, and then in Rome in 1961. The president, quite rightly, was Harold Cummins, the chair was José Pons of the Laboratorio de Antropologia, University of Barcelona, and the secretary Abhimanyu Sharma of New Delhi, with a committee including Mariassa Bat-Miriam of the Israeli Institute for Biological Research, anatomy professor Masaya Yasanuka of Nagasaki University, Japan, and geneticist Sarah B. Holt of the Galton Laboratory, University College London. Together they organized meetings "devoted to research on fingerprints, palms and soles," with the best known among them, Lionel Penrose, contributing to a 1971 conference in Warsaw, Dermatoglyphics in Human Genetics.[23]

Penrose was always clear that dermatoglyphics had a future. He

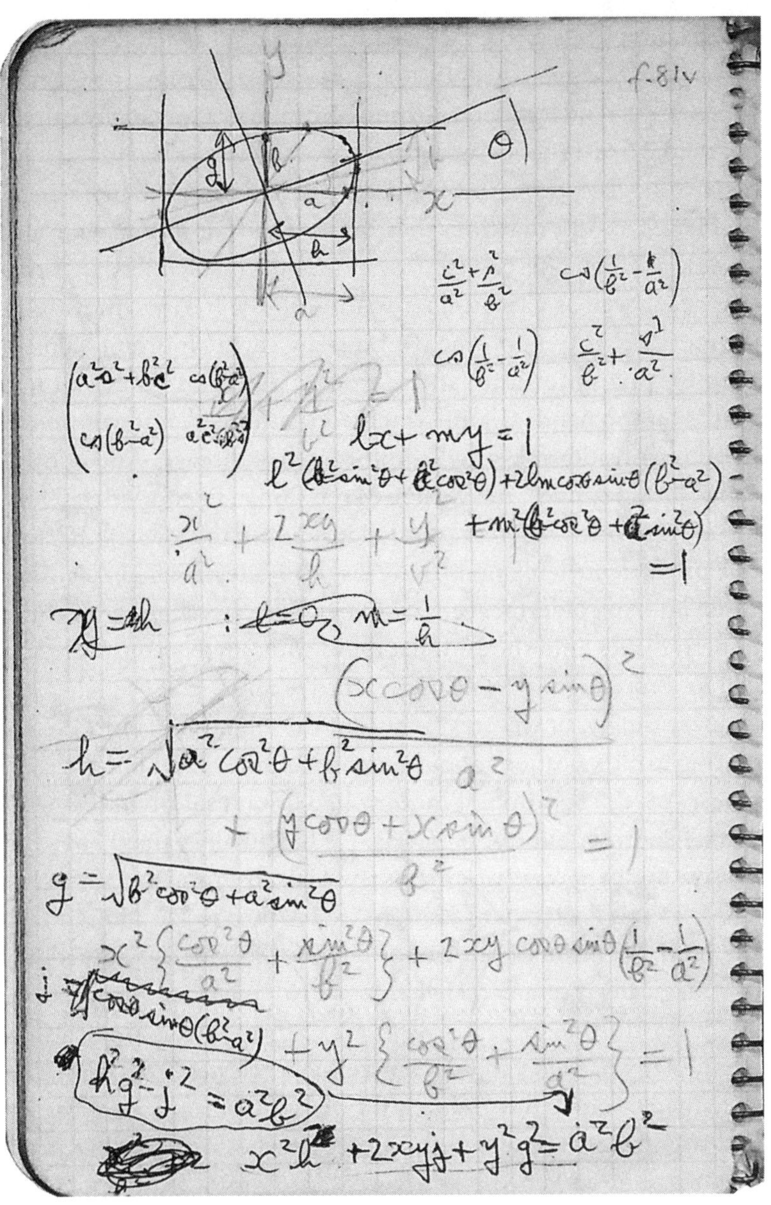

11.4 Lionel Penrose's mathematical language of the hand. Notebook on Dermatoglyphics, n.d. Lionel S. Penrose Papers, University College London, Special Collections, 2/42/1/2, f.81v. By kind permission, Professor Shirley Hodgson and University College London Library Services, Special Collections.

listed its applied significance as threefold: clinical diagnosis, cytological diagnosis, and genetical prognosis (for example, mosaicism in trisomy). And it did, for a time, continue as a way of investigating medical disorders. Autosomal trisomies were identifiable, not just trisomy 21, but trisomy 18, trisomy 13, and trisomy 8 mosaicism. The sex chromosomal links that Penrose had confirmed were also now dermatoglyphic canon: Turner's syndrome, Klinefelter's syndrome, as well as Cri-du-chat syndrome, Wolf-Hirschhorn syndrome, de Lange syndrome, Rubinstein-Taybi syndrome, Smith-Lemli-Optiz syndrome, and Cerebral Giantism. These all had marks on the hand. Claims emerged that linked patterns on the hand to intrauterine rubella and to celiac disease. And the so-called "Sydney line," the version of the simian line to which Charlotte Wolff took exception, was not just linked to Down syndrome, but also to acute leukemia. There was even a team who in 1978 cultured human epidermal cells from newborn foreskins, grew them in petri dishes, and showed that they "formed dermal patterns by cell movement going from arches into whorls."[24]

Increasingly, however, students of the engraved hand were reverting to old questions about human variation and human typologies: "race." This line of inquiry had always sat obliquely with Lionel Penrose, but not remotely. He had finally gained approval in 1963 to change his title from the Galton Professor of Eugenics to Galton Professor of Human Genetics, but that didn't make the UCL tradition of anthropometry and biological anthropology magically disappear. He belonged to a context, a generation, and an institution increasingly aware of the political implications of a biology of race, but which was nonetheless a generation that remained fascinated with human variation. One of his colleagues and friends at UCL was anthropology professor N. A. Barnicot, who considered Penrose as a kind of biological anthropologist, although he wondered why he so rarely published in anthropological journals. In truth their work was closely aligned, but it also differed in important ways. Barnicot was still interested in how to classify "mankind into races," but knew that Penrose was disinclined to use the term "race" at all, in the sense of human subspecies, preferring to think about the whole human race and its subpopulations. As Barnicot put it: "Penrose disliked the term race and even more the opinions of racists." And yet, if anything, this understated Penrose's position. He was far more than disinclined; he was unequivocal: "The use of the term 'race' must be discontinued altogether." It was inexact, archaic, and belonged to an unscientific era.

> At present there is no evidence that man can be divided biologically into subspecies . . . Unless and until such evidence is forthcoming, the term "race" has only one rational use in anthropology, namely to apply to the whole human race as distinct from other species.[25]

So Lionel Penrose told the population geneticists, sociologists, and anthropologists who gathered to work up an anti-racist UNESCO statement on race in the early 1950s. The term "population" might well be used instead, he suggested, groups of people who might be defined geographically, genealogically, linguistically, or culturally, according to the needs of any given investigation.[26]

Penrose stood apart from the UCL provenance on eugenics, distinct from Galton, who had invented the term, and from Ronald Fisher, who had applied it strongly to a class-based theory of fertility decline and to his advocacy for sterilization of the so-called feebleminded. Yet, paradoxically, it was interwar population geneticists (using Fisher's own theories) who had done the most to undo claims for any clear biology of race, whose statistical work made nonsense of the categorization of human variation by biological sign or phenotype. That generation showed that there were no patterns that held, statistically speaking; "the retreat of scientific racism," it has been called. And so, the kind of project that Francis Galton had attempted with fingerprints and that Harold Cummins and others continued to pursue with palms and soles was by the 1950s more than old-fashioned; it had simply been shown to be wrong. Galton already knew that, even if Cummins did not. The ambition to correlate particular characters to particular groups had long since come undone.[27]

And yet, both before and after Penrose's death in 1972, all manner of dermatoglyphic studies were undertaken that continued to try to fix "race" biologically. The opposite of Penrose's ambition of whole-world studies of the human species, these were typically micro-studies of the hands of this and that group, the more apparently isolated or self-contained the better. Before Penrose died, for example, the dermatoglyphic patterns of the Parsis of India were examined (1963), as were fingerprints of four central Australian Aboriginal communities (1967), and of "Australians of European Ancestry" (1971) as a control group. The year Penrose died, dermatoglyphic studies were pursued of "Iranians of African descent," and of "Caucasians of South Iran," and of "four-finger furrows" in Czechs and Gipsies (1972). After Penrose died, dermatoglyphic variation was studied among "Five Eskimo groups from North-western Alaska" (1978), "the Population of the USSR" (1978), and simian crease patterns among "Iranian endogamous groups" (1985).[28]

But why, in the age of a completely different understanding of human genetics, did such phenotype-based research—in this case on the hand—continue? One clue lies in the then-marginal and ill-funded institutional contexts of the researchers, often clustering in South Asia and in Eastern Europe. Dermatoglyphics was simple research in technical and technological terms, barely needing more than Richard Saunders's measuring compass and a suite of willing or unwilling hands. To put this into another context, it has been argued that one reason that Lejeune's team in early 1950s Paris pursued dermatoglyphics was because they lacked the equipment to do much else. After the war, there was barely a laboratory in Paris, still less a microscope adequate to any complex work. Even later, pursuing Lejeune's hunch about chromosomes was only just achievable in practice, using discarded and malfunctioning equipment. The team's breakthrough *was* possible in these crude postwar conditions, but going forward, cytology needed funding, increasingly specialized microscopes, as well as personnel trained in the karyotyping of cells. And it needed specific disciplinary training too, less in genetics than in pathology or animal or plant cytology. The kind of austerity that was the context of postwar Paris perhaps explains the healthy afterlife of dermatoglyphics, of palm and sole studies, that thrived in marginal, underfunded research locations, among small teams or even individual researchers, and in physical anthropology more than in biological sciences.[29]

At a meeting of the International Conference on Dermatoglyphics held in Athens in 1981, one contributor looked forward to dermatoglyphics in the twenty-first century. He knew that the population and taxonomic ambitions of dermatoglyphics were inadequate. In what sense was "German" or "Hungarian" or "British" a biological breeding group? One generation on from the UNESCO statement on race, it was now popularly as well as scientifically commonplace to state the obvious: "It is erroneous to conclude that the pink-skinned populations of the USA are a homogenous mass" was his example, though he still made exceptions: for Hindu castes "which change only slowly," studies might be plausible. But the next two decades, he wrote in 1981, will convince those in the field whether "populational variability" can legitimately be marked on the hand or not. His colleagues were still plugging away, presenting work on digital ridge count variations in "some castes in India," digital and palmar dermatoglyphics in Greeks, and finger ridge counts "among the five Turkman Groups in Iran." (The Persian inclination seems to have endured, Edward Heron-Allen would be pleased to see.) Lionel Penrose might have imagined a future without "race" and a future for dermatoglyphic reading of the hand that lay with medical genetics, diagnostics, and prognostics, but its actual destiny unfolded as a strange persistence of race science. Had he lived into the next millennium, our own twenty-first century, Penrose would have been astounded and troubled to read of attempts to biologically differentiate "Australian Aborigines" by a set of fingerprints supplied by a police station.[30]

Reading the hand in such a study was not about singular identity; rather, it was a continuation of Galton's (failed) search for ways to identify types of humans. This cadre of minor biologists and physical anthropologists were enacting Francis Galton's research wish, after his own failure to find "Englishmen" and "Jews" via their fingerprints. When scholars suggest that hand-based typological classification ended with Galton's failed attempts in the 1890s, they need to look again at the afterlife of dermatoglyphics in which precisely that was pursued; scientific palmists of the new millennium.[31]

FINGERPRINTS AND PALMISTRY: AN ENDPOINT

Like Galton and Cummins, Lionel Penrose did make a point of separating his work out from palmistry. "Dermatoglyphics is the scientific study of ridges in the skin of the palmar and plantar surfaces," he lectured in Oxford in the winter of 1965. "It has to be distinguished from palmistry, as astronomy from astrology."[32] Unlike Galton and Cummins, however, he put little effort into these professional demarcations: fine ridges for the scientists; crude crease lines for the palmists. Not only had he himself long been drawn to the significance of palm lines, he took the time to clarify that palmists' observations were not limited to creases in any case; they also observed "eminences, folds and grooves" of the whole hand. The main palmar creases had clearly been the object of inquiry for scientific as well as "mystical" study, especially, he noted, the transverse creases, the lines of head and heart, he named them. In the hands of subhuman primates, these were usually fused, he explained, and when this fusion occurred in human primates, this is called "the simian or four-finger crease." But he also corrected anatomists: these should not be called "flexure lines"—the common anatomical term—since they are formed by the twelfth week of gestation, before there is flexion in utero. In other words, Lionel Penrose engaged with both chiromantic and anatomical knowledge of the hand.[33] He and Isaac Newton would have had much to discuss.

Journalistic accounts of Penrose's work made the link too: "sophisticated, scientific finger- and palm-reading is only now beginning to play a role in the field of human genetics." In so many ways, "diagnosis by palm-reading" was precisely what Penrose was doing. And although his own intellectual genealogy lay with Galton's fingerprints, with Fisher's mathematics, and with Reginald

Langdon Down's clinical work, he himself became more interested toward the end of his life in that other species of hand reader, the palmists. He sometimes explained matters by palmistry terms: "the five-finger and three-finger creases . . . the lines of head and heart." He included images that showed anatomical terms—the thumb crease, the three-finger crease (or distal transverse), and the five-finger crease (or proximal transverse)—but also set them alongside "ancient names" for palmar creases and mounts: the lines of fate, head, and heart; the mounts of Venus, Luna, and Mars; the fingers of Jupiter, Saturn, and Apollo (fig. 11.5).[34]

In 1963, close to retirement, Lionel Penrose published "Fingerprints, Palms and Chromosomes" in *Nature*. The "transverse

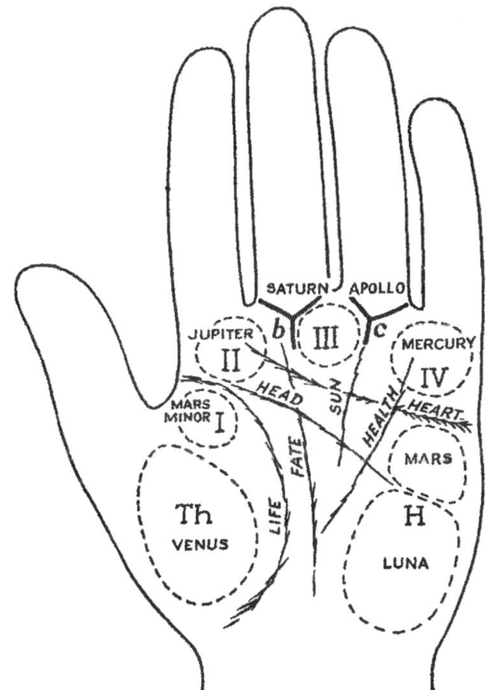

Fig. 4—Ancient names for palmar creases and areas.

11.5 Palmistry for *The Lancet*. "Ancient names for palmar creases and areas." Reprinted from L. S. Penrose, "Fingerprints and Palmistry," *The Lancet* 301, no. 7814 (1973): 1241, with permission from Elsevier.

palmar crease" had long since interested clinicians, playing an important part in diagnosis, but he noted also that students of chiromancy spoke of this crease as the fusion of the lines of the head and the heart. They usually interpreted it as "a sign of concentrative mental power." He was interested in traditional readings and noted that when the simian line appears in "normal" people, it is often taken to indicate "a firm and decisive character." And he was intrigued too that fortune tellers appreciated that creases become more complex over time, that they change with age. And so, "the lines of fortune and health may appear and then disappear again." Like any chiromancer, and especially like Desbarrolles, who had found and illustrated a letter *M* in the middle of the hand, Penrose also described how the four main crease lines on a palm intersect at the center "to form a kind of letter M." The crease lines "show interesting variations, but the ancient interpretations have to be revised." This aging hand expert's exception to palmistry was less about its truth claims than its technical accuracy. He thought it curious that palmists typically speak of a mound attached to each finger—the mounts of Jupiter, Saturn, Apollo, and Mercury—since this is plainly "erroneous." Actually, the mounts are only three in number and are interdigital.[35]

A gracious medical elder in the 1970s, Lionel Penrose looked backward in time, and well beyond his own twentieth-century life. He wondered about ancient hands, delighting in the fact that a ridge pattern could be discerned in 2,000-year-old Egyptian mummies. And he was interested in the history of hand reading as an expertise, dating from the India Veda period, he thought, and perhaps earlier in China. Repeating a convention found in almost every palmistry book, he presented his own short history of the hand, from medieval European applications of astrology, to Henry VIII's move to render palmistry illegal, to the 1735 "Witch Act," which kept it illegal. This was not repealed, he noted correctly, until 1951, with the Fraudulent Mediums Act. Penrose even referred to Cheiro's *You and Your Hand* (1932), published the same year as his own preliminary results from the Colchester Survey. This was all gathered in preparation for a book on the hand, but Lionel Penrose died before it was completed, on May 12, 1972.

Instead, a précis was published posthumously in the *Lancet* as "fingerprints and palmistry." The psychiatrist, mathematician, and geneticist died a historian as much as a geometrician of the hand.[36]

Lionel Penrose (1898–1972) lived the same broad vital dates as Charlotte Wolff (1897–1986) and Harold Cummins (1893–1976). Theirs were twentieth-century lives, joined by the hand. They were each medically trained, but their disciplinary trajectories were as divergent as their personal histories. Cummins was a conventional and conservative anatomist in the US South; Wolff was a lesbian Jewish emigré, schooled in erudite interwar European philosophy and psychology; Penrose was a Quaker, a Cambridge-trained mathematician-geneticist drawn to psychology, to psychiatry, and to those humans who fall beyond the normal, statistically speaking. There was notable common ground. For Charlotte Wolff and Lionel Penrose, one commonality was early 1920s German and Austrian psychology. The mind and the body, if not the self, was fascinating for both of them. They shared schooling and interest in gestalt psychology, the German word meaning in this context "pattern" or "configuration." Little wonder Penrose was interested. Wolff and Penrose also shared a motivation toward medical research in asylums. Penrose had the best possible authorized access, the responsibility of his first appointment with the Colchester Survey, and thereafter in Ontario, through the Galton Laboratory, and through his active retirement at the Harperbury Hospital. Wolff worked at a far harder and lower level, hampered by her refugee status, perhaps her gender, her sexuality, and her training. But for years, she was able to access some asylums and to gather data on hands, palms, lines, creases, diagnoses. She pushed this crudely through her basic calculations, seeking correlations, well above the training of Katharine St. Hill, who also looked at palms in asylums, but far below the mathematical expertise of Lionel Penrose, able to apply and test the latest in population genetics and then to link complicated topology to the curves and lines on fingers and palms. Penrose and Wolff were both members of the Society of Friends, Wolff perhaps more actively, Penrose as part of his inheritance, his family's cultural genes.

For Lionel Penrose and Harold Cummins, there was an important crossover in the classification of humankind. Cummins was a firm believer in race; Penrose was not. But they both spoke, researched, and engaged with the human, with anthropology, and with variation, within and without eugenics in its highest moment. Wolff and Cummins, in this regard, both lived in highly divided, highly racist polities—Nazi Germany on the one hand, Louisiana on the other. Wolff was a Jewish victim; Cummins was a white beneficiary. And yet they crossed over strongly in their research on primatology, both taking prints of, and interpreting the hands of, gorillas, chimpanzees, and gibbons, wondering how they and humans were related. But if Cummins the anatomist was interested in the physicality of the human body, Wolff the analyst was interested in the mind and the primate self.

Each of these medical readers of hands worked in and through the high modern era; the age of genetics, the age of racial states, and the age of psy. They were joined by hands, decoders of signs, interpreters of patterns, revealers of secrets. They each spoke the language of the hand, in different dialects perhaps, but in a manner mutually comprehensible through palmistry, and through the expansion of Galton's fingerprints into dermatoglyphics. The medical and scientific palmistry of Edward Heron-Allen, Katharine St. Hill, and even the Ellis family lurked behind these developments, and not far in the shadows. Nineteenth-century naturalists and anatomists of various orders, including the clinicians—the phrenologists—were in sharper focus. Before all of them, early modern physicians set out the language of the hand in the language of the heavens. They were themselves trying to make sense of the ancients, the foundational readers of bodily signs.

CONCLUSION

Occult Medicine and the Lines of Fate

For early moderns like Richard Saunders and Isaac Newton, the "occult" meant that which is invisible or hidden. In this sense, the word has crossed over into contemporary medical terminology, signaling symptomatic presentation whose cause is not organically discernible. Pain physicians tell me that "occult" is a term used constantly in referrals between specialists, meaning "unseen, unknown, unexplained." Hand specialists use it that way too. An "occult malignancy" causing pain in the thumb turned out to be bronchogenic carcinoma, in one case. In another, hand surgeons finally uncovered the reasons for "occult" pain: dorsal wrist ganglion. Subtle fractures had been "radiographically occult" in that case, and magnetic resonance imaging (MRI) was needed for the secret to be disclosed. The modern magic that applies magnetic field and radio waves now reveals the inside of our bodies, visualizing skin and flesh as well as cartilage and bone, making the hidden, the interior, visible and readable, bringing occult marks "engraven by nature" to light, where they can be decoded by radiological experts.[1]

With such extraordinary techniques and technologies, the antecedently incredible has indeed become true, as biologist Alfred Russel Wallace predicted, or, in this context (I might take the liberty of saying), that he divined. Of course, the MRI that renders visible otherwise occult signs is the outcome of a spectacular recent history of medical engineering, nuclear physics, clinical radiology, and more. But it is also connected to the strange work of eighteenth- and nineteenth-century physicists interested in animal magnetism, and to the Society for Psychical Research, who held magnets near bodies, experimentally looking for the different effect of the South Pole or the North Pole. More remotely, it is connected to hermetic Robert Fludd's "magnetical medicine," and Richard Saunders's discussion of sympathies and antipathies related to the "Magnet of the Sun, which is the subject of all admiration." Even earlier, the magnetism of Galen and Avicenna was an "insensible quality" that exceeded that of the four sensible qualities; heat, coldness, wetness, and dryness. Imagine what they would all perceive if they were time-traveling witnesses to MRI diagnostics that revealed otherwise occult ganglions in the wrist? It is the kind of wonder that perhaps only time-traveling historians imagine and indulge.[2]

A few modern chiromancers and palmists were enchanted by the visual magic of their own time. They reproduced new "X-rays" of the hand, which showed in life a deep skeletal interior that became only too apparent in death. Indeed, the very first X-ray of a human was of the hand of Anna Röntgen, taken in 1895. It was of her left hand, perhaps by happenstance, but chiromantically correct nonetheless. She reportedly exclaimed: "I have seen my death." Her husband, physicist Wilhelm Röntgen, immediately appreciated the significance of the magic he had stumbled across. He called it "x," the scientific symbol for an unknown. Reading my way through a thousand books and manuscripts, I have found the occasional X-ray to be the most curious. X-rays miss all that exterior flesh on which meaningful signs were engraved, in which the palmist was so expert. Unlike like modern magnetism—the MRI—X-rays also miss the soft tissue on the inside. And yet all of these imaging techniques, this desire to look inside the human body, caught the larger questions posed by chiromancy, palmistry, and dermatoglyphics: What is inside, how does that show on the outside, and what does that explain?[3]

If Paracelsus asserted that "nothing is so secret in a human being that it does not have an outward sign," we have been able to follow how comprehensions of inside and outside have changed, in relation to understanding selves, souls, the past, and the future, and that across chiromancy, medicine, and a whole range of modern human and life sciences. Putting it slightly differently to Paracelsus, and anticipating our own psychological vocabularies, Richard Saunders tells us across the centuries that hands are "looking-glasses wherein we see the Soul and Affections."[4] In the end, a history of the hand turns out to be a history of the human interior: Aristotle's physiognomic aphorisms; Charles Darwin's "emotion" ex-pressed (pressed outwards); Ida Ellis's phrenological character revealed; Charlotte Wolff's psychoanalytic and endocrinological body that speaks; and Lionel Penrose's hand phenotype with its occult genotype, finally decoded.

Tracking hand-based human interiors also tells us something of the body's relationship to time, both the past and the future. I began a history of the hand thinking it might reveal secrets about how the future and fate and fortune might be imagined similarly and differently. Instead, I found a trajectory over the centuries from hand experts' investment in the future to an investment in the past. In the early modern period, as a rule, the future was especially significant. The most relevant past was information regarding one's natality: the position of the heavenly bodies at birth. Yet even that needed decoding in order to forecast, both as prognostication and as prognostics. Certainly, a history of future-seeking continued and continues. Antique ideas about fortune, fate, destiny, and the will (connected to the hand via the mind) were brought into late modernity in various forms. And yet, the balance has shifted. Over the eighteenth, nineteenth, and twentieth centuries, the human interior, and practices of reading the human body, became increasingly about the past, which itself has extended backward into deeper time.

Nineteenth-century anatomists began to see in the human hand signs of a distant evolutionary past when comparing it to the extremities of other mammals and primates, in fossilized, skeletal, and living form. Darwin's anatomical and anthropological descendants took this in all kinds of directions. Francis Crookshank asserted a spurious species past in the hands of some humans, an "atavistic" recurrence that was evidence for his thesis about particular and distant common ancestors. Looking at the so-called simian lines, more reputable primatologists turned similar ideas into palm-based evidence of "the descent of man." It is perhaps this epic species story that served as the materialist substitute for God in a secularizing world. This retained within modernity a kind of antique magic in which signs on the body, if read with the correct expertise, could tell of long ago and predict how an ancient primate connection could be, and was, reproduced into the future. In a different political biology altogether, Friedrich Engels took the same idea and saw the great human capacity to hold and wield a tool and turned that into his long history of human labor and capital. For them all, the body spoke—its signs revealed meaning just as surely as the lines, crosses, stars, quadrangles, and triangles were decoded by Richard Saunders in 1663.

Palmists, psychiatrists, and geneticists alike looked for and found a more proximate, generational past in the hand as well: parentage, inheritance, genetics. Actually, Down syndrome is not inherited, but Langdon Down, Penrose, and Lejeune all thought it might be when they compared palm lines and patterns of parents and siblings. At the high point of dermatoglyphics, Harold Cummins found signs of paternity in the hand. Charlotte Wolff looked for inherited patterns too, but along with Adlerian analyst Francis Crookshank she knew that the modern self was also shaped by a personal, experiential history that imprinted the hand. When she read the palms of Aldous and Julian Huxley, like Peter the freeborn chimpanzee, she saw the loss of the mother. And if Wolff looked to the manifestation of an id and an ego, Lionel Penrose found identity at a different kind of internal depth again. The visible phenotype of the simian line was linked theoretically to an occult genotype, the latter eventually becoming visible and known once electron microscopes could reveal the interior of DNA within a cell. Chromosomes and the DNA within hold evidence and instruction that make each human unique. Galton, in a different language of the hand, had revealed just that of the fingerprint.

Like Mok, Peter, and Daisy in the London Zoo in 1938, it turns out that former prime minister Tony Blair has a simian line on each hand, at least according to Britain's current Chirological Society. Contemporary palmists enjoy and benefit from online readings of "famous simian line holders," squarely in the tradition of Edward Heron-Allen's society palmistry, Katharine St. Hill's hands of celebrities, and indeed Charlotte Wolff's chirology of Europe's interwar cultural luminaries. Now we are party to the inner secrets of Robert de Niro, Hillary Clinton, and Benedict Cumberbatch, as told by their rare palms. They sit in the same tradition of revelation and curiosity as nineteenth-century Richard Beamish's reproduction of the palm of professor William Whewell, who coined the terms "scientist" and "physicist," and of the two ancient mummies from Thebes. Contemporary palmists seem inclined to read simian lines positively: they are now stigmata with some cultural capital. A double simian line may be a sign of considerable distinction. In Japan, its rarity is read as luck, and for an online palmist who authors "God Given Glyphs," "many simian line owners pour their emotional energy into their work resulting in positive success in their vocational arena."[5] Obviously these palmists take crease lines, including the simian line, entirely seriously, but so do clinicians in the world's most reputable health institutions; the more surprising and perhaps more important point. New York's Mount Sinai Hospital, for example, also offers online instruction on the meaning of the simian crease, forming in the twelfth week of gestation in one in thirty people. It may be "normal" and mean nothing, but it may be associated with conditions that affect "mental and physical growth." If a neonate has the simian crease, parents are forewarned, the doctor may ask, "Is there a family history of Down syndrome? Does anyone else in the family have a single palmar crease without other symptoms?"[6]

It is another indulgence, but often enough I speculatively cast historical actors into our own times. The early nineteenth-century anatomist Charles Bell, who saw God's design in the human hand, would have endorsed and understood a contemporary palmist's description of the simian line as a God-given glyph. He would have understood and even expected the imprinting of the mind on the hand. He would certainly have expected Mount Sinai's clinical advice based on a basic anatomy of the hand. A century later, explicitly placing herself within Charles Bell's intellectual line, Charlotte Wolff would be able to discuss in detail these God-given glyphs, but as a believer in the eloquence of hands, not a believer in God. Along with Edward Heron-Allen and especially Katharine St. Hill, she would readily have comprehended hand reading's ongoing viability as a "healing therapy" that focuses on inner lives. Ida Ellis would have pretended to, and convincingly so. And as for simian lines and parental health advice, the Langdon Downs, Francis Crookshank, Charlotte Wolff, and Lionel Penrose would all simply see Mount Sinai Hospital as sensibly implementing their own recommendations.

Of what present has this been a long history? The history of the hand is a history of ways of finding the modern self, tantalizingly sitting between identity and identification. It is in part the backstory to carnivalesque palmistry, and in part to a history of health and medicine. Yet it is not to be reduced to an account of "pseudo-science" or "quackery." Rather, this has been a history of the vast range of more and less orthodox ways in which diagnosis, prognosis, and therapeutics have been conceptualized and implemented in the past. Put another way, this isn't (just) a history of fun and quirky medicine, it *is* the history of medicine. To be sure, some of the practices that Francis Crookshank, Harold Cummins, and Charlotte Wolff pursued at the high point of dermatoglyphics have become increasingly marginal; they were even then. But we should remember that they each extended the work of canonical Purkinje and Bell, of Cambridge anatomist of the hand Humphry, of neurologists Vaschide and Wallon, and of their contemporaries who literally wrote the textbooks, not just the updates of *Quain's Anatomy*, but also Frederic Wood Jones's *Principles of Anatomy as*

Seen in the Hand, for example. And we should appreciate that the brilliant Lejeune and Penrose were at the center of dermatoglyphic research too: there could hardly be two more distinguished genetic researchers of their era. And in any case, dermatoglyphics continues. Analysis of the hand's furrows is claimed to be "predictive" of breast cancer, prognostication again. Lionel Penrose's *atd* angle is put to twenty-first-century use. Dermatoglyphic patterns are linked to hypertension, to dental caries, even to Post-Traumatic Stress Disorder. And in *Human Biology*, hand reading is strangely returned to an original community with the dermatoglyphic analysis of "Hungarian and Gypsy populations."[7] Often these studies appear in obscure journals, but as recently as 1993 the *British Medical Journal* explained the relation of fingerprints and the shape of the palm to fetal growth and adult blood pressure.[8] This is hardly the "soul of the world," which seventeenth-century kabbalistic chiromancers sought and found. Nonetheless, the uptake of palm reading by mid-twentieth-century geneticists makes a history from chiromancy to genetics not just a whimsical comparison, but literally, actually, part of a connected history of health and medicine; all along a true story of magic and science.

Palmistry at the pier endures with its riotous mix of oriental and occidental occult, or sometimes in refusal of that in the tradition of scientific aspirant Katharine St. Hill. The chiromancy and physiognomy of *Hast Samudrika* flourishes, faces and hands read for near-future fortunes. But so, too, does medical fortune-telling. The extraordinary current popularity of DNA testing is derivative of this long history, a bodily reading of the future and the past. In the most explicit way, it also discloses otherwise occult knowledge of our singular selves, secrets contained within and decoded by experts. In the same way that thousands presented themselves to the Anthropometry Laboratory and provided Galton with data, so too the curious now send bio-samples and payment to AncestryDNA and 23andMe, to receive code readings not just of their own deep histories and proximate family histories, but also to be forewarned of their future health risks. It is as individualized and commercialized as any reading offered by Keiro or Katharine St. Hill or the Ellis family, who, we might say, invented diagnosis-by-mail.

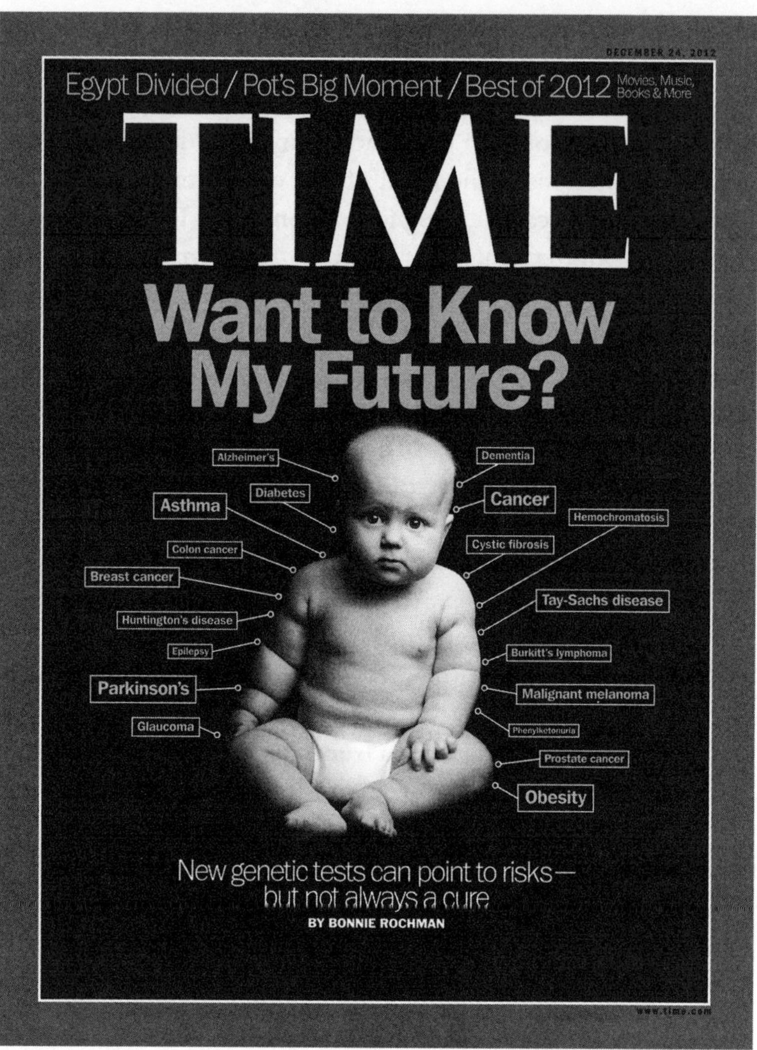

C.1 DNA testing for ancestry, traits, and health risks: a twenty-first-century endpoint for Richard Saunders's *Physiognomie and Chiromancie*. "Want to know my future?" Cover of *Time*, December 24, 2012.

"Welcome to you," entices the packaging of 23andMe, a twenty-first-century promise of identity that lies directly in the (linked) traditions of Cheiro's *You and Your Hand* and Galton's fingerprinting. 23andMe even trades on the weird and the wonderful, diagnosing traits like asparagus odor detection, fear of heights, hatred of the sound of chewing, and photic sneeze reflex. By comparison and for all their commercial expedience, the Ellis family's phrenological "character diagnosis" approaches the profound and is far closer to current analytic psychiatry, and especially to more everyday psychology and counseling. It is a caution not to just sneer or be amused by the seemingly antiquated: phrenology, physiognomy, and chiromancy. Just as Katharine St. Hill indicated in her *Medical Palmistry*, the body holds the signs of our future health. We still want to know the future story that our bodies tell. We still want those secrets to be disclosed by experts who are literate in a new kind of code altogether. *Time* magazine gathered this up for a cover in 2012: "Want to know my future?" (fig. C.1).[9]

This story has been an unlikely one, revealing a continuous history from chiromancy to genetics. My research here as a historian of modern biological sciences both started, and has unfolded, in a completely unexpected manner, from the electric moment that sparked my inquiry in the first place. In 1938, Charlotte Wolff—zoologist, psychologist, physician, seer, and hand reader—took Mok the gorilla's massive but lifeless hand, rolled ink onto it, pressed his mark onto paper, and after her own death, left the print behind in her disheveled papers. Eighty years later, I opened that archive folder and Mok's palm was revealed. But it was occult to me. I couldn't decode its context, its reason, its meaning, its history. I was not yet party to its hidden signatures (figs. C.2 and C.3). Now we can discern the chiromancy, physiognomy, primatology, palmistry, anatomy, evolutionary biology, neurology, endocrinology, and psychology therein. Now we can read that palm.

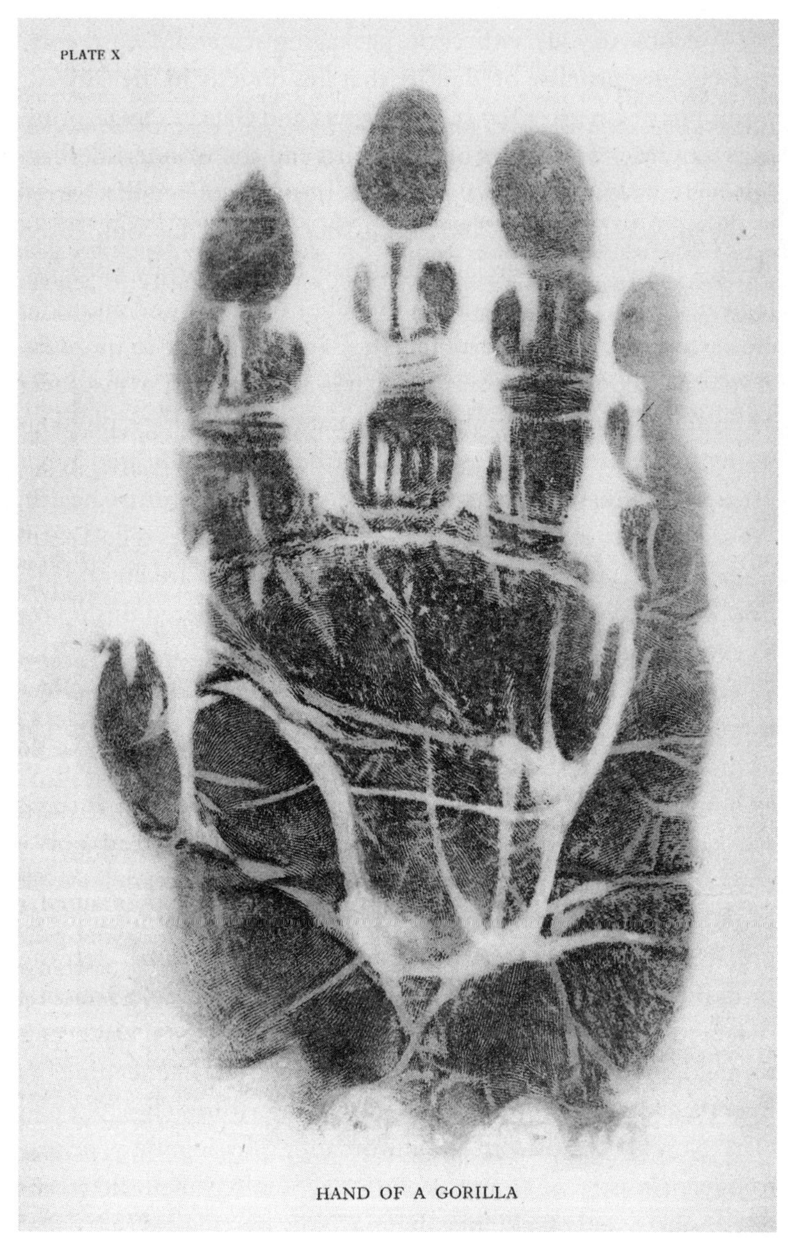

C.2 Hand of a gorilla. From Charlotte Wolff, *The Human Hand* (Methuen, 1942), plate X. Copy in author's possession.

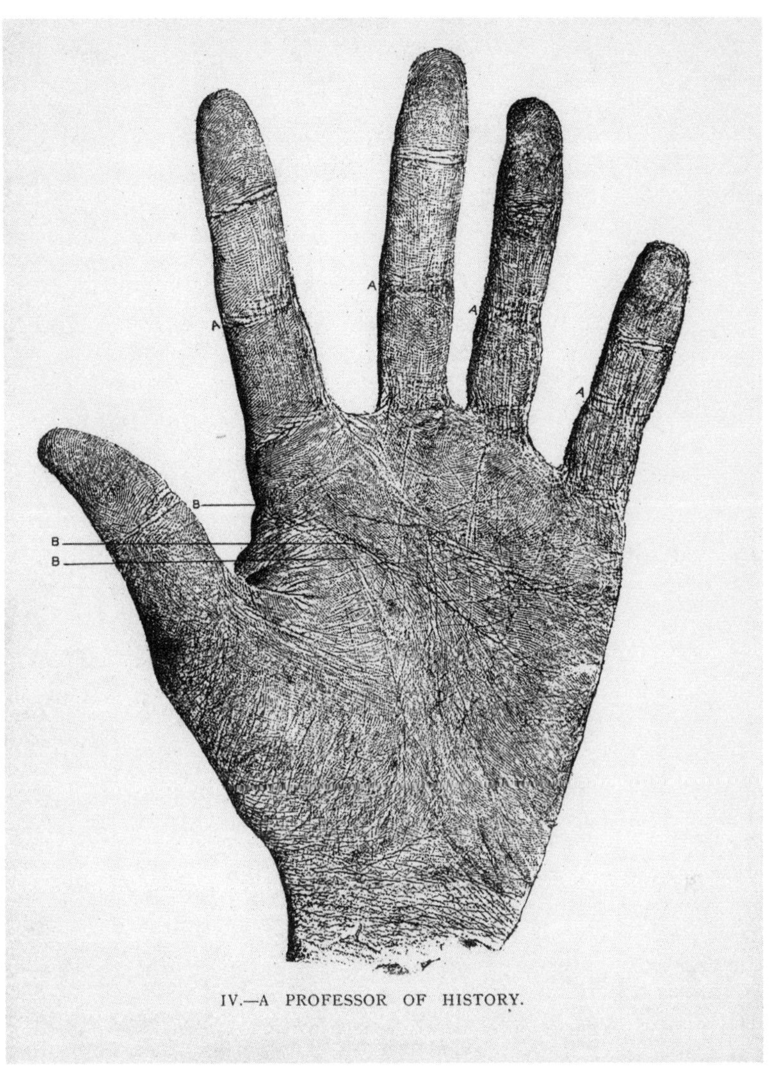

C.3 Palm print of "A Professor of History." "The Head Line shows extraordinary observation in some matters, but oblivion in others." From Ina Oxenford and Anna MacDowel Cosgrave, *Life Studies in Palmistry* (L. Upcott Gill, 1899), plate IV. Copy in author's possession.

Acknowledgments

My thanks to archivists and librarians in the Bodleian Library, Oxford, the British Library, the Houghton Library, Harvard University, the University of Liverpool Archive, and Cambridge University Library. I'm especially grateful for the skilled assistance of Rafa C. Siodor at UCL Library Services, Special Collections, and Nichola Court, archivist at the West Sussex Record Office, Chichester. Permissions for reproduction are detailed below each image, but in particular I thank Professor Shirley Hodgson, daughter of Lionel Penrose, for permission to reproduce from her father's papers, the British Psychological Society for permission to reproduce from the Wolff papers, and Sam Jones for permission to reproduce from the Edward Heron-Allen papers, in the Estate of Ivor Jones. Researching Edward Heron-Allen put me in touch with a range of willing experts. Joann Fletcher readily answered my queries on that Egyptian mummy. John Mahony of the Sette of Odd Volumes generously sent me a personally inscribed Opusculum, along with much-valued copies of Desbarrolles and other Heron-Allen materials. Thank you.

I was delighted to present some of this work as the Alison Winter Memorial Lecture, University of Chicago, 2023: Winter's *Mesmerized* has been on my shelves since 1998. I am grateful for the suggestions and corrections of colleagues and friends at home and abroad: Andrew Antoniou, David Armitage, Graham Bedford, Joanna Bourke, Khadija von Zinnenberg Carroll, Annie

Clarke, Michael Dunn, Rebecca Flemming, John Gascoigne, Nick Hopwood, Kasia Jesowska, Lauren Kassell, Emily Kern, Projit Mukharji, Cody Nitschke, Margaret Pelling, James Poskett, Sumathi Ramaswamy, Peter Godfrey Smith, Kenneth Zysk, and my research group over recent years: Tiarne Barratt-Young, Adam Bobbette, Michelle Bootcov, Jarrod Hore, Chi Chi Huang, Rimi Nandy, Stephen Pascoe, Aprajita Sarcar, Emma Thomas, Katrien Verbeke, Joel Wing-Lun. UNSW Sydney supports with both funding and freedom in a manner that is incomparable. Chelsea Barnett, Matthew Kidd, and Marie McKenzie have delivered immaculate research assistance. I have published on the hand in "Lionel Penrose and the Geometry of the Hand," *The Lancet* 403, no. 10440 (2024): 1978–79; and "The Disenchantment of Chiromancy: Reading Modern Hands from Palmistry to Genetics," *Past & Present* 263, no. 1 (May 2024): 125–69.

My thanks to Karen Darling and Fabiola Enríquez Flores at the University of Chicago Press and to Jin Auh and James Pullen at the Wylie Agency, in New York and London, respectively. I am enormously grateful for the expertise and time of readers for the Press: attentive and anonymous colleagues, my thanks.

Chris Brunton and Oscar Brunton are more useful than they probably realize as I try out ideas: thanks for your valued input. I appreciate also talks with my physician-brother Guy. In one passing conversation I stopped in my tracks when he used the term "occult" with regard to pain of unknown origin. Now there's a story. Thanks to the wonderful folk at Exeter Café where draft upon draft was marked up and mused upon. Likewise to the Barista of Bundanoon, Monsieur Justin François Crébert, who shared detailed historical knowledge of swordsmanship and hand proportions, as well as of Richard Francis Burton, fencer and orientalist. The closest of companions, Nerida Little, Kylie Hitchman, and Barbara Hatten, walk, talk, and cycle with me, constantly exchanging news and views. This book is for Barbara; author, neighbor, friend.

Alison Bashford, "Alegria," 2025

Notes

SECRETS DISCLOSED

1. Richard Saunders, *Palmistry, the secrets thereof disclosed* (London, 1663); D. T. Whiteside and M. A. Hoskin, eds., *The Mathematical Papers of Isaac Newton*, vol. 1: 1664–1666 (Cambridge University Press, 1967), 15–19; Martin Porter, *Windows of the Soul* (Clarendon Press, 2005), 27.

2. Matt Goldish, "Newton on Kabbalah," in *The Books of Nature and Scripture*, ed. James E. Force and Richard H. Popkin (Springer, 1994), 89–103; Allison P. Coudert, "Leibniz, Locke, Newton and the Kabbalah," in *The Christian Kabbalah*, ed. Joseph Dan (Harvard College Library, 1997), 149–79.

3. Lionel Penrose, "Fingerprints and Palmistry," *The Lancet* 301, no. 7814 (June 2, 1973): 1239–42.

4. "Simian Crease," Health Library, Icahn School of Medicine at Mount Sinai, https://www.mountsinai.org/health-library/symptoms/simian-crease.

5. F. G. Crookshank, "The Importance of a Theory of Signs and a Critique of Language in the Study of Medicine," in *The Meaning of Meaning: A Study of the Influence of Language Upon Thought and of the Science of Symbolism* (1923), ed. C. K. Ogden and I. A. Richards, 7th ed. (Harcourt, Brace and Co., 1945), 354; "Examination of the Hand (The Hand in Diagnosis)," Stanford Medicine, https://stanfordmedicine25.stanford.edu/the25/hand.html.

6. The Body as Text: Physical Examination Course for Medical Students, Stanford Medicine, https://stanfordmedicine25.stanford.edu/about/body-as-text-physical-exam-course.html.

7. Paracelsus, *Von den natürlichen Dingen* (1582) trans. Marke Ahonen, in "Medieval and Early Modern Physiognomy," *Sourcebook for the History of the Philosophy of Mind*, ed. Simo Knuuttila and Juha Sihvola (Springer, 2014), 636.

8. Eli Zaretsky, *Secrets of the Soul* (Knopf, 2005); Darian Leader, *Hands: What We Do with Them* (Hamish Hamilton, 2016).

9. V. S. Kumaraswami Mudaliar, *A Complete Treatise on the Science of Chirography and Chiromancy* (Victoria Press, 1918).

10. Alex Owen, *The Place of Enchantment* (University of Chicago Press, 2004); Egil Asprem, *The Problem of Disenchantment* (Brill, 2014); Lorraine Daston and Katharine Park, *Wonders and the Order of Nature 1150–1750* (Zone Books, 1998).

11. Joshua Landy and Michael Saler, eds., *The Re-Enchantment of the World* (Stanford University Press, 2009); Thomas Laqueur, "Why the Margins Matter," *Modern Intellectual History* 3, no. 1 (2006): 111–35; Michael Saler, "Modernity and Enchantment," *American Historical Review* 111, no. 3 (2006): 702. See also Jason Ā. Josephson Storm, *The Myth of Disenchantment* (University of Chicago Press, 2017).

12. Editor, "The Simian Crease," *The Lancet* 283, no. 7339 (April 25, 1964): 921.

13. Marta E. Hanson, "Hand Mnemonics in Classical Chinese Medicine," *Asia Major* 21, no. 1 (2006): 325–47; Katharine St. Hill, *Medical Palmistry, or, The Hand in Health and Disease* (Rider, 1927), 11.

14. Charles Darwin, *The Descent of Man and Selection in Relation to Sex* (1871) (John Murray, 1913), 76–77. This is the core insight extended by Raymond Tallis, *The Hand: A Philosophical Inquiry into Human Being* (Edinburgh University Press, 2003). Friedrich Engels, *The Part Played by Labour in the Transition from Ape to Man* (1896) (Foreign Languages Publishing House, 1952).

CHAPTER ONE

1. Johannes Rothman, *Xeipomantia, or the Art of Divining by the Lines and Signatures Engraven in the Hand of Man, by the Hand of Nature* (1652), in *The Works of the Late Most Excellent Philosopher and Astronomer Sir George Wharton*, trans. George Wharton, ed. John Gadbury (London, 1683), 528; Richard Saunders, *Palmistry, The Secrets thereof Disclosed* (London, 1663), 10.

2. Gershom Scholem, "Chiromancy," *Encyclopedia Judaica*, vol. 4 (Macmillan, 2007), 652–54. Zysk suggests that Nidarnar is likely the sage Nārada. Personal communication, June 2025. See Kenneth G. Zysk, *The Indian System of Human Marks*, 2 vols. (Brill, 2016), vol. 1, 123, 132–33, 136, 190–92.

3. "Physiognomonica," in *The Works of Aristotle, Volume VI, Opuscula*, trans. T. Loveday and E. S. Forster (Clarendon Press, 1913), 806a, 806b, 807b, 808a, 808b; Roger A. Pack, "On the Greek Chiromantic Fragment," *Transactions and Proceedings of the American Philological Association* 103 (1972): 367–80. See Porter on the terms "physiognomy," "physiognomony," and the vernacular "fisnomy" in *Windows of the Soul*, 18–35.

4. "Physiognomonica," 806b, 811a, 807a, 809b, 814b, 810a, 808a.

5. *Historia Animalium*, trans. D'Arcy Wentworth Thompson, in *The Works of Aristotle* (Clarendon Press, 1910), vol. IV, book I, part 15, 493; *De Partibus Animalium*, trans. William Ogle, in *The Works of Aristotle* (Clarendon Press, 1911),

book IV, part 10, 686; Baldasar Heseler, *Andreas Vesalius' First Public Anatomy at Bologna, 1540*, ed. Ruben Erikkson (Almqvist & Wiksells, 1959), 55–56.

6. K. C. Sen, *Hast Samudrika Shastra* (D. P. Taraporevala, 1960).

7. Zysk distinguishes the Indian system from Greek physiognomy, but suggests that they nonetheless share the idea that "the outer reveals the inner." Zysk, *The Indian System of Human Marks*, vol. I, ix.

8. [Anon.], *Palmistry, Chirognomancy, and other Prognostics (From Marathi)* (Bombay: Family Printing Press, 1888), 2, 4, 8, 15–17.

9. Sen, *Hast Samudrika Shastra*, xi.

10. Joseph Ziegler, "On the Various Faces of Hebrew Physiognomy as a Prognostic Art in the Middle Ages," in *Unveiling the Hidden—Anticipating the Future*, ed. Josefina Rodríguez-Arribas and Dorian Gieseler Greenbaum (Brill, 2021), 292–93.

11. *The Zohar*, cited in Ron Margolin, "Physiognomy and Chiromancy," in *Prognostication in the Medieval World*, 2 vols., ed. Matthias Heiduk, Klaus Herbers, and Hans-Christian Lehner (de Gruyter, 2020), vol. 2, 919; Michael D. Swartz, *The Signifying Creator* (New York University Press, 2012), 2; Scholem, "Chiromancy," 652–54.

12. Gersham Scholem, *Origins of the Kabbalah* (1962) (Princeton University Press, 2018); David B. Ruderman, *Kabbalah, Magic and Science* (Harvard University Press, 1988); Irmi Dubrau, "Physiognomy among Medieval Jews," in Heiduk et al., *Prognostication in the Medieval World*, vol. 2, 908–14; Ithamar Gruenwal, "Physiognomy, Chiromancy and Metoposcopy," in *Apocalyptic and Merkavah Mysticism* (Brill, 2014), 249–54; Margolin, "Physiognomy and Chiromancy," 915–24.

13. Porter, *Windows of the Soul*, 64–65; Stephanie Leitch, "Getting to How-To: Chiromancy, Physiognomy, Metoscopy and Prints in Secrets' Service," in *Quid est secretum?*, ed. Ralph Dekoninck, Agnės Guiderdoni, and Walter S. Melion (Brill, 2020), 635–40; William Eamon, *Science and the Secrets of Nature* (Princeton University Press, 1994). [Aristotle], *The secrete of secretes of Arystotle* (London, 1528), 45; Ahonen, "Medieval and Early Modern Physiognomy," 633–37; Steven J. Williams, *The Secret of Secrets* (University of Michigan Press, 2003). See Elaine Leong and Alisha Rankin, eds., *Secrets and Knowledge in Medicine and Science, 1500–1800* (Routledge, 2011).

14. Felix Klein-Franke, "The Geomancy of Ahmad B. 'Ali Zunbul," *Ambix* 20, no. 1 (1973): 26–35; Matthew Melvin-Koushki, "Geomancy in the Islamic World," in Heiduk et al., *Prognostication in the Medieval World*, vol. 2, 791; Joseph Ziegler, "Philosophers and Physicians on the Scientific Validity of Latin Physiognomy, 1200–1500," *Early Science and Medicine* 12, no. 3 (2017): 285–312; Emilie Savage-Smith, ed., *Magic and Divination in Early Islam* (Ashgate, 2004); Xing Wang, *Physiognomy in Ming China* (Brill, 2020), ch. 4.

15. Thomas Aquinas, *Summa theologiae* 2.2.95.1, discussed in Samuel P. Gillis Hogan, "Stars in the Hand," MA thesis, University of Saskatchewan, 2018,

45; Lynn Thorndike, *A History of Magical and Experimental Science*, 2 vols. (Columbia University Press, 1923), vol. 2, 606.

16. Chiromantia delineata, Kraków, BJ 551, fol. 117, reproduced in Benedek Láng, *Unlocked Books* (Pennsylvania State University Press, 2008), 127; Johann Hartlieb, *Buch aller verboten Kunst* (Augsburg, c. 1465); Johann Hartlieb, *Die Kunst Ciromantia* (Augsburg, c. 1480).

17. Hogan, "Stars in the Hand," 52, 210. See also Paola Zambelli, *The Speculum Astronomiae and Its Enigma* (Springer Dordrecht, 1992).

18. For judicial and natural astrology, see Charles Webster, *From Paracelsus to Newton* (Cambridge University Press, 1982); Hilary M. Carey, "Astrological Medicine and the Medieval English Folded Almanac," *Social History of Medicine* 17, no. 3 (December 2004): 345–63.

19. The Eadwine Psalter, Trinity College, Cambridge, MS R.17.1, fol. 282; see Charles F. Burnett, "The Earliest Chiromancy in the West," *Journal of the Warburg and Courtauld Institutes* 50 (1987): 189–95; MS. Digby Rolls 4, Bodleian Library, Oxford; Derek J. Price, *An Old Palmistry, being the earliest known book of palmistry in English* (W. Heffer, 1953), 1.

20. Johann Hartlieb, *Die Kunst Ciromantia*, John Rylands Library, Manchester, Blockbook Collection, 17261 (Augsburg: Jörg Schapf, [about 1480]).

21. Bartolemmeo della Rocca [Cocles], *Physiognomiae et chiromantiae compendium* (Strasbourg: J. Albertus, 1536), 59; [Anon.], *Chiromancy, or the art of divination by the hands* (1543) (London: G. Walker, c. 1800), 28.

22. Bartolommeo della Rocca [Cocles], *A brief and most pleasau[n]t epitomye of the whole art of phisiognomie . . . englished by Thomas Hyll* (London, 1556). See Porter, *Windows of the Soul*, ix; Thomas Hill, *A pleasant history: declaring the whole art of phisiognomy, orderly uttering all the speciall parts of man, from the head to the foot* (London, 1613), 166–73; Johannes ab Indagine, *Briefe introductions . . . unto the art of chiromancy*, trans. Fabian Withers (London, 1558), 3, 13, conclusion. A century later this translation was published as *The Book of Palmestry and Physiognomy* (London, 1656).

23. Indagine, *Briefe introductions*, book II, n.p.; Torreblanca (1678), trans. George Borrow, *The Zincali*, 2 vols., vol. I (John Murray, 1841), 125–26; *Chiromancy, or the art of divination by the hands* (1543) (G. Walker, c. 1800), 31. See Valentin Groebner, "Complexio/Complexion: Categorizing Individual Natures, 1250–1600," in *The Moral Authority of Nature*, ed. Lorraine Daston and Fernando Vidal (University of Chicago Press, 2004), 361–83.

24. Frances A. Yates, *Giordano Bruno and the Hermetic Tradition* (Routledge and Kegan Paul, 1964); B. P. Copenhaver, "Jewish Theologies of Space in the Scientific Revolution," *Annals of Science* 37, no. 5 (1980): 489–548; Rothman, *Xeipomantia*, title page.

25. Richard Saunders, *Physiognomie and Chiromancie* (London, 1671), Preface to Reader, n.p.

26. Saunders, *Physiognomie and Chiromancie* (1671), dedication to Ashmole and letter from William Lilly, n.p.

27. For the Society of Astrologers, see Michelle Pfeffer, "The Society of Astrologers (c. 1647–1684)," *British Journal for the History of Science* 54, no. 2 (2021): 133–53.

28. Sue Wiseman, Katharine Hodgkin, and Michelle O'Callaghan, eds., *Reading the Early Modern Dream* (Routledge, 2007); Saunders, *Physiognomie and Chiromancie* (1671), 1–2.

29. Saunders, *Physiognomie and Chiromancie* (1671), no page, "Catalogue of Authors," in Preface to Reader.

30. For *Kabbala Denudata* by Christian Knorr von Rosenroth, see Paul Kléber Monod, *Solomon's Secret Arts* (Yale University Press, 2013), 107–15.

31. Saunders, *Physiognomie and Chiromancie*, 43–50.

32. Saunders, *Physiognomie and Chiromancie*, 49–50.

33. Porter, *Windows of the Soul*, 25; Ian Maclean, *Logic, Signs and Nature in the Renaissance* (Cambridge University Press, 2002); Johannes Alsted, quoted in Porter, *Windows of the Soul*, 27, 11; Saunders, *Physiognomie and Chiromancie* (1671), 1; Porter, *Windows of the Soul*, 11; Leitch, "Getting to How-To," 635–59.

34. Paracelsus, *Von den natürlichen Dingen*, book I, trans. Ahonen, "Medieval and Early Modern Physiognomy," 636; Paracelsus, *Werke*, book I, 12, 174–77. See Michael Doan, "Paracelsus on Erfahrung and the Wisdom of Praxis," *Analecta Hermeneutica* 1 (2009): 168–85; Charles Webster, *Paracelsus* (Yale University Press, 2008).

35. Thomas Browne, *Religio Medici* (1682) (Sampson Low, Son & Marston, 1869), Second Part, Section One; Amadeo Murase, "The Homunculus and the Paracelsian *Liber de imaginibus*," *Ambix* 67, no. 1 (2020): 47–61.

36. Eadwine Psalter, Trinity College, Cambridge, MS R.17.1, f. 282. Burnett, "The Earliest Chiromancy in the West," 191, 194; Monod, *Solomon's Secret Arts*, 107–8.

37. [Anon.], *Palmistry, Chirognomancy, and other Prognostics*, 2–5.

38. MS. Digby Rolls 4, Bodleian Library, Oxford; Price, *An Old Palmistry*, 7–9; *The Conjurer's Guide!* (Glasgow, 1800).

39. John Bulwer, *Chirologia* (London, 1644); Saunders, *Palmistry* (4th ed., 1676), 38. See Justin E. H. Smith, "'A Corporall Philosophy,'" in *The Body as Object and Instrument of Knowledge*, ed. Charles T. Wolfe and Ofer Gal (Springer, 2010), 169–84.

40. John Bulwer, *Chirologia: Or The Naturall Language of the Hand* (London, 1644), 1, 2.

41. [Anon.], *An Old Egyptian Fortune-Teller's Last Legacy* (London, 1775).

42. [Anon.], *The High German Fortune-Teller* (London: Printed in Bow Church Yard, 1785); *The Conjurer's Guide!*, 12–13, 14–15, 21.

43. [Anon.], *The Conjurer's Guide!*, 20; [Anon.], *The High German Fortune-Teller*, 11.

44. [Anon.], *Wit's Cabinet* (London, c. 1700); Jacob ben Mordecai of Fulda, *Sefer Shoshanat Yaacov: . . . hokhmat ha-yad be-shoreshoten ve-hokhmat he-partsuf* (Amsterdam, 1706). Bookplate, Cambridge University Library 7816.d.89; Isaac Ḥayyim Cantarini, Ḥaye beśarim, 1775. A Tunisian manuscript of kabbalistic predictions, with a short treatise on chiromancy. BL Or 10357; [Anon.], *The Conjurer's Guide!*

45. Michael Saler, "Modernity and Enchantment," *American Historical Review* 111, no. 3 (2006): 702; Joshua Landy and Michael Saler, eds., *The Re-Enchantment of the World* (Stanford University Press, 2009).

CHAPTER TWO

1. George Hall, *The Gypsy's Parson: His Experiences and Adventures* (Sampson, Law, Martson, 1900), 298; Charles Godfrey Leland, *The English Gypsies and their Language* (Trübner & Co, 1874), ch. VI; John Crawfurd, "On the Origin of the Gypsies," *Transactions of the Ethnological Society of London* 3 (1865): 29; E. H. Palmer, ed., *Oriental Mysticism: A Treatise on the Sufistic and Unitarian Theosophy of the Persians, compiled from native sources* (Deighton, Bell, 1867), 74.

2. Thomas Browne, *Religio Medici* (1682) (Sampson Low, Son & Marston, 1869), Second Part, Section One.

3. An act concerning outlandish people, calling themselves Egypcions 1530 (22 Hen. 8 c. 10); Egyptians Act 1554 (1 & 2 Ph. & M. c. 4); Vagabonds Act 1530 (22 Hen. 8. c. 12). See H. T. Crofton, "Early Annals of the Gypsies in England," *Journal of the Gypsy Lore Society* 1, no. 1 (1888): 9; David Mayall, *Gypsy Identities 1500–2000* (Routledge, 2004).

4. An act concerning outlandish people, calling themselves Egypcions, 1530 (22 Hen. 8 c. 10); Egyptians Act 1554 (1 & 2 Ph. & M. c. 4); An Acte for punishment of Rogues Vagabondes and Sturdy Beggars, 1597 (39 Eliz. 1 c. 4); Vagrancy Act 1714 (12 Anne c. 23); An Act to amend and make more effectual the laws relating to rogues, vagabonds, and other idle and disorderly persons, and to houses of correction, 1744 (17 Geo. II, c. 5). See Danielle Boaz, "Fraud, Vagrancy and the 'Pretended' Exercise of Supernatural Powers in England, South Africa and Jamaica," *Law & History* 5, no. 1 (2018): 55–58; [Anon.], "Account of the Gipsies," *Chambers Miscellany of Useful and Entertaining Tracts*, vol. 16 (William and Robert Chambers, 1847), 7; John Morgan, "'Counterfeit Egyptians,'" *Romani Studies* 26, no. 2 (2016): 105–28; Nicholas Rogers, "Policing the Poor in Eighteenth-Century London," *Histoire Sociale/Social History* 24, no. 47 (May 1991): 127–47; Jennifer Mori, "Magic and Fate in Eighteenth-Century London," *Folklore* 129, no. 3 (2018): 254–77.

5. Vagrancy Act 1824 (5 Geo. 4 c. 83).

6. Heinrich Moritz Gottlieb Grellmann, *Dissertation on the Gipsies*, trans.

Matthew Raper (G. Bigg, 1787), ix, xviii; [Anon.], *Account of the Gipsies*, 5. See Klaus-Michel Bogdal, *Europe and the Roma: A History of Fascination and Fear* (Penguin, 2023).

7. John Hoyland, *A Historical Survey of the Customs, Habits, and Present State of the Gypsies* (York: John Hoyland, 1816), 80; [Anon.], *Account of the Gipsies*, 2.

8. Captain Thomas Newbold, "Gypsies in Egypt," *Journal of the Royal Asiatic Society of Great Britain and Ireland*, 16, part 2 (1856): 285; George Borrow, *The Zincali, Or An Account of the Gypsies of Spain* (John Murray, 1841), vol. 1, 160–62; Grellmann, *Dissertation on the Gipsies*; Hoyland, *A Historical Survey*; [Anon.], *Account of the Gipsies*, 3; Hoyland, *A Historical Survey*, 128.

9. [Anon.], *Account of the Gypsies*; Hoyland, *A Historical Survey*, 126, 142–43. This was a summary of the translation of Grellmann's *Die Zigeuner* (1783), Grellmann, *Dissertation on the Gipsies*, 167.

10. Grellmann, *Dissertation on the Gypsies*, 167; Charles Godfrey Leland, *The English Gypsies and Their Language* (Trübner & Co, 1873), xiii.

11. [Anon.], *Account of the Gipsies*, 29–30; George Borrow, *Romano Lavo-Lil. Word-Book of the Romany Or, English Gypsy Language* (John Murray, 1874).

12. Crofton, "Early Annals of the Gypsies in England," 2.

13. Charles Leland, *The Gypsies* (Houghton Mifflin, 1882), 182, 194.

14. Leland, *The Gypsies*, 337–38.

15. Leland, *The Gypsies*, 337–38.

16. Palmer, *Oriental Mysticism*, x; Richard F. Burton, "The Late E. H. Palmer: Personal Reminiscences," *Academy*, no. 574 (May 5, 1883): 311.

17. Richard F. Burton, *A Plain and Literal Translation of the Arabian Nights' Entertainments, Now Entituled* [*sic*] *The Book of The Thousand Nights and a Night* (Benares: Kamashastra Society, 1885); Richard F. Burton, *The Jew, The Gypsy and El Islam* (Hutchinson & Co., 1898), 256.

18. Grellmann, *Dissertation on the Gipsies*, 109.

19. "persons . . . wandering and p'tending themselves to be Egipcyans or wandering in the Habite Forme or Attyre of counterfayte Egipcians." An Acte for punishment of Rogues Vagabondes and Sturdy Beggars, 1598 (39 Eliz. 1 c. 4). See also Mori, "Magic and Fate in Eighteenth-Century London," 257; Owen Davies, *Popular Magic* (Hambeldon, 2007), 27.

20. Borrow, *The Zincali*, 125–30; Torreblanca, translated and quoted in Borrow, *The Zincali*, 127–28.

21. Borrow, *The Zincali*, 128, 129, 132.

22. Charles Godfrey Leland, *Gypsy Sorcery and Fortune Telling* (T. Fisher Unwin, 1891), xi, 1, 177, 189; Leland, *The Gypsies*, 62–65. See Ken Lee, "Orientalism and Gypsylorism," *Social Analysis* 44, no. 2 (2000): 129–56; Deborah Nord, *Gypsies and the British Imagination: 1807–1930* (Columbia University Press, 2006).

23. Leland, *Gypsy Sorcery*, 176.

24. Leland, *The Gypsies*, 62–65.

25. Charlotte Brontë, *Jane Eyre* (Smith, Elder, 1847).

26. George Eliot, *The Spanish Gypsy: A Poem* (William Blackwood & Sons, 1868); Leland, *The Gypsies*, 181. See Sarah Houghton-Walker, *Representations of the Gypsy in the Romantic Period* (Oxford University Press, 2014).

27. Joseph Glanvill, *The Vanity of Dogmatizing* (London, 1661), 197; Joseph Glanvill, *Scepsis Scientifica, or Confest ignorance, the way to science* (London, 1665), 147, 151.

28. Éliphas Lévi, *Dogme et rituel de la haute magie* (Baillière, 1861); Éliphas Lévi, *La science des esprits: révélation du dogme secret des kabbalistes, esprit occulte des Évangiles, appréciation des doctrines et des phénomènes spirites* (Baillière, 1865); Éliphas Lévi, *Transcendental Magic, its Doctrine and Ritual* trans. A. E. Waite (Redway, 1896), 3. See Christopher McIntosh, *Eliphas Lévi and the French Occult Revival* (State University of New York Press, 2011); Christopher Partridge, *The Re-Enchantment of the West Vol. 1* (T&T Clark International, 2004), 62–84; Nicholas Goodrick-Clarke, *The Western Esoteric Traditions* (Oxford University Press, 2008).

29. Lévi, *Transcendental Magic*, 22.

30. Lévi, *Transcendental Magic*, 351, 140–41; Adolphe Desbarrolles, *Les mystères de la main révélés et expliqués* (Dentu, 1859). This was partially abstracted into English in 1879 by an American, who borrowed the title of the book and made it his own: Robert Allen Campbell, *Philosophic Chiromancy: Mysteries of the Hand Revealed and Explained: the Art of Determining, From an Inspection of the Hands, the Person's Temperament, Appetites, Passions, Impulses, Aspirations, Mental Endowments, Character and Tendencies* (J. W. Campbell & Co., 1879).

31. Desbarrolles, *Les mystères de la main révélés et expliqué*, trans. Campbell, *Philosophic Chiromancy*, 20. Desbarrolles, translated by and quoted in Edward Heron-Allen, *A Manual of Cheirosophy* (Ward, Lock, 1885), 67, 65.

32. Rosa Baughan, *The Handbook of Palmistry* (Redway, 1883), 5–8.

33. Adda-Nari, "l'Isis des Indiens," in Adolphe Desbarrolles, *Les mystères de la main révélés et expliqué* (Dentu, 1859), 19. See Baughan, *The Handbook of Palmistry*, 7–8.

34. Helena Blavatsky, *The Secret Doctrine: The Synthesis of Science, Religion, and Philosophy*, 2 vols. (Theosophical Publishing Company, 1888).

35. William Q. Judge, *The Ocean of Theosophy* (Theosophical Publishing Society, 1893), ch. 5, "Body and Astral Body."

36. C. W. Leadbeater, "How clairvoyance is developed," *The Theosophist* 25, no. 5 (January 1904): 201–14; C. W. Leadbeater, *Clairvoyance* (Theosophical Society, 1899), 5, 10, 21–22, 25, 119; Judge, *Ocean of Theosophy*, ch. 5; Richard Noakes, *Physics and Psychics* (Cambridge University Press, 2019).

37. Judge, *Ocean of Theosophy*, ch. 16, "Psychic Forces, Laws and Phenomena."

38. Urs App, "Tibetan Theosophy," in *Sino-Tibetan Buddhism across the Ages*, ed. Ester Bianchi and Weirong Shen (Brill, 2021), 141–69.

39. J. B. Dale, *Indian Palmistry* (Theosophical Publishing Society, 1895), quoting Charles Coleman, *The Mythology of the Hindus* (Parbury Allen, 1832), 202.

40. Judge, "Chirognomy and Palmistry," 159–60; Dale, *Indian Palmistry*, 28.

41. Barad Kau Majumdar, "An Astrological Calculation Verified," *The Theosophist* 3, no. 4 (January 1882): 107.

42. Katharine St. Hill, *Medical Palmistry, Or the Hand in Health and Disease* (Rider, 1927), 91.

43. Qīmat Rā'i Vāsvā'ī, *Hamrāz majlis 'urfu qismata jī Rekha* (Karācī, Sindhu: Vishan Press, 1867), BL, SAC.1986.a.35114; Krishmanatha Raghuriathayi, *Palmistry, Chirognomancy, and Other Prognostics (from Marathi)* (Bombay, 1888); V. S. Kumaraswami Mudaliar & Son, *A Complete Treatise on the Science of Chirography and Chiromancy* (Victoria Press, 1918), 11, 72.

CHAPTER THREE

1. Charles Darwin, *The Descent of Man, and Selection in Relation to Sex* (1871) (John Murray, 1913), 76–77.

2. Joannes Evangelista Purkyně, *Commentatio de examine physiologico organi visus et systematis cutanei* (Breslau: Typis Universitatis, 1823). See Harold Cummins and Rebecca Wright Kennedy, "Purkinje's Observations (1823) on Finger Prints and Other Skin Features," *Journal of Criminal Law and Criminology* 31, no. 3 (1940): 343–56. This includes a translation of Purkinje's thesis from Latin to English, used here.

3. Purkyně, *Commentatio*, 348.

4. Charles Bell, *The Hand, Its Mechanism and Vital Endowments as Evincing Design*, 2nd ed. (William Pickering, 1833), v.

5. Bell, *The Hand*, 107.

6. Bell, *The Hand*, 107–8; George M. Humphry, *The Human Foot and the Human Hand* (Macmillan, 1861), 142.

7. Richard Owen, *On the Nature of Limbs* (John Van Voorst, 1849), 9; for example, Thomas Henry Huxley, *Evidence as to Man's Place in Nature* (Williams & Norgate, 1863), 85–94.

8. Charles Darwin, "Bell on the Hand [C. Bell 1833]," DAR 119: 5a, Darwin's reading notebooks, Darwin Correspondence Project; Charles Darwin, *On the Origin of Species by Means of Natural Selection* (John Murray, 1859), 434.

9. Humphry, *The Human Foot*, 3, 12–13; Darwin, *The Descent of Man*, 76–77.

10. Humphry, *The Human Foot*, 119, 126–27, 133, 146–47, 155; Edward Heron-Allen, *A Manual of Cheirosophy* (Ward, Lock, 1885), 39–40.

11. Charles Bell, *The Hand; Its Mechanism and Vital Endowments, as Evincing*

Design, 5th rev. ed. (John Murray, 1852), 303. Darwin drew from "Intermarriage" by Alex Walker, 1838. Darwin, *The Descent of Man*, 47–49.

12. R. B. Todd, "Abnormal Conditions of the Hand," in *Cyclopaedia of Anatomy and Physiology* (Longman, Brown, Green, Longmans & Robert, 1836–39), vol. II, 518; [Jones Quain], *Quain's Elements of Anatomy*, ed. William Sharpey, Allen Thompson, and John Cleland, 7th ed., 2 vols. (James Walton, 1867), vol. I, cciv; McDougall, "The Regions of the Hand," in *Cyclopaedia of Anatomy and Physiology*, 524.

13. McDougall, "The Regions of the Hand," 526; Heron-Allen, *A Manual of Cheirosophy*, 44–45; *Quain's Anatomy*, vol. 1, General Anatomy, cciv.

14. Charles Bell, "On the nervous circle which connects the voluntary muscles with the brain," *Philosophical Transactions of the Royal Society of London* 116 (December 31, 1826): 170; Bell, *The Hand*, 214.

15. Bell, "On the nervous circle"; Humphry, *The Human Foot*, 146–47; Bell, *The Hand*, 195.

16. Bell, *The Hand*, 180; Herbert Spencer, *The Principles of Psychology* (Longman, Brown, Green, and Longmans, 1855), 457. See also Mark Paterson, *Seeing with the Hands* (Edinburgh University Press, 2016).

17. Humphry, *The Human Foot*, 159; Charles Darwin, *The Expression of the Emotions in Man and Animals* (John Murray, 1872).

18. Humphry, *The Human Foot*, 188; Bell, *The Hand*, 180; Heron-Allen, *A Manual of Cheirosophy*, 46.

19. Darwin, *The Descent of Man*, 77; Humphry, *The Human Foot*, 156.

20. Humphry, *The Human Foot*, 206–7; V. S. Kumaraswami Mudaliar & Son, *A Complete Treatise on the Science of Chirography and Chiromancy* (Victoria Press, 1918), 10.

21. J. C. Lavater, *Physiognomy* (William Tegg, 1869), title page; J. C. Lavater and G. della Porta, *The Pocket Lavater, or The Science of Physiognomy* (Van Winkle & Wiley, 1817), 22. See Ludmilla Jordanova, "The Art and Science of Seeing in Medicine," in *Medicine and the Five Senses*, ed. W. F. Bynum and Roy Porter (Cambridge University Press, 1993), 122–33; Melissa Percival, *The Appearance of Character* (W. S. Maney & Son, 1999); Richard T. Gray, *About Face* (Wayne State University Press, 2004).

22. Lavater, "L'art de connaître les hommes par la physionomie," III, no. 1 (Paris: [L. Prudhomme], 1806–1809), in C. S. D'Arpentigny, *La Science de la Main*, trans. Edward Heron-Allen (Ward, Lock, 1885), 185; Charles Bell, quoted in R. B. Todd and William Bowman, "Locomotion," *The Physiological Anatomy and Physiology of Man* (John W. Parker, 1845), vol. I, 106; and in Edward Heron-Allen, *Manual of Cheirosophy*, 49. Ludmilla Jordanova, *Defining Features* (Reaktion, 2000).

23. Bell, *The Hand*, 217–18; Humphry, *The Human Foot*, 159; Heron-Allen, *A Manual of Cheirosophy*, 41; Darwin, *The Descent of Man*, 47–135.

24. Chart of mental geometry and synopsis and classification of the faculties. By Frederick Bridges, of the Musæum and School of Mental Geometry and Physiology, 16, Mount Pleasant, Liverpool, c. 1850, Wellcome Collection, 28445i.

25. [Anon.], *The Hand Phrenologically Considered: Being a Glimpse at the Relation of the Mind with the Organisation of the Body* (Chapman and Hall, 1848), 48, 53.

26. Richard Beamish Papers, MS/0332, Natural History Museum, London; Beamish cited Gall and Spurzheim's presentations to the French Institute, "On the Anatomy and Physiology of the Nervous System in General and the Brain in Particular," in 1808 and 1810; Beamish, *The Psychonomy of the Hand; or, The Hand An Index of Mental Development* (Frederick Pitman, 1865), 1; [Anon.], "Obituary. Richard Beamish, 1798–1873," *Minutes of the Proceedings of the Institution of Civil Engineers* 40, part 2 (1875): 246–51.

27. Richard Beamish to William Whewell, December 16, 1863, Trinity College Library, Cambridge, William Whewell Letters Received, Add. MS a/201/5; Beamish, *Psychonomy*, plate 23.

28. Beamish, *Psychonomy*, v, plates 1–28.

29. For his vanity, see Beamish, *Psychonomy*, vi; [Anon.], *The Hand Phrenologically Considered*, 3ff. English translations include *The Medical Times*; Craig, *The Book of the Hand*; Heron-Allen, *The Science of the Hand*; Beamish, *Psychonomy*, title page.

30. [Anon.], *Hand Phrenologically Considered*, 3–9.

31. Beamish, *Psychonomy*, plate 5, "Artistic Impulsive Type"; plates 13, 16, 21, 23, 64.

32. William Whewell, *The Philosophy of the Inductive Sciences, Founded Upon Their History*, 2 vols. (John Parker, 1847), vol. 1, 479ff.

33. Beamish, *Psychonomy*, 59; [Anon.], *The Hand Phrenologically Considered*, 70–71; D'Arpentigny, *Science of the Hand*, 199–207; Heron-Allen, *A Manual of Cheirosophy*, 139–40.

34. [Anon.], *Hand Phrenologically Considered*, 69–78.

35. Beamish, *Psychonomy*, 80–81.

36. [Anon.], *Hand Phrenologically Considered*, 73–74; Edward Heron-Allen, *Practical Cheirosophy: A Synoptical Study of the Science of the Hand* (G. P. Putnam's Sons, 1893), 67.

37. Beamish, *Psychonomy*, 60; Craig, *The Book of the Hand*, 107; D'Arpentigny, *The Science of the Hand*, 262; [Anon.], *Hand Phrenologically Considered*, 73–74.

38. Robert Knox, *The Races of Men* (H. Renshaw, 1850), v, 11, 49, 53. Knox, *Races of Men*, 11, 284, 319; D'Arpentigny, quoted in Knox, *Races of Men*, 196.

39. [Helix], Review of D'Arpentigny, *The Philosophy of the Hand*; and Robert Knox, *Races of Men*, published in *Westminster Review* 51–52, no. 1 (October 1849–January 1850): 1–21.

CHAPTER FOUR

1. Edward Heron-Allen, *A Manual of Cheirosophy: Being A Complete Practical Handbook of the Twin Sciences of Cheirognomy and Cheiromancy* (Ward, Lock, 1885), 90.

2. Catalogue of Printed Books and a Few Manuscripts, the property of the late E. Heron-Allen, Edward Heron-Allen Papers, West Sussex Record Office [WSRO], EHA MS 1/7/1.

3. "Studying People's Hands: Some of the Wonders of Palmistry," *New York Daily Tribune*, November 13, 1886. See also John P. Mahony and Barbara P. Mahony, "The Sir Vincent Horsley and Edward Heron-Allen Discussions," *Heron-Allen Society Newsletter*, no. 27 (Autumn 2015): 3–6; Edward Heron-Allen, Journals, vol. 1, 1881–1889, December 1, 1885. WSRO EHA 1/2/1/1.

4. Henry Frith and Edward Heron-Allen, *Chiromancy, or the science of palmistry* (Privately printed, 1882); Edward Heron-Allen, *Codex Chiromantiae* (Odd Volumes Opusculum: No. VII, 1883); Timothy J. McCann and John P. Mahoney, *A Brief Biography of Author and Polymath, Edward Heron-Allen, 1861–1943* (Privately printed for the Tallahassee Sette of Odd Volumes, no. III, 2023).

5. Review of Edward Heron Allen, *Codex Chiromantiae*, *Bibliographer* 5, no. 4 (March 1884): 114; "Form of Oath by which Vettius Valens, the Astronomer bound his disciples to secrecy." Foreword, Brother Edward Heron-Allen, Necromancer unto ye Sette of Odd Volumes (Privately printed, London, 1883); Membership of Ye Sette of Odd Volumes, July 13, 1883, in Heron-Allen, *Codex Chiromantiae*, end matter. A Book of Acquaintances, Edward Heron-Allen Papers, WSRO, EHA 1/5/1, manuscript entry for Powell.

6. Edward Heron-Allen, Journals, December 11, 1885; December 3, 1885; December 17 and 20, 1885; December 4, 1885; January 1, 1886, WSRO EHA 1/2/1/1.

7. Edward Heron-Allen, Journals, January 1, 1887, WSRO EHA 1/2/1/1; "The Fall of Cheirosophy," *Boston Globe*, February 16, 1888, 2. John P. Mahoney and Barbara Mahoney, "'Ship Stink, Immigrants and Hyperosmia,'" Heron-Allen Society Symposium, July 2, 2016. Typescript in author's possession, courtesy of John Mahoney.

8. For earlier learned magi, see Anthony Grafton, *Magus* (Allen Lane, 2023).

9. "Main gauche de Victor Hugo," in Adolphe Desbarrolles, *Les mystères de la main révélés et expliqué* (Chez L'Auteur, Boulevard Saint-Michel: Paris, n.d.). Plate opposite 922; Edward Heron-Allen, Chyromantia: manuscript, 1884–1893, Harvard University, Houghton Library, MS Eng 1624.

10. Edward Heron-Allen, "Inaugural Address of his Oddship Edward Heron-Allen," October 25, 1927, privately printed, 14–15; Joan Navarre, "Edward Heron-Allen," *Encyclopedia Iranica*, online, 2006; Oscar Wilde to Edward Heron-Allen, c. June 12, 1885, BL Eccles Bequest, ADD 81690; Heron-Allen influenced Wilde, including his *Lord Arthur Savile's Crime: A Story of*

Cheiromancy (1887). See Joan Navarre, "Oscar Wilde, Edward Heron-Allen, and the Palmistry Craze of the 1880s," *English Literature in Transition, 1880–1920* 54, no. 2 (2011): 174–84.

11. Edward Heron-Allen, *The Science of the Hand* (Ward, Lock, 1889); Heron-Allen published *Practical Cheirosophy: A Synoptical Study of the Hand* in 1887 (G. P. Putnam's Sons, 1887), an American précis published in 1899 as *The Language of the Hand*. Henry Frith and Edward Heron-Allen, *The Language of the Hand* (David McKay, 1899); Edward Heron-Allen, "Preface (1886)," in *The Science of the Hand*, 9. See also Robert Allen Campbell, *Philosophic Chiromancy: Mysteries of the Hand Revealed and Explained* (J. W. Campbell & Co., 1879), 22; Heron-Allen, *A Manual of Cheirosophy*, 67–68.

12. Heron-Allen, *A Manual of Cheirosophy*, 8, 18, 286–87.

13. List of London fortune tellers and remarks after caution from police, September 1912; Statement of "Professor Melini" (Thomas Morgan) and Response from Mr. E. W. Wallace, Secretary, Spiritualist Alliance and editor of "Light." TNA, Mepo 2/1323 (Fortune Telling and Palmistry: Prosecutions under the Vagrancy Act, 1824).

14. John P. Mahoney and Barbara P. Mahoney, "The Sir Vincent Horsley and Edward Heron-Allen Discussions," *Heron-Allen Society Newsletter* 27 (Autumn 2015): 8.

15. Quain, *Elements of Anatomy*, 8th ed., vol. ii, 214, cited in Heron-Allen, *Science of the Hand*, 70; Heron-Allen, *A Manual of Cheirosophy*, 45–48, 54.

16. Heron-Allen, *A Manual of Cheirosophy*, 71.

17. Heron-Allen, *A Manual of Cheirosophy*, 63, 65–69, 103.

18. Edward Heron-Allen, "Naming the Mounts," *Palmist's Review* 1, no. 1 (April 1899): 15.

19. Richard Noakes, *Physics and Psychics* (Cambridge University Press, 2019); Alison Winter, *Mesmerized* (University of Chicago Press, 2000).

20. Heron-Allen, *A Manual of Cheirosophy*, 64–65.

21. Rosa Baughan, *The Handbook of Palmistry* (Redway, 1883), 6; Johann Gottfried von Herder, "Idées sur la Philosophie de L'Histoire de l'Humanité," 1827, trans. in Heron-Allen, *A Manual of Cheirosophy*, 52; J. Abercombie, *Inquiries concerning the Intellectual Powers and the Investigation of Truth*, 8th ed. (John Murray, 1838); Johannes Müller, cited in Heron-Allen, *A Manual of Cheirosophy*, 51–52; Julius Bernstein, *The Five Senses of Man* (D. Appleton, 1883), 19.

22. Noakes, *Physics and Psychics*; Alfred Russel Wallace, *The Scientific Aspect of the Supernatural* (F. Farrah, 1866), 3–4.

23. "New Members and Associates," *Journal of the Society for Psychical Research* 1, no. 13 (February 1885): 258; Gustave Geley, "Materialized Hands," *Scientific American* 129, no. 5 (November 1923): 316–17, 373–74, discussing the molds made by Franek Kluski; "General Meeting," *Journal of Psychical Research* 13 (February 1885): 264.

24. "Lectures," *Journal of the Society for Psychical Research* 1, no. 1 (February

1884): 9; "Report of the General Meeting," *Journal of the Society for Psychical Research* 1, no. 18 (July 1885): 452–53; [W. F. Barrett], "On the Existence of a 'Magnetic Sense,'" *Journal of the Society for Psychical Research* 3 (April 1884): 41–44; Charles Daubeny, Presidential Address, *Report of the Twenty-Sixth Meeting of the British Association for the Advancement of Science* (John Murray, 1857), lii.

25. G. Hagopian to EHA, October 21, 1885, WSRO EHA 1/1/1/2, f 44; Edward Heron-Allen, *The Ruba'iyat of Omar Khayyam* (H. S. Nichols, 1898); Denis Wright, *The Persians Amongst the English* (London, 1985), 152–66; EHA to Shah of Persia, July 28, 1897, response "From Téhéran," August 23, 1897, WSRO, EHA 1/1/1/4 Correspondence.

26. Edward Heron-Allen, *The Science of the Hand* (Ward, Lock, 1889), 36.

27. Heron-Allen, *The Science of the Hand*, 31–41; Heron-Allen, *A Manual of Cheirosophy*, 27, 31.

28. Heron-Allen, "Naming the Mounts," 15.

29. Galen translated and cited by John Kidd, *On the adaptation of external nature to the physical condition of man* (1833), cited in Heron-Allen, *A Manual of Cheirosophy*, 20–21.

30. Heron-Allen, *A Manual of Cheirosophy*, 60.

31. Edward Heron-Allen, *The "Nefer" Sign* (Selsey Bill, Sussex: Printed for the Author, 1941); Joann Fletcher, "Edward Heron-Allen, His Friends and Their Mummies," *Opusculum VIII* (2005): 17–44.

32. James Poskett, *Materials of the Mind* (University of Chicago Press, 2019), 33–36, 41; *Catalogue of the Egyptian Antiquities in the Museum of Hartwell House* (W. M. Watts, 1858), 70; L. S. Penrose, "Fingerprints and Palmistry," *The Lancet* 301, no. 7814 (June 2, 1973): 1240–41.

33. Travel diary, Egypt 1903, vol. X, March 28, 1903, WRSO, EHA 1/2/1/10; Fletcher, "Edward Heron-Allen, His Friends and Their Mummies," 17–44. Personal communication with Joann Fletcher, Department of Archaeology, University of York, April 11, 2024.

34. Edward Heron-Allen, "Naming the Mounts," *Palmist's Review* 1, no. 1 (April 1899): 15.

35. Heron-Allen, *Codex Chiromantiae*; Heron-Allen, *A Manual of Cheirosophy*.

36. Arthur Conan Doyle, "A Study in Scarlet," *Beeton's Christmas Annual* (Ward, Lock, 1887).

37. Alfred Russel Wallace, *My Life: A Record of Events and Opinions*, 2 vols. (Chapman and Hall, 1905), vol. 1, 236; Edward Heron-Allen, *The Purple Sapphire and Other Posthumous Papers* (Philip Allan & Co., 1921).

CHAPTER FIVE

1. "Milestones," *Time*, October 19, 1936, 88; Cheiro, *Cheiro's Memoirs: The Reminiscences of a Society Palmist* (Rider, 1912), xiii, 29.

2. *Cheiro's Memoirs*, xiii; Keiro, *Keiro's Palmistry, Clairvoyance & Psychometry*

(George Routledge, 1905), 3, 19–20; Cheiro, *Read Your Past, Present & Future* (London Publishing Co., 1927), 27, 34.

3. "Fortune Teller's Ashes Buried after 9 Years," *The Mail* (August 4, 1945), 2; *Cheiro's Memoirs*, 3–5.

4. *Cheiro's Memoirs*, xv, 54, 198; Cheiro, *Cheiro's Language of the Hand* (Herbert Jenkins, 1900), v.

5. *Cheiro's Memoirs*, 14–15; Cheiro, *Read Your Past*, 19–20, 189; Cheiro, *Mysteries and Romances of the World's Greatest Occultists* (Herbert Jenkins, 1935), 207–9; Swami Vivekananda, plate XLII, Annie Besant, plate XXXVI, in *Cheiro's Language of the Hand*; Prajni Surabhi, "Swami Vivekananda's Palm Reading," in "All about Indian astrology, Vedanta, Yoga, Tantra, Martial Arts, Occult and Alternative Medicine," January 12, 2012, http://prajnasurabhi.blogspot.com/2012/01/swami-vivekanandas-palm-reading.html; Cheiro, *Bacharera pratiṭi dina a āpanāra bhāgya / Kiro; anubādaka Parīkshiṯ* (Kakātā: Ji. Nātha, 1981).

6. *Cheiro's Memoirs*, 167, 196ff, 218; Max Müller, *Lectures on the Science of Language* (Charles Scribner, 1862); *Cheiro's Language of the Hand*, 142–43; Cheiro, *Mysteries and Romances*, 173–78, 179.

7. *Cheiro's Language of the Hand*, 97; *Cheiro's Memoirs*, 158–59; Cheiro, *Mysteries and Romances*, 228–36.

8. Press Notice of Lectures, *New York Herald*, February 25, 1900; see entries for *Boston Herald* dated April 19 and June 16, in appendix to *Cheiro's Language of the Hand*. See British Film Archives, *The Language of the Hand*, a film on palmistry, British Film Institute, Reuben Library, 215035.

9. *Cheiro's Memoirs*, 62; Joseph O. Baylen, "Mark Twain, W. T. Stead and 'The Tell-Tale Hands,'" *American Quarterly* 16, no. 4 (Winter 1964): 606–12.

10. *Cheiro's Memoirs*, 217–18.

11. From *Cheiro's Language of the Hand*, "Appendix: Opinions of the Press and Public," n.p.; *Cheiro's Language of the Hand*, 156.

12. *Cheiro's Memoirs*, 59–61.

13. Charles Leland, *Gypsy Sorcery and Fortune Telling* (T. Fisher Unwin, 1891), 185; Cheiro, *You and Your Hand*, rev. Louise Owen (Jarrolds, 1969), xvii; *Cheiro's Memoirs*, 132.

14. Leland, *Gypsy Sorcery*, 180; *Cheiro's Memoirs*, 153, 167–68; E. S. D'Odiardi, *Medical Electricity: What Is It? And How Does It Cure?* (Swann Sonnenschein, 1893).

15. Baron Karl von Reichenbach, *Odic-Magnetic Letters*, trans. John S. Hittell (Calvin Blanchard, 1860), 35–36; *Cheiro's Language of the Hand*, v, vi, 162.

16. Cheiro, *Read Your Past*, 33.

17. *Cheiro's Language of the Hand*, 156.

18. *Cheiro's Memoirs*, 57, 203–4.

19. Cheiro, *Read Your Past*, 34; Cheiro, *If We Only Knew and Other Poems by Cheiro* (Chatto & Windus, 1895).

20. Report on Countess Cardelli, New Scotland Yard, October 25, 1892, National Archives, London (TNA), HO 144/541/A51669 (Vagrancy: Method of Dealing with Cases of Palmistry, Fortune Telling, etc.).

21. "War on Fortune Tellers," *The Standard*, September 17, 1912. News clipping in TNA, Mepo 2/1323 (Fortune Telling and Palmistry: Prosecutions under the Vagrancy Act, 1824); Cathcart Wason, clippings from Hansard, May 15 and November 22, 1911: TNA, Mepo 2/694 (Palmistry: Direction as to when Proceedings Should Be Taken); Alpheus Morton, clipping from Hansard, June 16, 1893: TNA, HO 144/541/A51669 (Vagrancy).

22. General Russell, Hansard clipping, July 13, 1900; and *Truth*, news clipping, June 12, 1890, TNA, HO 144/541/A51669 (Vagrancy); Cathcart Wason, Palmistry Advertisements (Mayfair), news clipping from Parliamentary Debates, House of Commons, June 26, 1912, TNA, Mepo 2/694 (Palmistry).

23. An Act for the Punishment of idle and disorderly Persons, and Rogues and Vagabonds, 5 Geo. 4 c. 83, Section IV; Fortune Telling, news clipping, TNA, Mepo 2/694 (Palmistry).

24. "War on Fortune Tellers," and List of London fortune tellers and remarks after caution from police, September 1912, TNA, Mepo 2/1323 (Fortune Telling and Palmistry).

25. List of London fortune tellers, TNA; Police Court report, Bournemouth, September 19, 1900, TNA, HO 144/541/A51669 (Vagrancy: Method of Dealing with Cases of Palmistry, Fortune Telling, etc.); Alexander Davies to the Right Hon. Sir White Ridley, October 10, 1900, TNA, HO 144/541/A51669 (Vagrancy); "Definition of the 'Occult Art,'" *Bournemouth Observer & Chronicle*, September 22, 1900, TNA, HO 144/541/A51669 (Vagrancy).

26. Crime—Palmistry, 1892, Metropolitan Police, TNA, Mepo 2/694 (Palmistry).

27. Report, Metropolitan Police, Great Marlboro Street Station, October 21, 1910, TNA, Mepo 2/694 (Palmistry); "War on Fortune Tellers," TNA; List of London fortune tellers, TNA.

28. Mr. Masterman, cited in "Fortune Telling," news clipping, c. 1911, Mepo 2/1323 (Fortune Telling and Palmistry); Extract, *Votes & Proceedings of the House of Commons*, Mr. Martin, Liberal, June 10, 1912, TNA, Mepo 2/694 (Palmistry); "War on Fortune Tellers," TNA.

29. "War on Fortune Tellers," TNA; Police Inspector and Superintendent Report, December 11, 1912, TNA, Mepo 2/1323 (Fortune Telling and Palmistry); Report in Mepo 2/1323 (Fortune Telling and Palmistry).

30. Fortune telling, December 14, 1912, TNA, Mepo 2/1323 (Fortune Telling and Palmistry); "Crystal Gazing in London: An Egyptian Fined £25," *The Morning Advertiser* (December 7, 1912), TNA, Mepo 2/1323 (Fortune Telling and Palmistry).

31. "Crystal Gazing in London," TNA.

32. List of London fortune tellers, TNA.

33. Louise Owen, preface to Cheiro, *You and Your Hand* (1969), xvi.

34. Letter to Home Secretary, October 9, 1912, TNA Mepo 2/1323 (Fortune Telling and Palmistry); Criminal Investigation Department, notes, October 14, 1912, TNA, Mepo 2/1323 (Fortune Telling and Palmistry).

35. "The Crusade against Palmistry," *Evening Star*, November 21, 1904, 6; Law Report, County of London Sessions, *The Times*, October 4, 1904, news clipping, TNA, HO 144/541/A51669 (Vagrancy); *The Times*, October 6, 1904, HO 144/541/A51669 (Vagrancy); Madame Keiro was Martha Stephenson, born Martha Faircloth, who married Charles Yates Stephenson in Ormskirk in 1888. *The Times*, October 6, 1904, TNA, HO 144/541/A51669 (Vagrancy); "The Crusade against Palmistry," 6; Charles and Martha Stephenson, Criminal Case, Palmistry. Reports of Hearing of Case at North London Sessions, 1904, HO 144/541/A51669 (Vagrancy); "Keiro, the World-Renowned Palmist and Madame Keiro, the World-Renowned Clairvoyante, will give Free Readings to Customers," *Westminster Gazette*, October 24, 1904. See also Owen Davies, "Newspapers and the Popular Belief in Witchcraft and Magic in the Modern Period," *Journal of British Studies* 37, no. 2 (April 1998): 139–65.

36. *The Times*, October 4, 1904, Law Report, County of London Sessions, news clipping, TNA, HO 144/541/A51669 (Vagrancy); *The Times*, October 5, 1904, news clipping, TNA, HO 144/541/A51669 (Vagrancy).

37. *The Times*, October 4, 1904, Law Report, TNA.

38. Owen Davies, *Witchcraft, Magic and Culture, 1736–1951* (Manchester University Press, 1999).

39. *The Times*, October 4, 1904, Law Report, TNA; *The Times*, October 7, 1904, Law Report, County of London Sessions, news clipping, TNA, HO 144/541/A51669 (Vagrancy).

40. *The Times*, October 7, 1904, Law Report, TNA; "Law's View of Palmistry," *Daily Mirror*, September 21, 1904; *The Times*, October 5, 1904, TNA, HO 144/541/A51669 (Vagrancy).

41. *The Times*, October 6, 1904, TNA, HO 144/541/A51669 (Vagrancy); *The Times*, October 5, 1904, TNA.

42. *The Times*, October 5, 1904, TNA. "Palmists on Trial," *Daily News* (London), October 5, 1904, 12. See also Emily Ogden, *Credulity* (University of Chicago Press, 2018), 9–18.

43. *The Times*, October 7, 1904, County of London Sessions, TNA, HO 144/541/A51669 (Vagrancy).

44. Keiro, *Keiro's Palmistry*, 17.

45. Police report, November 4, 1912, TNA, Mepo 2/1323 (Fortune Telling and Palmistry); "General Cable News, London," *Sunday Times*, March 4, 1917, 1.

46. *The Press*, New York, quoted in *Cheiro's Language of the Hand*, "Appendix: Opinions of the Press and Public."

47. Ida Ellis, *A Catechism of Palmistry* (Redway, 1898).

CHAPTER SIX

1. The Ellis Family, *Guide to Health* (Blackpool: Ellis Family, c. 1916), back-cover book descriptions signed as "The Ellis Family, Health Specialists, Promenade, Blackpool."

2. *Blackpool Gazette & Herald*, November 7, 1902; "Life Sketch," *Know Thyself* 1, no. 6 (February 1892): 49.

3. *Batley News*, August 7 and 21, 1891; *Know Thyself* 1, no. 5 (January 1892): 38.

4. *Know Thyself* 1, no. 1 (September 1891): 1; *Batley News*, March 11, 1892.

5. "Life Sketch," *Know Thyself* 1, no. 6 (February 1892): 49; Advertisement for Sexual Science, *Know Thyself* 1, no. 1 (September 1891), 8; Ida Ellis, *The Essentials of Conception and How to Prevent It* (Batley: Ida Ellis, 1891); *Know Thyself* 1, no. 2 (October 1891): 13; *Know Thyself* 1, no. 6 (February 1, 1892): 1.

6. *Batley News*, December 9, 1892; *Blackpool Gazette & Herald*, November 7, 1902; *Batley News*, February 15, 1895; *Shields Daily News*, January 10, 1896; *Stroud News and Gloucestershire Advertiser*, December 20, 1895.

7. J. K. Walton, "The Demand for Working-Class Seaside Holidays in Victorian England," *Economic History Review* 34, no. 2 (1981): 249–65; Tamara West, "Marginality and Modernity on the South Shore," *European History Quarterly* 52, no. 4 (2022): 572–92; Michael Saler, "Modernity and Enchantment," *American Historical Review* 111, no. 3 (June 2006): 702; "Doreena," Scott Macfie Gypsy Collection, University of Liverpool Special Collections, SMGC 1/2/PY.

8. "Four Hands" manuscript reading by Miss E. Smyth, *Palmist*, September 1872, University of Liverpool Special Collections, MS RPXXV.5.6.

9. *Fleetwood Chronicle*, November 17, 1893; for example, *Northern Echo*, December 18, 1894; *Blackpool Gazette & Herald*, November 17, 1893.

10. *Blackpool Times*, November 5, 1902; *Blackpool Gazette & Herald*, November 7, 1902. For a summary of Albert Ellis's views on local political matters, see *Blackpool Times*, October 22, 1902; *Blackpool Gazette & Herald*, November 7, 1902; *Blackpool Times*, July 2, 1904, and October 22, 1902; *Blackpool Gazette & Herald*, October 11, 1907.

11. 33 South Beach, Blackpool, Lancaster, Census Record, 1901, Public Record Office, RG 13/3972, TNA; 81/82 Central Beach, Blackpool, Census of England and Wales, 1911, TNA; 82 Central Beach, Blackpool, Census of England Wales, 1921, TNA.

12. Roger Cooter, *The Cultural Meaning of Popular Science* (Cambridge University Press, 1984); John Van Wyhe, *Phrenology and the Origins of Victorian Scientific Naturalism* (Routledge, 2004); James Poskett, *Materials of the Mind* (University of Chicago Press, 2019).

13. "How we arrange the faculties," *Know Thyself* (ed. Professor Ida Ellis) 1, no. 1 (September 1891): 4; Albert Ellis, *Temperament and Character*, 7th ed. (Blackpool: Ellis Family, 1909), preface, 10–11, 15.

14. Ellis, *Temperament and Character*, 9.

15. Ellis, *Temperament and Character*, 32–36.

16. Ellis, *Temperament and Character*, 20–29.

17. Albert Ellis, *Herbal Remedies for Diseases* (Blackpool: Ellis Family, c. 1917); Advice Charts in BL collected ephemera, [Tracts, charts, etc. on phrenology, palmistry, etc., Ellis, Albert], BL, 7383.cc.37.

18. Ellis Family, *Guide to Health*, 40–42; "Blackpool Palmistry," *Lancashire Evening Post*, August 23, 1904, 3.

19. Ellis Family, *Guide to Health*, 27, 44.

20. Ellis, *Temperament and Character*, 38–39.

21. *Palmistry Chart, Describing the Past, Present, and Future* (Blackpool: Ellis Family, c. 1910), passim; Ida Ellis, *A Catechism of Palmistry* (Redway, 1898); Ida Ellis, *Hand Physiognomy* (Blackpool: Ellis Family, 1899); Ida Ellis, *Key to Palmistry, with A Map of the Hand* (Blackpool: Ellis Family, 1924). See *Swindon Advertiser and North Wilts Chronicle*, August 26, 1904; Albert Ellis, *Matrimony: A book for those about to Marry* (Blackpool: Ellis Family, c. 1919).

22. "Crusade Against Palmists," *Manchester Evening News*, August 23, 1904, 2; "Blackpool Palmistry," *Lancashire Evening Post*, August 23, 1904, 3.

23. *Palmistry Chart*, 23–24.

24. There was a review in *Lloyd's Weekly Newspaper*, October 17, 1897; E. Clodd, *Fortnightly Review* 107 (May 1920): 765; Ida Ellis, *A Catechism of Palmistry: The Sciences of Chirognomy and Chiromancy Explained in the Form of Question and Answer*, 3rd ed. (Rider, 1917), preface.

25. Ida Ellis, *Catechism of Palmistry* (1917), 6–20.

26. *Batley Reporter and Guardian*, August 26, 1904, and September 19, 1891; *Know Thyself* 1, no. 2 (October 1891): 13; *Know Thyself* 1, no. 6 (February 1892): 1; *Dewsbury Chronicle and West Riding Advertiser*, October 17, 1891.

27. *Blackpool Gazette & Herald*, November 3, 1905, and October 31, 1905.

28. *Batley Reporter and Guardian*, August 26, 1904; *Manchester Evening News*, August 23, 1904; *Lancashire Evening Post*, August 23, 1904; *Bristol Times and Mirror*, August 24, 1904; *Blackpool Gazette & Herald*, November 15, 1904; *Morning Post*, December 17, 1901; *Brighouse News*, December 5, 1902; *Daily Telegraph & Courier* (London), September 24, 1904; *Torquay Times and South Devon Advertiser*, December 7, 1900; *Leeds Mercury*, February 10, 1903; *Northern Daily Telegraph*, September 11, 1900; *Sheffield Evening Telegraph*, August 29, 1901; "Crusade Against Palmists," *Manchester Evening News*, August 23, 1904, 2.

29. *The Times*, Tuesday, October 4, 1904, Law Report, October 3, County of London Sessions, news clipping, TNA, HO 144/541/A51669 (Vagrancy: Method of Dealing with Cases of Palmistry, Fortune Telling, etc.).

30. *Palmistry Chart*, 4.

31. *Fleetwood Chronicle*, May 10, 1912; *Lancashire Daily Post*, April 14, 1902.

32. *Belfast Telegraph*, November 28, 1912; *Birmingham Mail*, November 5, 1912; *The Manchester Guardian*, March 8, 1913, and July 20, 1916.

33. *Blackpool Gazette & Herald*, November 15, 1904; March 22, 1910; December 11, 1914; February 2, 1915; November 19, 1907; Ellis 1921 Census Records; *Fleetwood Chronicle*, July 26, 1912; *Blackpool Times*, July 9, 1902, and March 16, 1904.

34. *Lancashire Daily Post*, April 14, 1909; West, "Marginality and Modernity," 588–91.

35. Albert Ellis, England & Wales Government Probate Death Index 1858–2019, 1934; *The Salisbury Times*, November 13, 1908; *Blackpool Gazette & Herald*, October 28, 1910; *Croydon Chronicle and East Surrey Advertiser*, February 19, 1910; *Bucks Advertiser & Aylesbury News*, February 12, 1910; *Lewisham Borough News*, November 11, 1910; Ellis, Electoral Register 1930, Polling District Blackpool (Waterloo Road), 62; Frank Ellis, England & Wales Government Probate Death Index 1858–2019, November and December 1939; Ida Ellis, England & Wales Government Probate Death Index, 1858–2019, 1940. Their son, Frank Raphael Ernest, died on May 14, 1943, also at Trewinnard Manor, England & Wales Deaths 1837–2007, 1943.

36. Dora Yates, Blackpool Gypsy Notebook, 1957, Dora Yates Papers, University of Liverpool Archives, GLS D/3/6.

CHAPTER SEVEN

1. Katharine St. Hill, *The Book of the Hand: A Complete Grammar of Palmistry for the Study of Hands on a Scientific Basis* (Rider, 1927), 14–15; Katharine St. Hill, *The Grammar of Palmistry* (Redway, 1889), 6.

2. "The Gentle Science of Palmistry: A Chat with Mrs Katherine [*sic*] St. Hill," *The Northern Champion* (Taree), July 29, 1922, 7. There is some ambiguity surrounding her birth year. New Zealand Law Reports, Cases Determined by the Supreme Court and Court of Appeal of New Zealand; *New Zealand Times*, October 10, 1906; see short biography of St. Hill in *The Sphere*, August 13, 1927.

3. *The Queen*, July 3, 1909.

4. Max Weber, *The Theory of Social and Economic Organization*, trans. A. M. Henderson and Talcott Parsons (Oxford University Press, 1947). An "indignation meeting" was held in London by fortune tellers to protest the "exposure of their frauds." See *Daily Mirror*, July 11, 1904; *The Observer*, July 10, 1904; Katharine St. Hill, *The Palmist's Review* I, no. 1 (January 15, 1899): 3.

5. *The Palmist's Review* I, no. 1 (January 15, 1899): 1; [Anon.], "Literary Gossip," *The Colonies and India*, May 21, 1892, 18; Katharine St. Hill, *The Palmist's Review* I, no. 1 (January 15, 1899): 2.

6. *The Glasgow Herald*, February 2, 1893; Katharine St. Hill, Editorial, *The Palmist's Review* I, no. 1 (1899): 3. She lists "author" as her occupation for 1901 and 1921 census. A heavily notated copy of Katharine St. Hill, *The Book of the*

Hand (Rider, 1927). This copy was owned by Lady Atkinson of 74 Oakwood Court, Kensington. Copy in author's possession.

7. St. Hill, *Grammar*, iii; Edward Palmer, *The Arabic Manual: Comprising a Condensed Grammar of Both the Classical and Modern Arabic* (W. H. Allen, 1881); Henry Sweet, *A New English Grammar* (Clarendon Press, 1892); St. Hill, *The Book of the Hand*, 3–4. Original italics; St. Hill, *Grammar*, iii.

8. "The Gentle Science of Palmistry," 7.

9. St. Hill, *The Book of the Hand*, 1–2; Cocles, *Physiognomiae et chiromantiae compendium* (Strasbourg, 1653).

10. "The Gentle Science of Palmistry," 7; Maud Wheeler, *Moles or Birthmarks, and their Significance to Man and Woman* (Roxburghe Press, 1894). The Roxburghe Press had also published Langdon Taylor, *Precious Stones and Gems, with their reputed Virtues* (Roxburghe Press, 1895).

11. George Wilde, *Chaldean Astrology Up to Date: How to Cast the Horoscope and Read the Future in the Stars* (E. Marsh-Stiles, 1901); O. Hashnu Hara, "Editorial Notices," *Wings of Truth* I, no. 1 (1900): 1.

12. O. Hashnu Hara, "Early Lessons in Clairvoyance," *Wings of Truth* I, no. 1 (1900): 2; O. Hashnu Hara, "Practical Lessons in Theosophy," *Wings of Truth* IV, no. 1 (1903): 10; St. Hill, *Palmist's Review* I, no. 1 (January 15, 1899), 3.

13. St. Hill, *Palmist's Review* I, no. 1 (January 15, 1899): 3; J. J. Spark, *Scientific and Intuitional Palmistry* (Roxburghe Press, 1896); St. Hill, *The Palmist's Review* I, no. 1 (January 15, 1899): 3; Katharine St. Hill, *Medical Palmistry, or The Hand in Health and Disease* (Rider, 1927), 80; "The Gentle Science of Palmistry," 7.

14. "The Gentle Science of Palmistry," 7; [Anon.], "Literary Gossip," *The Colonies and India*, May 21, 1892, 18; "Palmists on Trial," *Daily News* (London), October 5, 1904.

15. St. Hill, *Book of the Hand*, 5; "Chirological Society's Examination List, 1898," *Palmist's Review* I, no. 2 (April 1899): 13.

16. Charles F. Rideal, "Preface," Katharine St. Hill, *Hands of Celebrities, or, Studies in Palmistry* (Roxburghe Press, 1896), 12, 13–14, 16; "Character Sketches of Celebrities," *Palmist and Chirological Review* 4, no. 50 (July 1896).

17. St. Hill, *Book of the Hand*, 11–13.

18. St. Hill, *Book of the Hand*, 7–8, 10.

19. St. Hill, *Medical Palmistry*, 83; St. Hill, *Grammar*, 28–30, 55–57. See also "On Time," in St. Hill, *Book of the Hand*, 287–96.

20. Obituary of Dr. Ralph Richardson of Holywell, *Flintshire Observer*, December 29, 1898.

21. Obituary of Dr. Ralph Richardson.

22. St. Hill, *Palmist's Review* I, 1 (January 15, 1899): 2–3; St. Hill, Editorial, *Palmist's Review* I, no. 1 (January 15, 1899): 2; "The Gentle Science of Palmistry," 7; St. Hill, *Book of the Hand*, 5.

23. St. Hill, *Medical Palmistry*, 20, 39, 41, 50–51, 62.

24. St. Hill, *Medical Palmistry*, 31–32, 51.

25. St. Hill, *Medical Palmistry*, 22, 40; "The Gentle Science of Palmistry," 7.

26. St. Hill, *Medical Palmistry*, 48–52; St. Hill, *Book of the Hand*, 184; St. Hill, *Grammar*, 52, 44. See also Aviva Briefel, *The Racial Hand in the Victorian Imagination* (Cambridge University Press, 2015).

27. St. Hill, *Medical Palmistry*, 6–8, 56.

28. St. Hill, *Medical Palmistry*, 3, 42, 85.

29. St. Hill, *Medical Palmistry*, 58.

30. The US Centers for Disease Control and Prevention's current list of signs and symptoms of Hansen's disease read entirely like Katharine St. Hill's: "Discolored patches of skin, usually flat, that may be numb and look faded (lighter than the skin around); growths (nodules) on the skin; thick, stiff or dry skin." Centers for Disease Control and Prevention, "Hansen's Disease (Leprosy), Signs and Symptoms," https://www.cdc.gov/leprosy/symptoms/index.html.

31. St. Hill, *Medical Palmistry*, 58–59.

32. St. Hill, *Medical Palmistry*, 13, 57; "The Gentle Science of Palmistry," 7.

33. "The Gentle Science of Palmistry," 7; St. Hill, *Medical Palmistry*, 34, 49–50.

34. St. Hill, *Medical Palmistry*, 10–12, 15–17.

35. St. Hill, *Medical Palmistry*, 10, 16–17, 41.

36. St. Hill, *Medical Palmistry*, v, vi, 4–5.

37. St. Hill, *Medical Palmistry*, 72, 87; "The Gentle Science of Palmistry," 7; St. Hill, *Medical Palmistry*, 66.

38. St. Hill, *Medical Palmistry*, 75, 78–79.

39. *Cheiro's Language of the Hand* (Herbert Jenkins, 1900), 152, 155; Ina Oxenford and Anna MacDowel Cosgrave, *Life Studies in Palmistry* (L. Upcott Gill, 1899), 63, 22, 69; "Congenital Idiot, 35 years of age," in Richard Beamish, *The Psychonomy of the Hand* (Frederick Pitman, 1865), plate 6.

40. "The Gentle Science of Palmistry," 7; St. Hill, *Medical Palmistry*, 84–88.

41. St. Hill, *Medical Palmistry*, 86–88. St. Hill raised money for animal charities and institutions including the Home of Rest for Horses, the Royal Society for the Prevention of Cruelty to Animals (RSPCA), the Home for Lost and Starving Dogs (Battersea), and Our Dumb Friends League (ODFL). See *Cheltenham Examiner*, December 9, 1896.

42. "Grand Fancy Fair at Pontypridd," *South Wales Star*, July 24, 1891, 7; *Western Mail*, July 17, 1891.

CHAPTER EIGHT

1. Francis Galton, "Composite Portraits, Made by Combining Those of Many Different Persons into a Single Resultant Figure," *Journal of the Anthropological Institute of Great Britain and Ireland* 8 (1878): 132–44. See Paul Rabinow, "Galton's Regrets," in *Essays on the Anthropology of Reason* (Princeton Univer-

sity Press, 1997), 112–28; Miss Baden Powell's outlines, Galton Papers, University College London Special Collections [UCLSC] GALTON/2/9/4/2/1; Profile of a bearded man, UCLSC GALTON 2/9/6/2/4/2. For portraiture, see Ludmilla Jordanova, *Defining Features* (Reaktion Books, 2000).

2. Francis Galton, *Fingerprint Directories* (Macmillan, 1895), 9–10.

3. [Anon.], "The Wonders of a Finger-Print," *The Sketch*, November 20, 1895, 188; Francis Galton, *Finger Prints* (Macmillan, 1892), 1; Systems of Ridges and the Creases in the Palm, in Francis Galton, *Finger Prints* (Macmillan, 1892), plate 3.

4. Keith Breckenridge, *Biometric State* (Cambridge University Press, 2014); Francis Galton, *Anthropometric Laboratory: Arranged by Francis Galton, F.R.S.* (William Clowes and Sons, 1884), for the International Health Exhibition; Michel Foucault, "Governmentality," in *The Foucault Effect*, ed. G. Burchell, C. Gordon, and P. Miller (University of Chicago Press, 1991), 87–104.

5. "Physiognomonica," trans. T. Loveday and E. S. Forster, in *The Works of Aristotle*, vol. VI, Opuscula (Clarendon Press, 1913), 807a.

6. Friedrich Engels, *The Part Played by Labour in the Transition from Ape to Man* (Foreign Languages Publishing House, 1952). Written in 1876, originally published 1896, *Anteil der Arbeit an der Menschwerdung des Affen*. Original italics.

7. Bertillon, *Signes Caractéristiques des Professions Manuelles*, 189, UCLSC GALTON/2/9/7.

8. Galton, *Finger Prints*, 197.

9. See Joseph Jacobs, "The Jewish Type, and Galton's Composite Photographs," *Photographic News* 29, no. 1390 (April 24, 1885): 268–69.

10. Francis Galton, *Narrative of an Explorer in Tropical South Africa* (John Murray, 1853).

11. Galton, *Finger Prints*, 192–93, 195, 196.

12. Lavater, "L'Art de Connaitre les Hommes par La Physionomie," III, no. 1 (Paris: [L. Prudhomme], 1806–1809), trans. Edward Heron-Allen, *The Science of the Hand* (Ward, Lock, 1885), 185.

13. Edward Henry, *Classification and Uses of Fingerprints* (George Routledge, 1900); Francis Galton, "The Patterns in Thumb and Finger Marks" (November 3, 1890), uncorrected proof, for meeting of the Fellows of the Royal Society, UCLSC GALTON 2/9/6/1; Galton evidence, Commission of Inquiry, quoted in *Finger Print Directories*, 31–33; Francis Galton, "Method of Indexing Finger-Marks," *Proceedings of the Royal Society*, received April 30, 1891, offprint from vol. 49, UCLSC GALTON 2/9/6/1 f. 5; Galton evidence, Commission of Inquiry, quoted in *Finger Print Directories*, 33, 40. See Chandak Sengoopta, *Imprint of the Raj* (Macmillan, 2003); Chandak Sengoopta, "Treacherous Minds, Submissive Bodies," in *Locating the Medical*, ed. Rohan Deb Roy and Guy N. A. Attewell (Oxford University Press, 2018); Simon A. Cole, *Suspect Identities* (Harvard University Press, 2002).

14. Translation of Purkinje's *Commentatio* from Latin to English, n.d., UCLSC GALTON/2/9/6/2/2 ff. 14 r-v; Francis Galton, "Method of indexing Finger-Marks," *Proceedings of the Royal Society*, received April 30, 1891, offprint from vol. 49, UCLSC GALTON 2/9/6/1 f. 5v, 7r, 6v; Galton's early categorization of fingerprint types, UCLSC GALTON/2/9/6/2/2.

15. Galton, *Finger Print Directories*, 10; Twins . . . portraits exhibited together with finger prints at Conversazione, Royal Society, June 12, 1895, typescript, UCLSC GALTON/2/9/6/9.

16. Commission of Inquiry, quoted in *Finger Print Directories*, 12; H. H. Wilder, "Palm and Sole Studies," *Biological Bulletin* 30 (February 1916): 135; [Anon.], "The Wonders of a Finger-Print," *The Sketch*, November 20, 1895, 188; Karl Pearson, *The Life, Letters, and Labours of Francis Galton*, vol. 3A (Cambridge University Press, 1930), 216.

17. Pearson, *The Life, Letters, and Labours of Francis Galton*, 216; "Fingerprints in Wax and Ink," UCLSC, GALTON/2/9/6/11/7.

18. Francis Galton, *Decipherment of Blurred Finger Prints* (Macmillan, 1893), 5; Francis Galton, "Method of indexing Finger-Marks," *Proceedings of the Royal Society*, received April 30, 1891, offprint from vol. 49, UCLSC, GALTON 2/9/6/1 f. 5v; "Directions for Making Palm and Sole Prints," typescript, n.d., UCLSC, GALTON/2/9/6/2/2 f.10; Fingerprints of the author, Galton, *Finger Prints*, title page.

19. Ina Oxenford, *Characteristic Hands* (F. N. Fowler, 1912), 11; Galton, *Finger Print Directories*, 60–61, 63–64, 111.

20. H. H. Wilder to Francis Galton, December 26, 1904, UCLSC, GALTON 3/3/22/31 ff.11–12; Galton, *Finger Print Directories*, 4, 9.

21. Arthur Conan Doyle, "The Sign of the Four," *Lippincott's Monthly Magazine* (February 1890): 145–223.

22. Richard Saunders, *Palmistry, the secrets thereof disclosed* (London, 1676), 59; "The Marks of Murder," *Daily Examiner*, December 30, 1888, 12.

23. Cheiro, *Cheiro's Language of the Hand*, 15th ed. (Herbert Jenkins, 1900).

24. Cheiro, *Cheiro's Memoirs: The Reminiscences of a Society Palmist* (Rider, 1912), 18, 186–88.

25. "The Gentle Science of Palmistry: A Chat with Mrs Katherine [*sic*] St Hill," *The Northern Champion* (Taree), July 29, 1922, 7; Charles F. Rideal, the publisher who co-edited the *Palmist and Chirological Review* with Katharine St. Hill, had a connection with Scotland Yard. Maurice Moser and Charles F. Rideal, *Stories from Scotland Yard* (George Routledge, 1890); Katharine St. Hill, *Medical Palmistry, or The Hand in Health and Disease* (Rider, 1927), 111–23; "Crusade Against Palmists," *Manchester Evening News*, August 23, 1904, 2.

26. "Blackpool Palmistry," *Lancashire Evening Post*, August 23, 1904, 3.

27. Katharine Furse to Philip Game, July 29, 1936, and Katharine Furse to Sir Harold Scott, Scotland Yard, March 12, 1951, TNA, Mepo 2/9979.

CHAPTER NINE

1. J. L. H. Down, "Observations on an ethnic classification of idiots," *Clinical Lecture Reports, London Hospital* 3 (1866): 259–62; G.E. Shuttleworth, "Mongolian Imbecility," *BMJ* 12 (1909): 664; R. Langdon Down, Comment, "Mongolian Imbecility," *BMJ* 12 (1909): 665; R. Down, "Notes and News," *Journal of Mental Science* 52 (1905): 188–89; Notes on patients suffering from mental handicap or mental illness seen by Dr. Reginald Langdon Down, some of whom were subsequently admitted to Normansfield, 1907–1910, London Metropolitan Archives, H29/NF/B/5/1, f. 1.2.

2. Richard Beamish, *The Psychonomy of the Hand* (Frederick Pitman, 1865), plate 1.

3. Paul Broca, "Le pli transversal du singe dans la main de l'homme," *Bulletins de la Société d'anthropologie de Paris*, II° Série, xii (1877): 431–32.

4. "John L. H. Langdon Down," *BMJ* 2, no. 1868 (October 17, 1896): 1170–71.

5. F. G. Crookshank, *The Mongol in Our Midst: A Study of Man and His Three Faces*, 3rd ed. (Kegan Paul, Trench, Trübner & Co., 1931), 178. See David Wright, "Mongols in our Midst," in *Mental Retardation in America*, ed. Steven Noll and James W. Trent Jr. (New York University Press, 2004), 92–119; Daniel Kevles, "'Mongolian Imbecility,'" in *Mental Retardation in America*, 120.

6. Henry Faulds, who anticipated Galton's fingerprinting, used the term "dermatographs." Henry Faulds, *Fingerprint Identification* (Wood, Mitchell, 1905), 14ff.

7. "Francis Graham Crookshank," *Journal of Nervous and Mental Disease* 79, no. 1 (January 1934): 122–23. The invention of *encephalitis lethargica* as a "disease" was analyzed in F. G. Crookshank, "The Importance of a Theory of Signs and a Critique of Language in the Study of Medicine," in C. K. Ogden and I. A. Richards, *The Meaning of Meaning: A Study of the Influence of Language Upon Thought and of the Science of Symbolism* (1923), 7th ed. (Harcourt, Brace and Co., 1945), 350–53.

8. Crookshank wrote an essay that prefaced Adler's book, "Individual Psychology: A Retrospect (and a Valuation)," in *Alfred Adler: Problems of Neurosis: A Book of Case Histories*, ed. Philippe Mairet (K. Paul, Trench, Trübner, 1929), vii–xxxvii; Crookshank, *Mongol*, 207.

9. Crookshank, "The Importance of a Theory of Signs," 341–54.

10. Ogden and Richards, eds., *The Meaning of Meaning* (1923), 14; F. G. Crookshank, *Diagnosis, and Spiritual Healing* (Kegan Paul, Trench, Trübner, 1927), 15; Crookshank, "The Importance of a Theory of Signs," 339; "Crookshank, MD [Obituary]," 122–23. See Carlo Ginzburg, "Morelli, Freud, and Sherlock Holmes," in *The Sign of Three*, ed. Umberto Eco and Thomas A. Sebeok (Indiana University Press, 1983).

11. Shuttleworth, "Mongolian Imbecility," 661; Crookshank, *Mongol*, 47–51; Daniel Pick, *Faces of Degeneration* (Cambridge University Press, 1989).

12. [Robert Chambers], *Vestiges of the Natural History of Creation* (John Churchill, 1844), 307–9; Crookshank, *Mongol*, 6–7.

13. Crookshank, *Mongol*, 52–53, 62, 404–5, 426–47.

14. F. G. Crookshank, "Hand-Prints of Mongolian and Other Imbeciles," *Transactions of the Medical Society of London* 44 (1921): 155–60; F. G. Crookshank, "Handprints of Mental Defectives," *The Lancet* 1, no. 5084 (February 5, 1921): 274; F. G. Crookshank, "The Hand-Markings of Mongolian Imbeciles: To the Editor of The Lancet," *The Lancet* 1, no. 5085 (February 12, 1921): 350. See also "Handprint in Relation to Mental Development," *Psyche* 4 (1923–24): 333; Bronislaw Malinowski, "The Problem of Meaning in Primitive Language," supplementary essay in Ogden and Richards, *The Meaning of Meaning*, 296–336; "Palmistry," abstract of a lecture delivered by Dr. Arthur Keith, January 15, 1901, *Man* 1, nos. 18–20 (1901): 26; Frederic Wood Jones, *Arboreal Man* (Arnold, 1918); Frederic Wood Jones, *The Principles of Anatomy as Seen in the Hand* (J. & A. Churchill, 1920); Crookshank, *Mongol*, 56.

15. Crookshank, *Mongol*, 206–7.

16. Crookshank, *Mongol*, 209–11.

17. Crookshank, *Mongol*, 217–18; *Cheiro's Language of the Hand* (Herbert Jenkins, 1900), 151, 96.

18. Crookshank, *Mongol*, 217, 219.

19. From the first edition, F. G. Crookshank, *The Mongol in Our Midst* (E. P. Dutton, 1924), 16.

20. W. H. L. Duckworth, *Morphology and Anthropology* (C. J. Clay, 1904); C. F. Sonntag, *The Morphology and Evolution of the Apes and Man* (John Bale, 1924), 96; Frederic Wood Jones, *Man's Place Among the Mammals* (Edward Arnold, 1929); Crookshank, *Mongol*, 220–22; G. M. V., "C. F. Sonntag, MD Edin.," *BMJ* 2, no. 3381 (October 17, 1925): 724.

21. Crookshank, *Mongol*, 219.

22. Ira S. Wile and S. Z. Orgel, "A Genetic Study of Mongolism," *Medical Journal and Record* 127 (April 1928): 431; and "A Study of the Physical and Mental Characters of Mongols," *International Clinics* 3 (October 1928): 145; "It is well shown on the left palm of the Virgin in one of the Buda Pesth Luinis." Crookshank, *Mongol*, 220. Bernardino Luini, *Virgin and Child with Saint Catherine of Alexandria and Barbara*, c. 1522–1525, oil on wood, Museum of Fine Arts, Budapest.

23. H. H. Wilder, "The Physiognomy of the Indians of Southern New England" (1912), offprint in Wilder Papers, Smith College, Massachusetts, Box 1059, folder 3; H. H. Wilder, "Scientific Palmistry," *Popular Scientific Monthly* 61 (1902): 41–54; Print of the Right hand of Harris Hawthorne Wilder, frontispiece to Harold Cummins and Charles Midlo, *Finger Prints, Palms and Soles* (Dover, 1943).

24. H. H. Wilder, "Palm and Sole Studies," *Biological Bulletin* 30, no. 2 (February 1916): 135.

25. H. H. Wilder, *The Pedigree of the Human Race* (George G. Harrap, 1926), 251; Sarah B. Holt, "Harold Cummins (1894–1976) Obituary," *Journal of Medical Genetics* 13 (1976): 540.

26. Harold Cummins, "The Configurations of Epidermal Ridges in a Human Acephalic Monster," *Anatomical Record* 26, no. 1 (August 1923): 1–13; Holt, "Harold Cummins," 540.

27. Roswell H. Johnson, "Pads on the Palm and Sole of the Human Fetus," *American Naturalist* 33, no. 393 (1899): 729–34; Cummins and Midlo, *Finger Prints, Palms and Soles*, 2nd ed. (New York: Dover, 1961), 185, photograph p. 179; Cummins and Midlo, *Finger Prints, Palms and Soles*, 185.

28. Charles Midlo and Harold Cummins, *Palmar and Plantar Dermatoglyphics in Primates* (Wistar Institute of Anatomy and Biology, 1942); Harold Cummins and S. D. S. Spragg, "Dermatoglyphics in the Chimpanzee: Description and Comparison with Man," *Human Biology* 10, no. 4 (December 1938): 457–510; Cummins and Midlo, *Finger Prints, Palms and Soles*, 156–73; Wilder, "Palm and Sole Studies," 135–72, 211–52; for Cummins, see Daniel Asen, "Secrets in Fingerprints," *Canadian Medical Association Journal* 190, no. 19 (2018): E597–99; H. H. Wilder, "Racial Differences in Palm and Sole Configuration," *American Anthropologist* New Series 6, no. 2 (April–June 1904): 244–93; H. H. Wilder, "Racial Differences in Palms and Sole Configuration. Palm and Sole Prints of Japanese and Chinese," *American Journal of Physical Anthropology* 5 (1922): 143–206; Harold Cummins et al., "Revised Methods of Interpreting and Formulating Palmar Dermatoglyphics," *American Journal of Physical Anthropology* 12, no. 3 (January–March 1929): 415–73. For ongoing studies, see F. A. Miller, "Dermatoglyphics and the Persistence of 'Mongolism,'" *Social Studies of Science* 33, no. 1 (2003): 75–94.

29. Holt, "Harold Cummins (1894–1976) Obituary"; Harold Cummins, MD, "The Cadaver," Lecture to the Tulane Medical History Society, December 17, 1973, Tulane University Archives, https://archive.org/details/01Track01_20151117; Harold Cummins, "Dermatoglyphics in Eskimos from Point Barrow," *American Journal of Physical Anthropology* 20, no. 1 (April–June 1935): 13–17; Harold Cummins, "Dermatoglyphics of Bushmen (South Africa)," *American Journal of Physical Anthropology* 13, no. 4 (December 1955): 699–709. For further "ethnic" studies of the simian line, see R. S. Bali, *Anthropology of Crease Morphogenesis* (New Delhi: Concept Publishing Company, 1994). See also Daniel Asen, "'Dermatoglyphics' and Race after the Second World War," in *Global Transformations in the Life Sciences, 1945–1980*, ed. Patrick Manning and Mat Savelli (University of Pittsburgh Press, 2018), 61–77. Studies by Hill and Abel, noted in Cummins and Midlo, *Finger Prints, Palms and Soles*, 259; H. H. Wilder to Francis Galton, November 28, 1894, University College London Special Collections, GALTON 2/9/6/2/2 ff. 32–33.

30. Cummins and Midlo, *Finger Prints, Palms and Soles*, 253.

31. Cummins and Midlo, *Finger Prints, Palms and Soles*, 271, 225–56, 273–

74. For German scholarship, see Gisela Meyer-Heydenhagen, "Die palmaren Hautleisten bei Zwillingen," *Zeitschrift für Morphologie und Anthropologie* 33, no. 1 (1934): 1–42; S. Bettmann, "Über die Vierfingerfurche," *Zeitschrift für Anatomie und Entwicklungsgeschichte* 98 (1932): 487–503. See also J. Jadin and J.A. Valšík, "Aperçu sur l'état sanitaire des Pygmées de l'Ituri/The Finger Prints of Central African Pygmies, Negroes and Their Crossbreeds," *Anthropologie* 16, nos. 1–4 (1948): 69–100.

32. Cummins and Midlo, *Finger Prints, Palms and Soles*, 251–52.

33. William M. Shanklin and Harold Cummins, "Dermatoglyphics in Rwala Bedouins," *Human Biology* 9, no. 3 (1937): 357–65; Cummins and Midlo, *Finger Prints, Palms and Soles*, 265–68.

34. Cummins and Midlo, *Finger Prints, Palms and Soles*, 276–79.

35. Jane Stafford, "New Palm Study Is Real Science," *Science News Letter* 22, no. 593 (August 20, 1932): 114.

36. Crookshank, *Mongol*; S. S. Sarkar, "The Simian Crease," *Zeitschrift für Morphologie und Anthropologie* 51, no. 2 (1961): 212, 217.

37. Cummins and Midlo, *Fingerprints, Palms and Soles*, 280–81; for example, Blanka Schaumann and Milton Alter, *Dermatoglyphics in Medical Disorders* (Springer, 1976).

38. H. K. Kumbnani, "Main Line Index and Transversality in Trisomy 21," in *Dermatoglyphics: An International Perspective*, ed. Jamshed Mavalwala (De Gruyter Mouton, 1978), 351–56.

CHAPTER TEN

1. Alfred Adler, *What Life Should Mean to You* (George Allen & Unwin, 1932), ch. 2, "Mind and Body"; Charlotte Wolff, *On the Way to Myself: Communications to a Friend* (Methuen, 1969), 90.

2. Charlotte Wolff, *Love Between Women* (Gerald Duckworth, 1971); Charlotte Wolff, *Psychologie der lesbischen Liebe: eine empirische Studie der weiblichen Homosexualität*, trans. Christel Buschmann (Rowohlt, 1977); Atina Grossmann, "German Women Doctors from Berlin to New York," *Feminist Studies* 19, no. 1 (Spring 1993), 65–88; Toni Brennan and Peter Hegarty, "Charlotte Wolff and Lesbian History," *Journal of Lesbian Studies* 14, no. 4 (2010): 338–58; Charlotte Wolff, *Bisexuality: A Study* (Quartet Books, 1977); Charlotte Wolff, *Magnus Hirschfeld: A Portrait of a Pioneer in Sexology* (Quartet Books, 1986).

3. Charlotte Wolff, "CV etc. 1928–1937," typescript, Wellcome Collection, London, British Psychological Society Archive, Charlotte Wolff Papers, PSY/WOL/5/1 [hereafter CWP]; Laurens Schlicht, "Graphology in Germany in the 1920s and 1930s," *NTM Zeitschrift für Geschichte der Wissenschaften, Technik und Medizin* (NTM Journal for the History of Science, Technology and Medicine) 28, no. 2 (June 2020): 149–79; Ludwig Klages, *Handschrift und Charakter; gemeinverständlicher Abriss der graphologischen Technik* (J. A. Barth, 1917); Wolff,

On the Way to Myself, 99; Bryan S. Turner, "Reflections on the Epistemology of the Hand," in *Regulating Bodies* (Routledge, 1992), 99–121.

4. Wolff, *On the Way to Myself*, 71, 182; James L. Jarrett, ed., *Nietzsche's Zarathustra: Notes of the Seminar given in 1935–39 by C. G. Jung*, vol. 2, part I (Routledge, 1989), 360; Julius Spier, "Psychochirologie eine neue Therapie," *Arztliche Rundschau* 40, no. 24 (1930): 339–40; Julius Spier, *The Hands of Children: An Introduction to Psycho-Chirology* (Kegan Paul & Co., 1944); Alexandra Nagel, "The Psychochirologist Julius Spier and the Art of Reading Hands During the Interbellum" (PhD, Leiden University, 2020); Alexandra Nagel, "Jung, Julius Spier, and Palmistry," *Jung Studies* 14, no. 1 (January 2020): 65–81; Alexandra H. M. Nagel, "From Chiromancy to Psychochirology," *Aries* 21, no. 2 (2021): 246–70.

5. Richard T. Gray, *About Face* (Wayne State University Press, 2004); Ian Verstegen, "The Politics of Physiognomic Perception," *Gestalt Theory* 44, nos. 1–2 (August 2022): 183–200.

6. Wolff, *On the Way to Myself*, 90, 96–97, 103; M. E. Warlick, "Palmistry as Portraiture," in *Surrealism, Occultism and Politics*, ed. Tessel M. Bauduin, Victoria Ferentinou, and Daniel Zamani (Routledge, 2018), 57–72; Charlotte Wolff, *The Human Hand* (Methuen, 1942), 91.

7. Docteur Lotte Wolff, "Les Révélations Psychiques de la Main," *Le Minotaure* 6 (Winter 1935), offprint, CWP, PSY/WOL/6/3/1; Michel Huteau, "A Star of French Psychology," *Bulletin de Psychologie* 494, no. 2 (2008): 173–99; Nicolae Vaschide, *Essai sur la psychologie de la main* (Marcel Rivière, 1909); Greta Plaitano, "Technologies de l'oeil, psychologie de la main: Nicolae Vaschide et la photographie médico-artistique," *Cinéma&Cie* 20, no. 35 (2020): 29–40; Charlotte Wolff, *The Hand in Psychological Diagnosis* (1951) (New Delhi: Sagar, 1972), 42; Handprints no. 10, Medicus et Psychologues, CWP, PSY/WOL/2/10; *Mains du Mongoliens* on rice paper, n.d., CWP PSY/WOL/3/4.

8. See letters between Wolff and Maria and Aldous Huxley, CWP PSY/WOL/1/3/5; Wolff, *On the Way to Myself*, 94–95.

9. Charlotte Wolff, *Studies in Hand Reading* (Chatto & Windus, 1936). Her "analysands" included Vaslav Nijinsky, Daisy Fellowes, Helen Grund, André Breton, Paul Elliard, René Crevel, Aldous Huxley, Julian Huxley, Julien Green, Gerald Heard, George Bernard Shaw, Virginia Woolf, Osbert Sitwell, T. S. Eliot, Lady Ottoline Morrell, André Derain, Max Ernst, Balthus, Maurice Ravel, Armande de Polignac, Vicomte de Noailles, F. M. Alexander, Cecil Beaton, Herman Schryver, Man Ray, Horst P. Horst, George Hoyningen-Huene, Ija Abdy Gaye, Anna May Wong, John Gielgud, and Antonin Artaud; Charlotte Wolff, "CV etc. 1928–1937," typescript, CWP PSY/WOL/5/1.

10. Undated news clipping from *News Chronicle; The Morning Post*, July 18, 1936; *The Listener*, August 5, 1936; *Spectator*, July 24, 1936; *New York Times*, July 4, 1937, in CWP PSY/WOL/6/3/4 Scrapbook.

11. "These Names Makes News: Paree Still Gay," *Daily Express*, January 7,

1936, CWP PSY/WOL/6/3/4 Scrapbook; Charles Bell, *The Hand*, quoted in Wolff, *The Hand in Psychological Diagnosis*, xiii.

12. Wolff, *Studies in Hand Reading*.

13. Wolff, *The Human Hand*, 121–22.

14. Aldous Huxley, typescript preface, CWP PSY/WOL/6/3/2; Wolff, *The Human Hand*, 33–34, 24.

15. Wolff, *On the Way to Myself*, 71; Wolff, *The Hand in Psychological Diagnosis*, x, 90; Wolff, *On the Way to Myself*, 80–85; Charlotte Wolff, "Les Principes de la chirologie," in *Encyclopédie française*, Tome VIII, *La vie mentale*, ed. Henri Wallon (Société de gestion de l'Encyclopédie francaise, 1938); Charlotte Wolff, *A Psychology of Gesture* (Methuen, 1945), ix–x.

16. Wolff, *The Human Hand*, 120–21; Katharine St. Hill, *The Book of the Hand* (Rider, 1927), 6, 9. Wolff, *On the Way to Myself*, 89. See also Abbie Garrington, *Haptic Modernism* (Edinburgh University Press, 2022), ch. 3.

17. Wolff, *The Human Hand*, 105, 61.

18. Wolff, *On the Way to Myself*, 187, 176. For the trance, see Mikkel Borch-Jacobsen, "Hypnosis in Psychoanalysis," *Representations* 27 (Summer 1989): 92–110.

19. E. Alix, "Recherches sur la disposition des lignes papillaires de la main et du pied," *Annales des Sciences Naturelles (Zoologie)* 5e sér., T. 8 (1868): 359; Paul Broca, "L'ordre des primates. Parallèle anatomique de l'homme et des singes," *Bulletins et Mémoires de la Société d'Anthropologie de Paris* (1869): 228–401; François Leuret, Pierre Louis Gratiolet, et al., *Anatomie comparée du système nerveux considéré dans ses rapports avec l'intelligence* (Baillière, 1839); Wolff, *The Human Hand*, 103, 105, 61, 95, 91; Wolff, *On the Way to Myself*, 84.

20. Wolff, *The Hand in Psychological Diagnosis*, ix–x; Adolph Friedemann, "Handbau und psychosis," *Archives für Neurologie und Psychiatrie* 82 (1928): 439–99; C. J. C. Earl, "The Primitive Catatonic Psychosis of Idiocy," *British Journal of Medical Psychology* 14 (October 1934): 230; and much later, Y. Haft-Pomrock and Y. Ginath, "Differences Between Schizophrenics and Normal Controls Using Chirological (Hand) Testing," *Israel Journal of Psychiatry and Related Sciences* 19, no. 1 (1982): 5–22; Charlotte Wolff and Henry R. Rollin, "The Hands of Mongolian Imbeciles in Relation to Their Three Personality Groups," *Journal of Mental Science* 88, no. 372 (1942): 415–18; Cyril Burt, *The Measurement of Mental Capacities: A Review of the Psychology of Individual Differences* (Oliver and Boyd, 1927); Charlotte Wolff, "Character and Mentality as Related to Hand-Markings," *British Journal of Mental Psychology* 18, nos. 3–4 (February 1941): 364–82; Charlotte Wolff, "The Hand of the Mental Defective," *British Journal of Medical Psychology* 20, no. 2 (1944): 147–60.

21. Harold Cummins, review of Wolff, *The Human Hand*, in *American Journal of Psychology* 57, no. 1 (1944): 126–27; Wolff, "The Hand of the Mental Defective," 148.

22. Wolff, "The Hand of the Mental Defective," 148 [original emphasis];

Lancelot Hogben, "The Scientific Basis of Organotherapy," *Journal of the Medical Association of South Africa* 3, no. 23 (December 14, 1929): 668; S. L. Simpson, *Major Endocrine Disorders*, 2nd ed. (Oxford University Press, 1948); Wolff, *The Hand in Psychological Diagnosis*, 13–14; see also review of *The Hand in Psychological Diagnosis*, in *The Lancet*, March 22, 1952; August Werner, *Endocrinology* (Lea & Febiger, 1937); Wolff, *The Hand in Psychological Diagnosis*, 14–17, 20.

23. Wolff, *The Hand in Psychological Diagnosis*, 139; Charlotte Wolff to the Editor, *British Journal of Psychiatry* 115, no. 519 (1969): 252–53; Wolff, *The Human Hand*, 95, 126.

24. Wolff, *The Hand in Psychological Diagnosis*, 10, xi; Bell, *The Hand*, quoted in Wolff, *The Hand in Psychological Diagnosis*, xi–xiii.

25. Wolff, *The Hand in Psychological Diagnosis*, 7; "The Hand and Intelligence," manuscript for ch. 1, *The Hand in Psychological Diagnosis*, CWP, PSY/WOL/6/3/2.

26. Handprints I, Aldous Huxley Family, CWP, PSY/WOL/2/1; Superintendent, Regent's Park to Charlotte Wolff, January 27, 1939, CWP, PSY/WOL/6/3/2.

27. Julian Huxley, "From Fin to Fingers: The Evolution of Man's Hand," *Illustrated London News* (December 1930): 1138–39. See Alison Bashford, *The Huxleys* (University of Chicago Press, 2022), ch. 6; Wolff, *The Human Hand*, 68.

28. Charlotte Wolff, "The Form and Dermatoglyphs of the Hands and Feet of Certain Anthropoid Apes," *Proceedings of the Zoological Society of London* A107, no. 3 (September 1937): 347.

29. Wolff, "The Form and Dermatoglyphs of the Hands and Feet of Certain Anthropoid Apes," 348.

30. Harold Cummins and Charles Midlo, *Finger Prints, Palms and Soles* (Dover, 1943), 299; Charlotte Wolff, "A Comparative Study of the Form and Dermatoglyphs of the Extremities of Primates," *Proceedings of the Zoological Society of London* A108, no. 1 (April 1938): 143–61; Charlotte Wolff, "The Form and Dermatoglyphs of the Hands and Feet of Certain Anthropoid Apes," 347.

31. Richard Beamish, *The Psychonomy of the Hand* (Frederick Pitman, 1865), plate 1, "Gorilla"; Raphael Weldon to Francis Galton, April 30, 1894, "Chimpanzee Prints," UCLSC, GALTON/2/9/6/3/4/5; Henry Faulds, *Guide to Finger-Print Identification* (Wood, Mitchell, 1905), 17. See also Henry Faulds, "On the Skin-Furrows of the Hand," *Nature* 22 (October 28, 1880): 605; Noel Jaquin, *The Hand of Man* (Faber & Faber, 1933), plate 1, 25.

32. Wolff, "A Comparative Study of the Form and Dermatoglyphs of the Extremities of Primates," 157; Wolff, *The Human Hand*, 94.

33. Wolff, "A Comparative Study of the Form and Dermatoglyphs of the Extremities of Primates," 157; Charlotte Wolff, typescript, "La forme et les dermatoglyphes des mains et pieds d'un gorilla," CWP, PSY/WOL/2/27; other prints from Mok are in the album CWP, PSY/WOL/3/7; Arthur Kollmann, *Der Tastapparat der Hand der menschlichen Rassen und der Affen in*

seiner Entwickelung und Gliederung (Leopold Voss, 1883), 27; Wolff, "A Comparative Study of the Form and Dermatoglyphs of the Extremities of Primates," 159.

34. Wolff, *The Human Hand*, 97; Wolff, "The Form and Dermatoglyphs of the Hands and Feet of Certain Anthropoid Apes," 348; Charlotte Wolff, Notes for the form and dermatoglyphs of the hands and feet of certain anthropoid apes, 1937–38: CWP, PSY/WOL/6/3/1.

35. "An experiment in science," with Charlotte Wolff and Alan Best, of London Zoo, October 27, 1937, news-clipping offprint, CWP, PSY/WOL/1/3/5.

36. For example, Mellin Novikova and Alexander Kuznetsov, "Palmar Flexion Creases and Finger Linkage Groups in New World Monkeys—Functional and Evolutionary Palmistry," *Biological Communications* 62, no. 3 (December 2017): 181–201.

37. S. G. Purvis-Smith, "The Sydney Line: A Significant Sign in Down's Syndrome," *Australian Paediatric Journal* 8, no. 4 (August 1972): 198–200; Charlotte Wolff to Dr. Eleanor Brown, March 31, 1974, and April 25, 1973, CWP, PSY/WOL/6/3/1; Wolff, *The Human Hand*, 116; Eleanor Brown to Charlotte Wolff, January 29, 1973, CWP/WOL/6/3/1; Charlotte Wolff to Eleanor Brown, April 24, 1974, CWP/WOL/6/3/1.

CHAPTER ELEVEN

1. Soraya de Chadarevian, *Heredity Under the Microscope* (University of Chicago Press, 2020).

2. Lionel Penrose, "Chromosomes and Dermatoglyphic Patterns," typescript, October 1968, Lionel Penrose Papers, University College London Special Collections [hereafter LPP] PENROSE/2/43/2/2.

3. Max Weber, "Science as a Vocation" (1917), *Daedalus* 87, no. 1 (Winter 1958): 117.

4. Harry Harris, "Lionel Sharples Penrose, 1898–1972," *Biographical Memoirs of Fellows of the Royal Society* 19 (December 1973): 521.

5. Colchester Survey, LPP PENROSE/2/19/2; Edmund Ramsden, "Remodelling the Boundaries of Normality," in *Human Heredity in the Twentieth Century*, ed. Bernd Gausemeier, Staffan Müller-Wille, and Edmund Ramsden (Pickering and Chatto, 2013), 39–54; Lionel S. Penrose, *Mental Defect* (Sidwick & Jackson, 1933).

6. N. A. Barnicot, "Penrose's Work and View in Relation to Anthropology," *British Journal of Psychiatry* 125 (1974): 553; Harris, "Lionel Sharples Penrose," 524–25.

7. Penrose, *Mental Defect*, 20ff; Harris, "Lionel Sharples Penrose," 525.

8. Lionel Penrose, Notes for Litchfield Lecture, Oxford, January 20, 1965, typescript, LPP, PENROSE 2/42/3/2; L. S. Penrose, "Fingerprints and Palmistry," *The Lancet* 301, no. 7814 (June 2, 1973): 1240–41.

9. List of Published Papers by Sarah Holt, typescript, LPP PENROSE 2/42/9/1.

10. Michael Smith, *Lionel Sharples Penrose* (Colchester, n.d.), 31; L. S. Penrose, "The Creases on the Minimal Digit in Mongolism," *The Lancet* 218, no. 5637 (September 12, 1931): 585; Penrose, *Mental Defect*, 99–101.

11. "Mongolian Imbecility," *Nature* 107, no. 2685 (April 14, 1921): 218–19; K. Codell Carter, "Early Conjectures That Down Syndrome Is Caused by Chromosomal Nondisjunction," *Bulletin of the History of Medicine* 76, no. 3 (Fall 2002): 537–39; L. S. Penrose, "Maternal Age, Order of Birth and Developmental Abnormalities," *Journal of Mental Science* 85 (1939): 1149.

12. Wolff, *The Human Hand*, 98, 103, 106; Doris Mann Snedeker, "A Study of the Palmar Dermatoglyphics of Mongoloid Imbeciles," *Human Biology* 20, no. 3 (September 1948): 146–55.

13. Carter, "Early Conjectures," 560; F. A. Miller, "Dermatoglyphics and the Persistence of 'Mongolism,'" *Social Studies of Science* 33, no. 1 (2003): 76.

14. Penrose, "Fingerprints and Palmistry," 1239–42; Miller, "Dermatoglyphics," 75–94; Gordon Allen et al., "Mongolism: Letter to the Editor," *The Lancet* 277, no. 7180 (April 8, 1961): 775. See Carter, "Early Conjectures," 530; Tony Booth, "Labels and Their Consequences," in *Current Approaches to Down's Syndrome*, ed. David Lane and Brian Stratford (Holt Reinhart and Winston, 1985).

15. R. Turpin and J. Lejeune, "Dermatoglyphic Study of the Palms of Mongolian Idiots and of Their Parents and Siblings," *La Semaine des Hôpitaux* 29, no. 76 (December 1953): 3955–67; Miller, "Dermatoglyphics," 80. See Daniel Kevles, *In the Name of Eugenics* (Harvard University Press, 1985), 245–47.

16. L. Beckman and A. Norring, "Finger and Palm Prints in Schizophrenia," *Acta Genetica et Statistica Medica* 13, no. 2 (1963): 170–77; "Dermatoglyphics Takes New Role in Medicine," December 25–26, 1965, LPP, PENROSE/2/42/9/I.

17. See microscopy photographs in Lionel Penrose, "Chromosomes and Dermatoglyphics," typescript, LPP, PENROSE/2/43/2/2.

18. Blanka Schaumann and Milton Alter, *Dermatoglyphics in Medical Disorders* (Springer-Verlag, 1976).

19. The Eadwine Psalter, Trinity College, Cambridge, MS R.17.1, fol. 282; Johannes ab Indagine, *Introductiones apotelesmaticae elegantes* (Strasbourg, 1522); [Anon.], *The Conjurer's Guide!* (Glasgow, 1800); L. S. Penrose, "Familial Studies on Palmar Pattern in Relation to Mongolism," *Proceedings of the International Congress of Genetic and Hereditas* 8, suppl. vol. (1949).

20. L. Beckman, K. H. Gustavson, and A. Norring, "Dermal Configurations in the Diagnosis of the Down Syndrome: An Attempt at a Simplified Scoring Method," *Acta Genetica et Statistica Medica* 15, no. 1 (1965): 3–12, referring also to Penrose, "Fingerprints, Palms and Chromosomes," 933.

21. Lionel Penrose, "Chromosomes and Dermatoglyphic Patterns," Octo-

ber 1968. Typescript. LPP PENROSE 2/43/2/2; Barnicot, "Penrose's Work," 556; L. S. Penrose, "Dermatoglyphic Topology," *Nature* 205 (1965): 544; L. S. Penrose and D. Loesch, "Topological Classification of Palmar Dermatoglyphics," *Journal of Mental Deficiency Research* 14, no. 2 (June 1970): 111–28; L. S. Penrose, "Fingerprints and Palmistry," 1239–40; L. S. Penrose, "Genetical Studies in Dermatoglyphics," typescript, October 2, 1971, presented at the Warsaw conference "Dermatoglyphics in Human Genetics," LPP PENROSE 2/42/7/1; Lionel Penrose, Proof of "Dermatoglyphic Topology," *Nature*, 1965, LPP PENROSE 2/42/2/2; L. S. Penrose, "Finger-Prints, Palms and Chromosomes," *Nature* 197, no. 4871 (March 9, 1963): 933.

22. Danuta Z. Loesch, *Quantitative Dermatoglyphics: Classification, Genetics, and Pathology* (Oxford University Press, 1983). Foreword by Cedric A. B. Smith.

23. "Conferences on Dermatoglyphics," *Oceania* 33, no. 1 (1962): 50; typescript of Penrose's paper "Genetical studies in dermatoglyphics" for a Conference on "Dermatoglyphics in Human Genetics," Warsaw, October 2, 1971, LPP PENROSE/2/42/7. See also Danuta Loesch, "The Contributions of L. S. Penrose to Dermatoglyphics," *Journal of Mental Deficiency Research* 17 (1973): 1–17.

24. Lionel Penrose, "Dermatoglyphics" manuscript, n.d., LPP PENROSE 2/42/3/1; Schaumann and Alter, *Dermatoglyphics in Medical Disorders*; Penrose, "Fingerprints and Palmistry," 1242; H. Green and J. Thomas, "Pattern Formation by Cultured Human Epidermal Cells: Development of Curved Ridges Resembling Dermatoglyphs," *Science* 200, no. 4348 (1978): 1385.

25. Penrose's ambition at one point, for example, was to work out the "frequencies of genes for the world as a whole." L. S. Penrose and N. McArthur, "World Frequencies of the O, A, and B Blood Group Genes," *Annals of Eugenics* 15 (1951): 302; N. A. Barnicot, "Biology and Human Variation," *Race* 1, no. 2 (1959): 40–50; Lionel Penrose, quoted in UNESCO, *The Race Concept: Results of an Enquiry* (UNESCO, 1952), 24.

26. N. A. Barnicot, "Penrose's Work and View in Relation to Anthropology," *British Journal of Psychiatry* 125 (December 1974): 555. Lionel Penrose, "Genetics of the Human Race," in *Genetics in the 20th Century*, ed. L. C. Dunn (Macmillan, 1951). See also Jenny Bangham and Soraya de Chadarevian, "Human Heredity after 1945," *Studies in History and Philosophy of Biological and Biomedical Sciences* Part A 47 (2014): 45–49.

27. Elazar Barkan, *The Retreat of Scientific Racism* (Cambridge University Press, 1996).

28. Jamshed Mavalwala, "The Dermatoglyphics of the Parsis of India," *Zeitschrift für Morphologie und Anthropologie* 54, no. 2 (1963): 173–89; M. K. Robson and P. A. Parsons, "Fingerprint Studies on Four Central Australian Aboriginal Tribes," *Archaeology and Physical Anthropology in Oceania* 2, no. 1 (1967):

69–78; Sardool Singh, "Quantitative and Qualitative Analysis of the Dermatoglyphics of the Palms of Australians of European Ancestry," *Human Biology* 43, no. 4 (1971): 486–92; Jan Benes, "A Contribution to the Occurrence of Four-Finger Furrows in Czechs and Gipsies," *Anthropologie* 10, no. 2/3 (1972): 77–82; Robert J. Meier, "Dermatoglyphic Variation in Five Eskimo Groups from North-Western Alaska," in *Dermatoglyphics: An International Perspective*, ed. Jamshed Mavalwala (Mouton, 1978), 145–52; H. L. Heet, "Dermatoglyphic Differentiation of the Population of the USSR," in *Dermatoglyphics: An International Perspective*, ed. Jamshed Mavalwala (Mouton, 1978), 177–94; M. S. Kamali, "Simian Crease Polymorphism among Fifteen Iranian Endogamous Groups," *Anthropologischer Anzeiger* 43 (September 1985): 217–25.

29. Kevles, *In the Name of Eugenics*, 245; Miller, "Dermatoglyphics," 79.

30. Jamshed Mavalwala, "Dermatoglyphics: Looking Forward to the 21st Century," in *Progress in Dermatoglyphic Research*, ed. Christos S. Bartsocas (Alan R. Liss, 1982), 13–23, foreword by Bartsocas, viii; C. Cho, "Finger Dermatoglyphics of Australian Aborigines in the Northern Territory of Australia," *Korean Journal of Biological Sciences* 4, no. 1 (2000): 91–94.

31. Galton, *Finger Prints*, 196; Paul Rabinow, "Galton's Regret," in *Essays on the Anthropology of Reason* (Princeton University Press, 1997), 112–28; For twenty-first-century studies, see Zhang, Hai-Guo, et al., "Dermatoglyphics from All Chinese Ethnic Groups Reveal Geographic Patterning," *PLoS One* 5, no. 1 (2010): e8783; Neeti Kapoor and Ashish Badiue, "Digital Dermatoglyphics: A Study on Muslim Population from India," *Egyptian Journal of Forensic Sciences* 5, no. 3 (2015): 90–95; Imene Namouchi, "Anthropological Significance of Dermatoglyphic Trait Variation: An Intra-Tunisian Population Analysis," *International Journal of Modern Anthropology* I (2011): 12–27; I. Singh and R. K. Garg, "Finger Dermatoglyphics: A Study of the Rajputs of Himachal Pradesh," *The Anthropologist* 6, no. 2 (2004): 155–56; B. Mohan Reddy et al., "Quantitative Dermatoglyphics and Population Structure in Northwest India," *American Journal of Human Biology* 12, no. 3 (2000): 315–26.

32. Notes for Litchfield Lecture, Oxford, January 20, 1965. Typescript, Lecture Notes, LPP PENROSE/2/42/3/2.

33. Penrose, "Fingerprints and Palmistry," 1239–42.

34. "Diagnosis by Palm-Reading," *Medical World News*, September 27, 1963, offprint in LPP PENROSE 2/42/9/4; Penrose, "Fingerprints and Palmistry," 1239–42. See also Miller, "Dermatoglyphics," 75–94.

35. From Penrose, "Finger-Prints, Palms and Chromosomes," 933, 935; and Penrose, "Fingerprints and Palmistry," 1240–42.

36. Penrose, "Fingerprints and Palmistry," 1240; L. S. Penrose, "On the Interaction of Heredity and Environment in the Study of Human Genetics (with Special Reference to Mongolian Imbecility)," *Journal of Genetics* 25 (1932): 407–22.

CONCLUSION

1. Personal communication with pain physician Dr. Guy Bashford and rehabilitation physician Professor John Speed, October 1, 2024; T. G. Abrahams, "Occult Malignancy Presenting as Metastatic Disease to the Hand and Wrist," *Skeletal Radiology* 24 (1995): 135–37; Phuc Vo et al., "Evaluating Dorsal Wrist Pain: MRI Diagnosis of Occult Dorsal Wrist Ganglion," *Journal of Hand Surgery* 20, no. 4 (1995): 667–70; S. Goldsmith et al., "Magnetic Resonance Imaging in the Diagnosis of Occult Dorsal Wrist Ganglions," *Journal of Hand Surgery* 33, no. 5 (20): 595–99; Mohamed Jarraya et al., "Radiographically Occult and Subtle Fractures," *Radiology Research and Practice* 1 (2013): 1–10.

2. Lauren Kassell, "Magic, Alchemy and the Medical Economy in Early Modern England: The Case of Robert Fludd's Magnetical Medicine," in *Medicine and the Market in England and Its Colonies, c. 1450–1850*, ed. Mark Jenner and Patrick Wallis (Palgrave, 2007), 88–107; Saunders, *Physiognomie and Chiromancie*, preface; Christoph Sander, "Tempering Occult Qualities: Magnetism and Complexio in Early Modern Medical Thought," *Early Science and Medicine* 28 (2023): 573–96.

3. Richard B. Gunderman, *X-Ray Vision: The Evolution of Medical Imaging and Its Human Significance* (Oxford University Press, 2013).

4. Richard Saunders, *Physiognomie and Chiromancie*, 75.

5. Jasper Gerard, "Palmed Off," *Times*, November 26, 1998; Modern Hand Reading Forum, https://www.modernhandreadingforum.com/t326-famous-simian-lines-which-famous-persons-have-a-simian-line; "Simian Line Two," God Given Glyphs, https://godgivenglyphs.com/chirology-articles/simian-lines-2/.

6. "Simian Crease," https://www.mountsinai.org/health-library/symptoms/simian-crease.

7. Azra Metovic et al., "Predictive Analysis of Palmar Dermatoglyphics in Patients with Breast Cancer for Small Bosnian-Herzegovinian Population," *Medical Archives* 72, no. 5 (2018): 357–61; A. S. Nagy et al., "Comparative Analysis of Dermatoglyphic Traits in Hungarian and Gypsy Populations," *Human Biology* 76, no. 3 (2004): 383–400.

8. K. M. Godfrey et al., "Relation of Fingerprints and Shape of the Palm to Fetal Growth and Adult Blood Pressure," *BMJ* 307, no. 6901 (1993): 405–9.

9. "What Are the Traits Covered in the 23andMe Health Report?," https://www.xcode.life/23andme-raw-data/23andme-health-report-risk-worth-alternative-cost/.

Select Bibliography of Secondary Sources

Ahonen, Marke. "Medieval and Early Modern Physiognomy." In *Sourcebook for the History of the Philosophy of Mind: Philosophical Psychology from Plato to Kant*, edited by Simo Knuuttila and Juha Sihvola, 633–37. Springer, 2013.

App, Urs. "Tibetan Theosophy: Helena Blavatsky's Tantric Connection." In *Sino-Tibetan Buddhism Across the Ages*, edited by Ester Bianchi and Weirong Shen, 141–69. Brill, 2021.

Asen, Daniel. "'Dermatoglyphics' and Race after the Second World War: The View from East Asia." In *Global Transformations in the Life Sciences, 1945–1980*, edited by Patrick Manning and Mat Savelli, 61–77. University of Pittsburgh Press, 2018.

Asprem, Egil. *The Problem of Disenchantment: Scientific Naturalism and Esoteric Discourse, 1900–1939*. Brill, 2014.

Bangham, Jenny, and Soraya de Chadarevian. "Human Heredity after 1945: Moving Populations Centre Stage." *Studies in History and Philosophy of Biological and Biomedical Sciences* Part A 47 (2014): 45–49.

Barkan, Elazar. *The Retreat of Scientific Racism: Changing Concepts of Race in Britain and the United States Between the World Wars*. Cambridge University Press, 1996.

Bashford, Alison. *The Huxleys: An Intimate History of Evolution*. University of Chicago Press, 2022.

Baylen, Joseph O. "Mark Twain, W. T. Stead and 'The Tell-Tale Hands.'" *American Quarterly* 16, no. 4 (Winter 1964): 606–12.

Berkowitz, Carin. "Charles Bell's Seeing Hand: Teaching Anatomy to the Senses in Britain, 1750–1840." *History of Science* 52, no. 4 (2014): 377–400.

Berland, Kevin. "Outward Sign and Inward Condition: Recent Studies in Physiognomy, Anthropometry, and Related Sciences." *Eighteenth-Century Studies* 36, no. 2 (Winter 2003): 266–69.

Boaz, Danielle. "Fraud, Vagrancy and the 'Pretended' Exercise of Supernatural Powers in England, South Africa and Jamaica." *Law & History* 5, no. 1 (2018): 54–84.

Bogdal, Klaus-Michel. *Europe and the Roma: A History of Fascination and Fear.* Penguin, 2023.

Booth, Tony. "Labels and Their Consequences." In *Current Approaches to Down's Syndrome*, edited by David Lane and Brian Stratford, 3–24. Holt Reinhart and Winston, 1985.

Borch-Jacobsen, Mikkel. "Hypnosis in Psychoanalysis." *Representations* 27 (Summer 1989): 92–110.

Boxer, Carly B. "Reading Palms, Handling Hands: Learning, Embodiment, and Images in a Late Medieval Chiromancy Roll." *Eolas* 14 (2022): 100–115.

Boyle, Marjorie O'Rourke. "God's Hand." In *Senses of Touch: Human Dignity and Deformity from Michelangelo to Calvin*, 172–256. Brill, 2021.

Breckenridge, Keith. *Biometric State: The Global Politics of Identification and Surveillance in South Africa, 1850 to the Present.* Cambridge University Press, 2014.

Brennan, Toni, and Peter Hegarty. "Charlotte Wolff and Lesbian History: Reconfiguring Liminality in Exile." *Journal of Lesbian Studies* 14, no. 4 (2010): 338–58.

Briefel, Aviva. *The Racial Hand in the Victorian Imagination.* Cambridge University Press, 2015.

Burnett, Charles S. F. "The Earliest Chiromancy in the West." *Journal of the Warburg and Courtauld Institutes* 50 (1987): 189–95.

Carey, Hilary M. "Astrological Medicine and the Medieval English Folded Almanac." *Social History of Medicine* 17, no. 3 (December 2004): 345–63.

Carter, K. Codell. "Early Conjectures That Down Syndrome Is Caused by Chromosomal Nondisjunction." *Bulletin of the History of Medicine* 76, no. 3 (Fall 2002): 528–63.

Cole, Simon A. *Suspect Identities: A History of Fingerprinting and Criminal Identification.* Harvard University Press, 2002.

Cooter, Roger. *The Cultural Meaning of Popular Science: Phrenology and the Organization of Consent in Nineteenth-Century Britain.* Cambridge University Press, 1984.

Copenhaver, Brian P. *Magic in Western Culture: From Antiquity to the Enlightenment.* Cambridge University Press, 2015.

Coudert, Allison P. "Leibniz, Locke, Newton and the Kabbalah." In *The Christian Kabbalah: Jewish Mystical Books and Their Christian Interpreters*, edited by Joseph Dan. Harvard College Library, 1997.

Daston, Lorraine, and Katharine Park. *Wonders and the Order of Nature 1150–1750.* Zone Books, 1998.

Davies, Owen. "Newspapers and the Popular Belief in Witchcraft and Magic in the Modern Period." *Journal of British Studies* 37, no. 2 (April 1998): 139–65.

Davies, Owen. *Popular Magic: Cunning-Folk in English History*. Hambeldon, 2007.

Davies, Owen. *Witchcraft, Magic and Culture, 1736–1951*. Manchester University Press, 1999.

de Chadarevian, Soraya. *Heredity Under the Microscope: Chromosomes and the Study of the Human Genome*. University of Chicago Press, 2020.

Dubrau, Irmi. "Physiognomy among Medieval Jews." In *Prognostication in the Medieval World: A Handbook*, vol. 2, edited by Matthias Heiduk, Klaus Herbers, and Hans-Christian Lehner, 908–14. De Gruyter, 2020.

Eamon, William. *Science and the Secrets of Nature: Books of Secrets in Medieval and Early Modern Culture*. Princeton University Press, 1994.

Fletcher, Joann. "Edward Heron-Allen, His Friends and Their Mummies: EHA and Ancient Egypt." *Proceedings of the 5th Heron-Allen Symposium 2005: Opusculum VIII* (2005): 17–44.

Forster, Regula. "The Pseudo-Aristotelian *Sirr al-asrār/Secretum secretorum*." In *Prognostication in the Medieval World: A Handbook*, vol. 2, edited by Matthias Heiduk, Klaus Herbers, and Hans-Christian Lehner, 965–70. De Gruyter, 2020.

Franklin, J. Jeffrey. *Spirit Matters: Occult Beliefs, Alternative Religions, and the Crisis of Faith in Victorian Britain*. Cornell University Press, 2018.

Garrington, Abbie. *Haptic Modernism: Touch and the Tactile in Modernist Writing*. Edinburgh University Press, 2022.

Gillis Hogan, Samuel P. "Stars in the Hand: The Manuscript and Intellectual Contexts of British Latin Medieval Chiromancy and Its Scholastic and Astrological Influences." Master of Arts thesis, University of Saskatchewan, 2018.

Ginzburg, Carlo. "Morelli, Freud, and Sherlock Holmes: Clues and Scientific Method." In *The Sign of Three: Dupin, Holmes, Peirce*, edited by Umberto Eco and Thomas A. Sebeok, 81–118. Indiana University Press, 1988.

Goldish, Matt. "Newton on Kabbalah." In *The Books of Nature and Scripture*, edited by James E. Force and Richard H. Popkin, 89–103. Springer, 1994.

Gonzalez-Crussi, Frank. *The Language of the Face*. MIT Press, 2023.

Goodrick-Clarke, Nicholas. *The Western Esoteric Traditions*. Oxford University Press, 2008.

Grafton, Anthony. *Magus: The Art of Magic from Faustus to Agrippa*. Allen Lane, 2023.

Gray, Richard T. *About Face: German Physiognomic Thought from Lavater to Auschwitz*. Wayne State University Press, 2004.

Groebner, Valentin. "Complexio/Complexion: Categorizing Individual Natures, 1250–1600." In *The Moral Authority of Nature*, edited by Lorraine Daston and Fernando Vidal, 361–83. University of Chicago Press, 2004.

Grossmann, Atina. "German Women Doctors from Berlin to New York:

Maternity and Modernity in Weimar and in Exile." *Feminist Studies* 19, no. 1 (Spring 1993): 65–88.

Gunderman, Richard B. *X-Ray Vision: The Evolution of Medical Imaging and Its Human Significance*. Oxford University Press, 2013.

Hanson, Marta E. "Hand Mnemonics in Classical Chinese Medicine: Texts, Earliest Images, and Arts of Memory." *Asia Major* 21, no. 1 (2006): 325–47.

Heiduk, Matthias, Klaus Herbers, and Hans-Christian Lenher, eds. *Prognostication in the Medieval World: A Handbook*. 2 vols. De Gruyter, 2020.

Hillman, David, and Carla Mazzio, eds. *The Body in Parts: Fantasies of Corporeality in Early Modern Europe*. Routledge, 1997.

Houghton-Walker, Sarah. *Representations of the Gypsy in the Romantic Period*. Oxford University Press, 2014.

Hoyland, Robert. "The Islamic Background to Polemon's Treatise." In *Seeing the Face, Seeing the Soul: Polemon's Physiognomy from Classical Antiquity to Medieval Islam*, edited by Simon Swain, 227–80. Oxford University Press, 2008.

Hoyland, Robert. "Physiognomy in Islam." *Jerusalem Studies in Arabic and Islam* 30 (2005): 361–402.

Huteau, Michel. "A Star of French Psychology: Nicolae Vaschide (1874–1907)." *Bulletin de Psychologie* 494, no. 2 (2008): 173–99.

Jordanova, Ludmilla. "The Art and Science of Seeing in Medicine: Physiognomy 1780–1820." In *Medicine and the Five Senses*, edited by W. F. Bynum and Roy Porter, 122–33. Cambridge University Press, 1993.

Jordanova, Ludmilla. *Defining Features: Scientific and Medical Portraits 1660–2000*. Reaktion Books, 2000.

Josephson-Storm, Jason Ā. *The Myth of Disenchantment: Magic, Modernity, and the Birth of the Human Sciences*. University of Chicago Press, 2017.

Kassell, Lauren. "The Economy of Magic in Early Modern England." In *The Practice of Reform in Health, Medicine, and Science, 1500–2000: Essays for Charles Webster*, edited by Margaret Pelling and Scott Mandelbrote, 43–57. Ashgate, 2005.

Kassell, Lauren. "Magic, Alchemy and the Medical Economy in Early Modern England: The Case of Robert Fludd's Magnetical Medicine." In *Medicine and the Market in England and Its Colonies, c. 1450–1850*, edited by Mark Jenner and Patrick Wallis, 88–107. Palgrave, 2007.

Kassell, Lauren. *Medicine and Magic in Elizabethan London: Simon Forman: Astrologer, Alchemist, and Physician*. Oxford University Press, 2005.

Kevles, Daniel. *In the Name of Eugenics: Genetics and the Uses of Human Heredity*. Harvard University Press, 1985.

Kevles, Daniel. "'Mongolian Imbecility': Race and Its Rejection in the Understanding of a Mental Disease." In *Mental Retardation in America: A Historical Reader*, edited by Steven Noll and James W. Trent Jr., 120–29. New York University Press, 2004.

Khalid, Muhammad Ali, and Tarif Khalidi. "Is Physiognomy a Science? Reflections on the *Kitāb al-Firāsa* of Fakhr al-Dīn al-Rāzī." In *The Occult Sciences in Pre-Modern Islamic Cultures*, edited by Nade El-Bizri and Eva Orthmann, 67–82. Ergon Verlag, 2018.

Klein-Franke, Felix. "The Geomancy of Ahmad B. 'Ali Zunbul: A Study of the Arabic Corpus Hermeticum." *Ambix* 20, no. 1 (1973): 26–35.

Landy, Joshua, and Michael Saler, eds. *The Re-Enchantment of the World: Secular Magic in a Rational Age*. Stanford University Press, 2009.

Láng, Benedek. *Unlocked Books: Manuscripts of Learned Magic in the Medieval Libraries of Central Europe*. Pennsylvania State University Press, 2008.

Laqueur, Thomas. "Why the Margins Matter: Occultism and the Making of Modernity." *Modern Intellectual History* 3, no. 1 (April 2006): 111–35.

Leader, Darian. *Hands: What We Do with Them*. Hamish Hamilton, 2016.

Lee, Ken. "Orientalism and Gypsylorism." *Social Analysis* 44, no. 2 (2000): 129–56.

Leitch, Stephanie. "Getting to How-To: Chiromancy, Physiognomy, Metoscopy and Prints in Secrets' Service." In *Quid est secretum? Visual Representations of Secrets in Early Modern Europe, 1500–1700*, edited by Ralph Dekoninck, Agnès Guiderdoni, and Walter Melion, 635–59. Brill, 2020.

Leitch, Stephanie. "Mapping the Hand and Scanning the Forehead: Embedding Knowledge in Astrological Images." In *Emblems in the Free Imperial City*, edited by Mara R. Wade, Christopher D. Fletcher, and Andrew C. Schwenk, 262–92. Brill, 2024.

Leong, Elaine, and Alisha Rankin, eds. *Secrets and Knowledge in Medicine and Science, 1500–1800*. Routledge, 2011.

Loesch, Danuta. "The Contributions of L. S. Penrose to Dermatoglyphics." *Journal of Mental Deficiency Research* 17, no. 1 (1973): 1–17.

Maclean, Ian. *Logic, Signs and Nature in the Renaissance: The Case of Learned Medicine*. Cambridge University Press, 2002.

Mahoney, John P., and Barbara P. Mahoney. "'Ship Stink, Immigrants and Hyperosmia': A Tale of Ex-Palmist Edward Heron-Allen in the Castle Garden Affair." 16th Annual Heron-Allen Society Symposium, Chichester, July 2, 2016. Typescript courtesy of John Mahoney.

Mahoney, John P., and Barbara P. Mahoney. "The Sir Vincent Horsley and Edward Heron-Allen Discussions." *Heron-Allen Society Newsletter* 27 (Autumn 2015): 8.

Margolin, Ron. "Physiognomy and Chiromancy: From Prediction and Diagnosis to Healing and Human Correction." In *Prognostication in the Medieval World: A Handbook*, vol. 1, edited by Matthias Heiduk, Klaus Herbers, and Hans-Christian Lehner, 915–24. De Gruyter, 2020.

Mayall, David. *Gypsy Identities 1500–2000: From Egipcyans and Moon-men to the Ethnic Romany*. Routledge, 2004.

McIntosh, Christopher. *Eliphas Lévi and the French Occult Revival*. State University of New York Press, 2011.

Melvin-Koushki, Matthew. "Geomancy in the Islamic World." In *Prognostication in the Medieval World: A Handbook*, vol. 2, edited by Matthias Heiduk, Klaus Herbers, and Hans-Christian Lehner, 788–93. De Gruyter, 2020.

Miller, F. A. "Dermatoglyphics and the Persistence of 'Mongolism.'" *Social Studies of Science* 33, no. 1 (2003): 75–94.

Monod, Paul Kléber. *Solomon's Secret Arts: The Occult in the Eighteenth Century*. Yale University Press, 2013.

Morgan, John. "'Counterfeit Egyptians': The Construction and Implementation of a Criminal Identity in Early Modern England." *Romani Studies* 26, no. 2 (2016): 105–28.

Mori, Jennifer. "Magic and Fate in Eighteenth-Century London: Prosecutions for Fortune-Telling, c. 1678–1830." *Folklore* 129, no. 3 (2018): 254–77.

Müller-Wille, Staffan. "Corners, Tables, Lines: Towards a Diagrammatics of Race." *Nuncius* 36, no. 3 (2021): 517–31.

Müller-Wille, Staffan. "Race and Kinship: Anthropology and the 'Genealogical Method.'" In *The Politics of Making Kinship: Historical and Anthropological Perspectives*, edited by Erdmute Alber, David Warren Sabean, Simon Teuscher, and Tatjana Thelen, 79–113. Berghahn Books, 2023.

Nagel, Alexandra. "From Chiromancy to Psychochirology: The Modern Transformation of a Mantic Art." *Aries: Journal for the Study of Western Esotericism* 21, no. 2 (2021): 246–70.

Nagel, Alexandra. "Jung, Julius Spier, and Palmistry." *Jung Studies* 14, no. 1 (January 2020): 65–81.

Nagel, Alexandra. "The Psychochirologist Julius Spier and the Art of Reading Hands During the Interbellum." PhD diss., Leiden University, 2020.

Navarre, Joan. "Oscar Wilde, Edward Heron-Allen, and the Palmistry Craze of the 1880s." *English Literature in Transition* 54, no. 2 (2011): 174–84.

Noakes, Richard. *Physics and Psychics: The Occult and the Sciences in Modern Britain*. Cambridge University Press, 2019.

Nord, Deborah. *Gypsies and the British Imagination: 1807–1930*. Columbia University Press, 2006.

Owen, Alex. *The Place of Enchantment: British Occultism and the Culture of the Modern*. University of Chicago Press, 2004.

Pack, Roger A. "Aristotle's Chiromantic Principle and Its Influence." *Transactions of the American Philological Association* 108 (1978): 121–30.

Pack, Roger A. "On the Greek Chiromantic Fragment." *Transactions and Proceedings of the American Philological Association* 103 (1972): 367–80.

Partridge, Christopher. *The Re-Enchantment of the West: Alternative Spiritualities, Sacralization, Popular Culture, and Occulture*, vol. 1. T&T Clark, 2004.

Paterson, Mark. *Seeing with the Hands: Blindness, Vision and Touch After Descartes*. Edinburgh University Press, 2016.

Pearl, Sharrona. *About Faces: Physiognomy in Nineteenth Century Britain*. Harvard University Press, 2010.

Percival, Melissa. *The Appearance of Character: Physiognomy and Facial Expression in Eighteenth-Century France*. W. S. Maney & Son, 1999.

Pfeffer, M. "The Society of Astrologers (c. 1647–1684): Sermons, Feasts and the Resuscitation of Astrology in Seventeenth-Century London." *British Journal for the History of Science* 54, no. 2 (2021): 133–53.

Pick, Daniel. *Faces of Degeneration: A European Disorder, c. 1848–1918*. Cambridge University Press, 1989.

Plaitano, Greta. "Technologies de l'oeil, psychologie de la main: Nicolae Vaschide et la photographie médico-artistique." *Cinéma&Cie* 20, no. 35, Mediatic Handology: Shaping Images, Interacting, Magicking (2020): 29–40.

Porter, Martin Henry. *Windows of the Soul: Physiognomy in European Culture 1470–1780*. Clarendon Press, 2005.

Porter, Roy. *Health for Sale: Quackery in England, 1660–1850*. Manchester University Press, 1989.

Poskett, James. *Materials of the Mind: Phrenology, Race, and the Global History of Science, 1815–1920*. University of Chicago Press, 2019.

Rabinow, Paul. "Galton's Regret: Of Types and Individuals." In *Essays on the Anthropology of Reason*, 112–28. Princeton University Press, 1997.

Ramsden, Edmund. "Remodelling the Boundaries of Normality: Lionel S. Penrose and Population Surveys of Mental Ability." In *Human Heredity in the Twentieth Century*, edited by Bernd Gausemeier, Staffan Müller-Wille, and Edmund Ramsden, 39–54. Pickering and Chatto, 2013.

Rogers, Nicholas. "Policing the Poor in Eighteenth-Century London: The Vagrancy Laws and Their Administration." *Histoire Sociale/Social History* 24, no. 47 (May 1991): 127–47.

Ruderman, David B. *Kabbalah, Magic and Science: The Cultural Universe of a Sixteenth-Century Jewish Physician*. Harvard University Press, 1988.

Saler, Michael. "Modernity and Enchantment: A Historiographic Review." *American Historical Review* 111, no. 3 (2006): 692–716.

Savage-Smith, Emilie, ed. *Magic and Divination in Early Islam*. Ashgate, 2004.

Scholem, Gershom. "Chiromancy." In *Encyclopaedia Judaica*, vol. 4, 652–54. Macmillan, 2007.

Scholem, Gershom. *Major Trends in Jewish Mysticism*. Schocken Books, 1946.

Sen, K. C. *Hast Samudrika Shastra, or, the Indian Science of Hand Reading* (D. P. Taraporevala, 1960).

Sengoopta, Chandak. *Imprint of the Raj: How Fingerprinting Was Born in Colonial India*. Macmillan, 2003.

Sengoopta, Chandak. "Treacherous Minds, Submissive Bodies: Corporeal Technologies and Human Experimentation in Colonial India." In *Locating the Medical: Explorations in South Asian History*, edited by Rohan Deb Roy and Guy N. A. Attewell. Oxford University Press, 2018.

Smith, Michael. *Lionel Sharples Penrose: A Biography*. Lavenham, 1999.

Swartz, Michael D. *The Signifying Creator: Nontextual Sources of Meaning in Ancient Judaism*. New York University Press, 2012.

Tallis, Raymond. *The Hand: A Philosophical Inquiry into Human Being*. Edinburgh University Press, 2003.

Thorndike, Lynn. "Chiromancy in Mediaeval Latin Manuscripts." *Speculum* 40 (1965): 674–706.

Thorndike, Lynn. *A History of Magical and Experimental Science*. 2 vols. Columbia University Press, 1923.

Turner, Bryan S. "Reflections on the Epistemology of the Hand." In *Regulating Bodies: Essays in Medical Sociology*, 99–121. Routledge, 1992.

Van Den Broek, Gerard J. "Signs and Signatures: Reading God's Herbal." *Semiotica* 63, nos. 1–2 (1987): 109–28.

Van Wyhe, John. *Phrenology and the Origins of Victorian Scientific Naturalism*. Routledge, 2004.

Wang, Xing. *Physiognomy in Ming China: Fortune and the Body*. Brill, 2020.

Warlick, M. E. "Palmistry as Portraiture: Dr. Charlotte Wolff and the Surrealists." In *Surrealism, Occultism and Politics: In Search of the Marvellous*, edited by Tessel M. Bauduin, Victoria Ferentinou, and Daniel Zamani, 57–72. Routledge, 2018.

Weber, Max. *The Theory of Social and Economic Organization*. Translated by A. M. Henderson and Talcott Parsons. Oxford University Press, 1947.

Webster, Charles. *From Paracelsus to Newton: Magic and the Making of Modern Science*. Cambridge University Press, 1982.

Webster, Charles. *Paracelsus: Medicine, Magic and Mission at the End of Time*. Yale University Press, 2008.

West, Tamara. "Marginality and Modernity on the South Shore: Blackpool's Fortune Tellers, Authenticity and Belonging." *European History Quarterly* 52, no. 4 (2022): 572–92.

Williams, Steven J. *The Secret of Secrets: The Scholarly Career of a Pseudo-Aristotelian Text in the Late Middle Ages*. University of Michigan Press, 2003.

Winter, Alison. *Mesmerized: Powers of Mind in Victorian Britain*. University of Chicago Press, 1998.

Wiseman, Sue, Katharine Hodgkin, and Michelle O'Callaghan, eds. *Reading the Early Modern Dream: The Terrors of the Night*. Routledge, 2007.

Wright, David. *Downs: The History of a Disability*. Oxford University Press, 2011.

Wright, David. "Mongols in Our Midst: John Langdon Down and the Ethnic Classification of Idiocy, 1858–1924." In *Mental Retardation in America: A Historical Reader*, edited by Steven Noll and James W. Trent Jr., 92–119. New York University Press, 2004.

Yates, Frances A. *Giordano Bruno and the Hermetic Tradition*. Routledge and Kegan Paul, 1964.

Yates, Frances A. *The Occult Philosophy in the Elizabethan Age*. Routledge and Kegan Paul, 1979.

Zambelli, Paola. *The* Speculum Astronomiae *and Its Enigma: Astrology, Theology and Science in Albertus Magnus and His Contemporaries*. Springer Dordrecht, 1992.

Zaretsky, Eli. *Secrets of the Soul: A Social and Cultural History of Psychoanalysis*. Knopf, 2005.

Ziegler, Joseph. "On the Various Faces of Hebrew Physiognomy as a Prognostic Art in the Middle Ages." In *Unveiling the Hidden—Anticipating the Future: Divinatory Practices Among Jews Between Qumran and the Modern Period*, edited by Josefina Rodriguez-Arribas and Dorian Gieseler Greenbaum, 286–310. Brill, 2021.

Ziegler, Joseph. "Philosophers and Physicians on the Scientific Validity of Latin Physiognomy, 1200–1500." *Early Science and Medicine* 12, no. 3 (2007): 285–312.

Zysk, Kenneth G. *The Indian System of Human Marks*. 2 vols. Brill, 2016.

Index

Page numbers in italics refer to figures.